Handbook of Museum Textiles

Scrivener Publishing
100 Cummings Center, Suite 541J
Beverly, MA 01915-6106

Publishers at Scrivener
Martin Scrivener (martin@scrivenerpublishing.com)
Phillip Carmical (pcarmical@scrivenerpublishing.com)

Handbook of Museum Textiles

Volume I
Conservation and Cultural Research

Edited by
Seiko Jose
Sabu Thomas
Pintu Pandit
Ritu Pandey
and
Vandana Gupta

Scrivener Publishing

WILEY

This edition first published 2023 by John Wiley & Sons, Inc., 111 River Street, Hoboken, NJ 07030, USA and Scrivener Publishing LLC, 100 Cummings Center, Suite 541J, Beverly, MA 01915, USA

For more information about Scrivener publications please visit www.scrivenerpublishing.com.

Wiley Global Headquarters
111 River Street, Hoboken, NJ 07030, USA

For details of our global editorial offices, customer services, and more information about Wiley products visit us at www.wiley.com.

Library of Congress Cataloging-in-Publication Data

ISBN 978-1-119-79170-6

Cover image: Pixabay.Com
Cover design by Russell Richardson

Set in size of 11pt and Minion Pro by Manila Typesetting Company, Makati, Philippines

Printed in the USA

Contents

Preface xix

1 Textile Museums: Its Importance, Objectives and Functions 1
Vandana Gupta and Mangesh S. Manekar
1.1 Introduction 1
1.2 Museum Definition and Its Existence 2
1.3 Textile Museums and Their History 4
1.4 Importance of Textile Museums 6
1.4.1 Connective Value 6
1.4.2 Collective Value 7
1.4.3 Educative Value 10
1.4.4 Economic Value 11
1.5 Objectives of Textile Museum 14
1.6 Function of Textile Museum 14
1.7 Textile Museum and Their Future Prospects 16
1.8 Conclusion 16
References 16

2 Textile Intangible Cultural Heritage of the World 19
Ritu Pandey, Vandana Gupta, Pintu Pandit, Kumar Rohit and Suruchi Pandey
2.1 Introduction 19
2.1.1 Ancient Textiles of the World 20
2.2 Ancient Textiles of the World: Techniques and Historical Perspective 24
2.2.1 Ancient Heritage Textiles—Asia and Australia 24
2.2.1.1 Chinese Textiles 24
2.2.1.2 Japanese Textiles 24
2.2.1.3 Indian Textile 26
2.2.1.4 Turkish Textiles 27
2.2.1.5 Persian Rugs 27
2.2.2 African Textile 28
2.2.3 Scandinavian Textile Design 29
2.2.4 American Textiles 30
2.3 Role of Museum: Textile as a Part of Heritage and Culture 32
2.4 Marketing Strategies for Museums and Conservation 33

2.5 Conclusion 35
References 35

3 Important Woven Textile Specimens in World Museums 39
Karthika Audinet
3.1 Introduction 39
3.2 Methodology 40
3.3 Prehistoric Textiles 42
3.3.1 Textile Specimen 1.3.1 44
3.3.2 Textile Specimen 1.3.2 44
3.3.3 Textile Specimen 1.3.3 45
3.3.4 Textile Specimen 1.3.4 45
3.3.5 Textile Specimen 1.3.5 46
3.3.6 Textile Specimen 1.3.6 46
3.4 Textiles from Peru 46
3.4.1 Textile Specimen 1.4.1 48
3.4.2 Textile Specimen 1.4.2 48
3.4.3 Textile Specimen 1.4.3 49
3.4.4 Textile Specimen 1.4.4 49
3.4.5 Textile Specimen 1.4.5 49
3.4.6 Textile Specimen 1.4.6 50
3.4.7 Textile Specimen 1.4.7 50
3.5 Chinese Textiles 50
3.5.1 Textile Specimen 1.5.1 52
3.5.2 Textile Specimen 1.5.2 52
3.5.3 Textile Specimen 1.5.3 53
3.5.4 Textile Specimen 1.5.4 54
3.5.5 Textile Specimen 1.5.5 54
3.5.6 Textile Specimen 1.5.6 55
3.5.7 Textile Specimen 1.5.7 55
3.6 Textiles from the Indian Subcontinent 56
3.6.1 Textile Specimen 1.6.1 57
3.6.2 Textile Specimen 1.6.2 58
3.6.3 Textile Specimen 1.6.3 58
3.6.4 Textile Specimen 1.6.4 58
3.6.5 Textile Specimen 1.6.5 59
3.6.6 Textile Specimen 1.6.6 59
3.6.7 Textile Specimen 1.6.7 59
3.7 Textiles from Sudan, Egypt, Mali, and Ghana (North and West African Textiles) 60
3.7.1 Textile Specimen 1.7.1 61
3.7.2 Textile Specimen 1.7.2 62
3.7.3 Textile Specimen 1.7.3 63
3.7.4 Textile Specimen 1.7.4 63
3.8 Textiles from Japan 64
3.8.1 Textile Specimen 1.8.1 65

3.8.2 Textile Specimen 1.8.2 65
3.8.3 Textile Specimen 1.8.3 66
3.8.4 Textile Specimen 1.8.4 66
3.8.5 Textile Specimen 1.8.5 66
3.8.6 Textile Specimen 1.8.6 67
3.9 Textiles from Iran and Central Asia 67
3.9.1 Textile Specimen 1.9.1 69
3.9.2 Textile Specimen 1.9.2 69
3.9.3 Textile Specimen 1.9.3 69
3.9.4 Textile Specimen 1.9.4 70
3.9.5 Textile Specimen 1.9.5 70
3.10 Textiles from Italy and France 71
3.10.1 Textile Specimen 1.10.1 73
3.10.2 Textile Specimen 1.10.2 73
3.10.3 Textile Specimen 1.10.3 73
3.10.4 Textile Specimen 1.10.4 73
3.11 Conclusion: Toward an Understanding of the Historic Foundations of Woven Structures 74
References 77
List of Museums 82
Glossary 83

4 Types of Conservation of Textiles in the Museum: Their Importance and Scope 87
K. B. Binita and B. Sunita
4.1 Introduction 87
4.2 Importance of Conservation 88
4.3 Principles of Conservation 89
4.3.1 Determining the Need and Laying the Foundation for Conservation 89
4.3.2 The Conservation Assessment and Process 89
4.3.3 Obtaining a Conservation Assessment 89
4.3.4 Selecting an Assessor or Conservator 89
4.3.5 The Assessment as a Planning Tool 90
4.3.6 Conservation Collection Condition Survey 90
4.3.7 Object Treatment 90
4.3.8 Risk Assessment and Management 90
4.4 Types of Textile Articles Conserved 90
4.5 Methods of Conservation 91
4.5.1 Preventive Conservation 91
4.5.1.1 Climate 92
4.5.1.2 Light 92
4.5.1.3 Insects 93
4.5.1.4 Microorganisms 93
4.5.1.5 Dust, Soil, and Other Contaminants 94
4.5.1.6 Disaster 94

4.5.2 Curative/Interventive Conservation 95
4.5.2.1 Surface Cleaning 95
4.5.2.2 Vacuuming 95
4.5.2.3 Wet Cleaning 95
4.5.2.4 Solvent or Dry Cleaning 96
4.5.2.5 Stabilization 96
4.6 Storage, Display, and Handling of Museum Textiles 96
4.7 Scope of Conservation 97
4.7.1 Education and Knowledge Dissemination 97
4.7.2 Lecture, Seminar, Workshops, and Research 98
4.7.3 Photography and Publication 98
4.8 New Approaches in Conservation 98
4.9 Conclusion 99
References 99
Webliography 100

5 Fashion and Textile Museums Across the Globe 101
Arpana Kamboj and Surabhi Mahajan
5.1 Introduction 101
5.2 Victoria and Albert Museum, London 103
5.2.1 History 103
5.2.2 Collection 104
5.3 Fashion Museum, Bath, UK 104
5.3.1 History 104
5.3.2 Collection 105
5.4 Metropolitan Museum of Art, New York City 106
5.4.1 History 106
5.4.2 Collection 107
5.5 Musée De La Mode Et Du Textile, France 108
5.5.1 History 108
5.5.2 Collection 109
5.6 Palais Galliera, France 109
5.6.1 History 110
5.6.2 Collection 110
5.6.2.1 Eighteenth Century Dress Office 110
5.6.2.2 Nineteenth Century Ensembles Division 111
5.6.2.3 Fashion of the Principal Half of 20th Century 111
5.6.2.4 Haute Couture 111
5.6.2.5 Contemporary Office 111
5.6.2.6 Extras Office 111
5.7 Kyoto Costume Institute, Japan 111
5.7.1 History 112
5.7.2 Collection 112
5.8 Museum of Fashion Institute of Technology, New York, USA 113
5.8.1 History 113
5.8.2 Collection 114

5.9 Museo Del Traje, Spain 115
5.9.1 History 115
5.9.2 Collection 116
5.10 Fashion Institute of Design & Merchandising, California 116
5.10.1 History 117
5.10.2 Collection 117
5.11 Kent State University Museum, USA 117
5.11.1 History 118
5.11.2 Collection 119
5.12 Conclusion 119
References 119

6 Documentation of Museum Textiles 123
Simmi Bhagat and Radhana Raheja
6.1 Introduction 123
6.2 Functions of Documentation 124
6.3 Features of Documentation System 125
6.4 Collection Management Policy 126
6.5 Assessment Standards 128
6.5.1 Collection Assessment 128
6.5.2 Assessment of Objects 129
6.6 Types of Documentation 130
6.6.1 Written Description 130
6.6.2 Photographic Records 131
6.7 Formats of Documentation 136
6.7.1 Styles of Written Documentation 136
6.7.2 Manual and Digitized Documentation 136
6.8 Case Study 137
6.9 Conclusion 141
References 141

7 Ideal Storage Conditions for Museum Textiles 143
Simmi Bhagat and Kanika Sachdeva
7.1 Introduction 143
7.2 Published Standards in Museum Storage 144
7.3 Storage Design and Architecture 145
7.3.1 Museum Storage Building and Space Allocation 146
7.3.2 Building Monitoring and Maintenance 146
7.4 Environmental Conditions 147
7.4.1 Temperature and Relative Humidity 147
7.4.2 Light 148
7.5 Storage Techniques 148
7.5.1 Accession and Labeling 149
7.5.2 Flat Storage 149
7.5.3 Rolled Storage 150
7.5.4 Hanging Storage 151

7.5.5 Special Storage 152
7.6 Safety Systems 153
7.6.1 Location, Structural, and Physical Protection 153
7.6.2 Perimeter Alarms 153
7.6.3 Invigilation 154
7.6.4 Key Security 154
7.7 Disaster Handling 154
7.7.1 Protecting from Fire 155
7.7.2 Protecting from Floods 155
7.7.3 Protecting from Pests 156
7.7.4 Day-to-Day Maintenance 156
7.8 Managing Dust and Dirt 157
7.9 Pollutants 157
7.10 Conclusion 159
References 159

8 Tools and Methods for Handling and Storage of Museum Textiles 161
Pratikhya Badanayak, Seiko Jose, Ragini Dubey and Ritu Pandey
8.1 Introduction 161
8.2 Care, Maintenance, and Handling of Museum Textiles 162
8.2.1 General Storage Factors 162
8.2.2 General Guideline in Handling 163
8.3 Ideal Conditions, Temperature, Humidity 163
8.4 Storage Units 163
8.5 Storage Materials 164
8.6 Tools Used in Maintenance of Museum Textiles 164
8.6.1 Equipping the Workspace 164
8.6.2 Housekeeping 164
8.6.2.1 Cleaning the Collection and Environment 166
8.6.2.2 Basic and Best Practices for Checking and Monitoring in Museum 166
8.6.3 Materials and Supplies 167
8.6.3.1 Handling 167
8.6.4 Packing and Unpacking 168
8.6.5 Moving 170
8.6.6 Rolling and Unrolling 170
8.7 Labeling 170
8.8 Cleaning 171
8.9 Dealing with Separations 171
8.10 Tools Used for Displaying Museum Textiles 172
8.10.1 Showcases and Galleries 172
8.10.2 Frames 172
8.10.3 Mannequins 173
8.10.4 Hangers 174
8.11 Handling During Transportation 175
8.11.1 By Road 175

8.11.2 By Rail 176
8.11.3 By Sea 176
8.11.4 By Air 176
8.12 Handling Techniques and Conservation Practices of Ancient Textiles in Museums 177
8.12.1 Egyptian Shroud 177
8.12.2 Jordanian Belt 177
8.12.3 Silk Textile 177
8.12.4 Coptic Tapestry 178
8.13 Conclusions 178
References 178

9 Roles and Responsibilities of Museum Professionals **181**
Kanika Sachdeva
9.1 Introduction 181
9.2 History of Museums Professionals Training in India 182
9.3 Roles in a Textile Museum 182
9.3.1 Conservator 185
9.3.2 Conservation Scientist 185
9.3.3 Curator 186
9.3.4 Collections Manager 187
9.3.5 Registrar/Documentalist 187
9.3.6 Historian 188
9.3.7 Exhibition Coordinator/Designer 188
9.3.8 Museum Education Officer 189
9.3.9 Photographer 189
9.3.10 Information Technologist 190
9.3.11 Health and Safety Officer 190
9.3.12 Security Officer 191
9.4 Conclusion 191
References 191

10 Ancient Weaving and Dyeing Techniques **193**
Hannah Dewey, Meghan Lord, Seonyoung Youn, Januka Budhathoki-Uprety and Kavita Mathur
10.1 Introduction to Weaving 193
10.2 Ancient Weaving by Geographical Region 194
10.2.1 In the Middle East and Central Eurasia 194
10.2.2 In Egypt 195
10.2.3 In Greece, Italy, and Romania 196
10.2.4 In India 198
10.2.5 In Southeast Asia and China 199
10.2.6 In The Americas 200
10.3 Conclusion on Weaving Techniques 203
10.4 Introduction to Dyes and Dyeing Technologies 203
10.5 Ancient Dyes, Pigments, and Dyeing Technologies 203
10.5.1 Indigoids (Indigo and Tyrian Purple) 203

10.5.2 Quinonoids (Madder) 204
10.5.3 Carotenoids (Saffron) 204
10.5.4 Flavonoids 205
10.5.5 Dihydropyran (Brazilwood and Logwood) 205
10.5.6 Tannins 205
10.6 Conclusion 205
References 205

11 Armours: Ancient Metallic Textiles **209**
Ritu Pandey, Ragini Dubey, Pintu Pandit, Suruchi Pandey, Mukesh Kumar Sinha and Amarish Dubey
11.1 Introduction 209
11.2 Parts of Armour and Accessories 210
11.2.1 Helmet 210
11.2.2 Coif 210
11.2.3 Ventail 213
11.2.4 Mail 213
11.2.5 Hauberk 213
11.2.6 Gauntlet and Pauldron 213
11.2.7 Sabatons and Greaves 214
11.3 Armour Designs 215
11.4 Armour Materials 215
11.5 Metallic Costume of King Tutankhamen 217
11.6 Conclusion 217
References 218

12 Textile Conservation in India: A Case Series **219**
Deepshikha Kalsi, Elizabeth-Anne Haldane and Lynda Hillyer
12.1 Introduction 219
12.2 Internship Training in Textile Conservation at the V&A 220
12.2.1 Condition Assessment 221
12.2.2 Case Study: Conservation of a Painted and Dyed Cotton Chintz Appliqué Panel 221
12.2.3 Condition Assessment 222
12.2.4 Conservation Treatment and Mounting 222
12.3 Setting Up a Textile Conservation Studio in India 224
12.4 Conservation of an 19th Century Jama 225
12.5 Case Study—Conservation of a Military Frock Coat 227
12.5.1 Historical Context 227
12.5.2 Documentation of Construction and Condition Assessment 228
12.5.3 Conservation Treatment 229
12.5.4 Customizing the Mannequin Mount 232
12.6 Developing Display and Mounting Solutions for Flat Textiles and Costumes for the Special Exhibition PRA-KASHI Silk, Gold and Silver from the City of Lights at the National Museum, New Delhi 232
12.7 Technical Analysis and Documentation 233
12.8 Training and Outreach 234

12.8.1 Case Study—Indian Museum, Kolkata 235
12.8.2 Case Study—The Registry of Sarees, Bangalore 235
12.9 Conclusion 236
Acknowledgments 237
References 237

13 Symbolism and Conservation of Indigenous African Textiles for Museums 239
Raphael Kanyire Seidu, Ebenezer Kofi Howard, Edward Apau and Benjamin Eghan
13.1 Introduction 239
13.2 Types of Indigenous African Textiles 240
13.2.1 African Weave Traditions 240
13.2.1.1 Smock Weaves/Fugu 240
13.2.1.2 Aso-Oke 242
13.2.1.3 Kente 243
13.2.1.4 Kete 244
13.2.1.5 Akwete 245
13.2.1.6 Berber Cloth 246
13.2.1.7 Shuka Cloth 247
13.2.1.8 Kuba Raffia Cloth 247
13.2.2 African Dye Traditions 248
13.2.2.1 Adire 248
13.2.2.2 Ukara 250
13.2.2.3 Mud Cloth 251
13.2.3 African Print Traditions 253
13.2.3.1 Adinkra Cloth 253
13.2.3.2 Kanga Cloth 255
13.2.3.3 Shweshwe 256
13.2.3.4 Ankara or African Wax Prints (West Africa) or Kitenge (East Africa) 256
13.2.4 Other African Traditions 257
13.2.4.1 Bark Cloth 257
13.2.4.2 Fon Appliqué Cloth 258
13.3 Indigenous African Textiles Techniques 259
13.4 Museums in African 259
13.4.1 Challenges of Museums in Africa 260
13.4.2 Contribution of Technology for African Museums 260
13.5 Conclusion 261
References 261
Appendix (Figure sources) 265

14 Conservation of Textile Immemorial: The Fading Past of Uttarakhand Museums 267
Pooja Singh and Alka Goel
14.1 Introduction 267
14.2 Materials and Methods 269

14.2.1 Selection of Locale 269
14.2.2 Tool Preparation and Data Collection 269
14.2.3 Data Collection 269
14.2.4 Statistical Analysis of the Data 270
14.2.4.1 Weighted Mean Score 270
14.3 Results and Discussion 270
14.3.1 General Information About the Museums 270
14.3.1.1 The Number of People Who Work at the Museums that Have Been Chosen 270
14.3.1.2 Conservation Laboratories 271
14.3.1.3 Acquisition of Textile Antiquities 272
12.3.1.4 Ageing of Textile Articles Placed in Different Museums 274
14.3.1.5 Air Circulation Facilities in Museums 274
14.3.1.6 Protective Measures Used to Protect the Windows/ Ventilators From Sunlight and Dust 275
14.3.1.7 Methods Used for Identification of Fibers 276
14.3.1.8 The Details of Temperature and Relative Humidity Ranges in a Variety of Museums 276
14.3.2 Types of Display Techniques Used for Textile Antiquities 276
14.3.2.1 Labeling Methods Carried Out for the Displayed Artifacts 279
14.3.3 Storage Equipments Used in Selected Museums 280
14.3.4 The Collection of Textile Artifacts Collections in Various Museums of Uttarakhand 281
14.3.4.1 Details of Stored Textile Materials 281
14.3.4.2 Govind Ballabh Pant Museum, Almora 281
14.3.4.3 Tribal Museum, Munsyari 282
14.3.4.4 Kumaon Regiment Museum, Ranikhet 282
14.3.4.5 Lok Sangrah, Folk Culture Museum, Bhimtal 283
14.3.4.6 Jim Corbett Museum 283
14.3.4.7 Gurney House Museum, Nainital 284
14.3.5 Various Methods of Prevention Used in Various Selected Museums 284
14.3.5.1 Covering Materials Used for Various Artifacts Displayed in Selected Museums 285
14.3.5.2 Special Kind of Lighting System in the Museum to Protect the Textiles/Garments From Fading/Ageing 286
14.3.5.3 Touching on Museum Antiquities 287
14.3.6 Conservation Techniques Used in the Museum 287
14.3.6.1 Pretreatments Given to Textile Antiquities and Display Boards 287
14.3.6.2 Methods of Reinforcing the Deteriorated Textile Antiquities 288
14.3.6.3 Backing Material Used in Conservation of Museum Textiles 288

14.4 Conclusion 289
References 289

15 The Conservation and Display of Indian Textiles at the Victoria and Albert Museum 291
Elizabeth-Anne Haldane, Lynda Hillyer and Deepshikha Kalsi
15.1 Introduction to the V&A and the Indian Textile Collections 291
15.2 Care of Collections 294
15.3 Conservation 295
15.3.1 Principles of Conservation 295
15.3.2 Assessing Condition, Causes of Deterioration 295
15.3.3 Preventive Conservation 297
15.3.4 Understanding the Object—Context and Scientific Investigation 297
15.4 Object Treatment 300
15.4.1 Object Treatment—Cleaning 300
15.4.2 Surface Cleaning and Humidification 301
15.4.3 Wet Cleaning 303
15.4.4 Solvent Cleaning 305
15.4.5 Stabilization and Support 306
15.5 Display 308
15.6 Conclusion 312
Acknowledgments 312
References 313

16 Between Science and Art: Activities of the Natural Dyeing Laboratory 315
Katarzyna Schmidt-Przewoźna
16.1 Introduction 315
16.2 Promotion of Antique Dyes, Pigments, and Prints 320
16.2.1 Projects 320
16.2.2 Workshop and Exhibitions 321
16.2.3 Color Catalog of Ancient Dye and Its Reproduction 323
16.2.4 Reconstruction of Ancient Dyeing Techniques 323
16.3 Analysis of Antique Polish Kontush Sash Dyeing Material: A Case Study 324
16.4 Conclusion 325
Acknowledgment 326
References 326

17 Visitor Interactions and Museum Textiles 327
Kanika Sachdeva and Divya Singhal Gupta
17.1 Introduction 327
17.2 Textile Exhibitions—Challenges in Display 328
17.2.1 Display Method 328
17.2.1.1 Open Display or Display Cases 329
17.2.1.2 Display Design 330
17.2.1.3 Points to be Considered While Planning a Textile Display 332

17.2.1.4 Level of Interaction Between the Visitors and the Objects on Display 333
17.2.2 Display Lighting 334
17.2.2.1 Hacks for Appropriate Lighting of Textile Exhibitions in Museums 335
17.3 Exhibition Protocols Followed by the Museum 335
17.3.1 Safety Guidelines—Visitor Safety, Conduct and Access 335
17.3.2 Safety of Museum Artifacts 336
17.3.3 Let Us Look at Some Examples of the Protocols Followed by the Museums and the Changes that Have Taken Place After the Pandemic 336
17.4 Photography and Memorabilia 336
17.5 Access Guidelines for Museum Storage 337
17.6 An Ideal Textile Exhibition 338
17.6.1 Case Study 1 338
17.6.2 Case Study 2 338
17.7 Conclusion 339
References 339

18 Educational Value of Clothing and Textile Museums 341
Sara Marcketti and Jennifer Gordon
18.1 Introduction 341
18.2 Importance of Conservation in Textiles and Clothing Collections 342
18.3 Frameworks for Material Culture Analysis in the Learning Process 344
18.4 The Value of Collections to Students' Education 344
18.4.1 The Collection at Iowa State University 345
18.5 Taxonomy of Significant Learning and Collections 345
18.5.1 Foundational Knowledge 346
18.5.2 Application 347
18.5.3 Integration 348
18.5.4 Human Dimensions 349
18.5.5 Caring 350
18.5.6 Learning How to Learn 351
18.6 Conclusion 352
References 352

19 Career in Textile Museum 355
Maanasaa Sethuraman, Suruchi Pandey and Ritu Pandey
19.1 Introduction 355
19.2 Sources of Textile Museum Collections 356
19.3 Scope of Careers in Textile Museum 358
19.3.1 Job Opportunities 358
19.3.1.1 Public Sector 359
19.3.1.2 Private Sector 361
19.3.1.3 Opportunities Offshores 362
19.3.2 Changing Hiring Trends 362

19.3.2.1 Work-Life Balance in Careers in Museum Textile 363
19.3.2.2 Job Description 363
19.4 Glimpses of Work in Progress on Museum Textile 366
19.5 Sourcing for Talent at Textile Museums 369
19.5.1 Private Job Sites 369
19.5.2 Museum Websites 370
19.5.3 Consultant Hiring 371
19.5.4 Social Media 371
19.5.5 Newspaper Advertisement 371
19.5.6 Word of Mouth 372
19.5.7 Campus Hiring 372
19.6 Educational Opportunities 374
19.6.1 School or Pre-University Level 374
19.6.2 University Level (Under Graduation) 374
19.6.3 Postgraduation 374
19.7 Sample Organization Structure 375
19.8 Limitations and Challenges in the Field of Textile Museum 375
19.9 Conclusion 375
Acknowledgment 379
References 379

Index **383**

Preface

Textile museums showcase textiles and costumes of our past civilization and history. The Museum's collection plays a significant part in connecting the historical context of ancient costumes. Textile museums open the records of textile raw materials, weaving techniques and tools used, dyes, pigments, and textile artifacts' texture.

Evaluation of textile fibers, dyes, and pigments through carbon dating and advanced chemical characterization is covered in detail in Volume II of this book. The current book (Volume I) focuses on textile museums, cultural heritage, artifacts, responsibilities of museum professionals, career opportunities in this field, and most importantly, conservation of antique textiles in museums. The book addresses the proper display, storage, upkeep, handling, and conservation technology of textile artifacts to ensure its safety for coming generations.

The book aims to promote textile museum education and meet the needs of textile students, professionals, and museum curators. The incorporation of several case studies to restore the lost glory of textiles is well elucidated. This will appeal to the textile conservators and the practicing curators.

Comprising 19 chapters, this book is a unique repository of theoretical and practical details of museum practices. Chapters on textile museums, importance, cultural heritage, conservation, and documentation of textiles are covered in-depth. Conservation case studies and examples have been highlighted in many chapters. Management practices and guidelines to pursue a career in the museum textile field have been given due attention. The respective authors of the various chapters of this book are all of international repute as researchers, academicians, conservators, and curators in this field. They have done justice to their topics.

We are very grateful to all the supporting institutions: Central Sheep and Wool Research Institute, Indian Council of Agricultural Research, Govt. of India, Avikanagar, Rajasthan, India, Mahatma Gandhi University, Kottayam, Kerala, India, National Institute of Fashion Technology, Ministry of Textiles, Govt, of India, Mithapur Farms, Patna, India, Department of Textiles & Clothing, Chandra Shekhar Azad University of Agriculture and Technology, Kanpur, India, and Chitkara Design School (CDS), Chitkara University, Punjab, India. Thank you for providing us all the support for conceptualizing and producing the entire book. We like to thank our family members and all well-wishers for standing with us all the time, as always. At the end, all the thanks to Almighty God for directing us with his omnipresence, power, and positive force.

Editors
Seiko Jose
Sabu Thomas
Pintu Pandit
Ritu Pandey
Vandana Gupta
September 2022

1

Textile Museums: Its Importance, Objectives and Functions

Vandana Gupta[1]* and Mangesh S. Manekar[2]

[1]Chitkara Design School, Chitkara University, Rajpura, Punjab, India
[2]Parul Institute of Fine Arts, Parul University, Vadodara, Gujarat, India

Abstract
The word museum is derived from the Latin word muses, meaning "a source of inspiration or to be absorbed in one's thought." Museums are institutions that bring past to life and hold the greatest artistic treasures of different periods and cultural history. The collection and restoration of artifacts and heirlooms of historic, artistic, and scientific importance is part of museum activities. Along with this, unique objects are made available for public viewing by organizing permanent and temporary exhibitions. Looking at the definition of museum, it has evolved with development in the society. Museums around the world have always been revered as store houses of information. Its space is used to conserve and display objects, materials of different time and era, which otherwise is fading from the lives of individuals. Both government funded and privately owned museums are established by each state, which are open for public viewing. These museums acquaint masses with the history and in turn contribute to their decision making process. Each museum has its own role to play, such as textile museums contribute in determining the trends of the past and establishing new trends on the basis of current research. This chapter will take you to a journey from museum establishment to their role in society, emphasizing textile museums.

Keywords: History, textile museum, collective, connective, educative, economic

1.1 Introduction

There is a tendency in mankind to collect materials for their personal use or store them for their next generation to acquaint them with their ancestral civilization and culture. It is taught in many design institutes that new discoveries and inventions can take place by understanding and digging into the past. The poet Thomas Stearns Eliot helps us reframe these ideas. He reminds us that tradition cannot be possessed; it must be reinvented and rediscovered by each generation [1]. Museums are used to retain objects from a certain age and period that are likely to disappear in the future. Objects from antique vehicles, textile, or extinct species can still be found and preserved for many observations and inspection.

**Corresponding author*: vandana.g178@gmail.com

Seiko Jose, Sabu Thomas, Pintu Pandit, Ritu Pandey and Vandana Gupta (eds.) Handbook of Museum Textiles: Volume I Conservation and Cultural Research, (1–18) © 2023 Scrivener Publishing LLC

Museums serve to keep history alive for certain period and ages. They encourage the public to familiarize themselves with their past, which essentially encourages them to make future choices. In order to assess the clothing and fashion patterns of recent years, textile museums play a significant role. Museums are a source of inspiration for fashion designers, who in the contemporary era are still searching for ancient types of art that might generate wonders to current time. Textile museums are store house of textiles of historic importance and also play a major role in bringing the small unit artisans; their skills and development process among masses, apart from sharing about high profile designers. They contribute to the country's overall economy, especially in the case of emerging economies. To understand the importance of museums, it is important to explore the reason of their existence, as well as how these museums are divided into different classification according to the artifacts. This chapter provides understanding about the importance of museum existence, emphasizing on the role of textile museums in our society.

1.2 Museum Definition and Its Existence

The word museum is derived from the Latin word muses, meaning 'a source of inspiration or to be absorbed in one's thought'. Museums are institutions that bring past to life and hold the greatest artistic treasures of different periods and cultural history. They collect and care for objects of scientific, artistic and historic importance making them available for public viewing through permanent and temporary exhibits. The definition of museum has evolved with development in society and is shared by International Council of Museums (1946-2007). In 1946 the word "museums" was defined as "all collections open to the public, of artistic, technical, scientific, historical or archaeological material, including zoos and botanical gardens, but excluding libraries, except in so far as they maintain permanent exhibition rooms" [2]. In 2001, museum was explained as a nonprofit making, permanent institution in the service of society, its development, made open to the public, acquires, conserves, researches, communicates, exhibits for purposes of study, education, enjoyment, material evidence of people and their environment. In continuation during 2007, museum was explained as a nonprofit, permanent institution in the service of society and its development, open to the public, which acquires, conserves, researches, communicates and exhibits the tangible and intangible heritage of humanity and its environment for the purposes of education, study and enjoyment; emphasizing more on the new role of museums for the betterment of society [3]. To understand how and why museums were established, it is important to know why people started collecting, who visited, and what do we learn from them. Among several reasons of collecting some reasons are given by Akin, 1996 which can be listed as:

a) Collectors wish to connect themselves with history
b) Thrill of the chase i.e.,

- The research involved in finding an object
- Competition with fellow collectors and
- Source of pride in finding a long sought after object are motivators [4].

Further, Alexander *et al.* [5] also stated that collections seems to be instinctive for human beings, and may be based on pursuit of knowledge, connoisseurship or simply obsessive collecting.

One such example is the exquisite and versatile collection of art objects, such as ivory, modern painting, bidri ware, textiles, and many more at the Salar Jung Museum established in 1951 and located on the southern bank of the river Musi in Hyderabad, Telangana state of India depicting the catholicity of Nawab Mir Yusuf Ali Khan, Salar Jung III after whom the museum is named. Most of the museum pieces displayed belonged to his collections, which he collected from well-known antique dealers from all corners of the world [6].

The desire to collect is there in human from the time immortal. Ancient civilizations like Greek, Roman, medieval Europe collected art and antiques as private collections. In 16th century, with developing interest in human and history, specialized collections came into existence in the form of natural history, archeological collections, manuscripts, portraits. During the 17th century, learned societies, such as Academy of Sciences in Paris & Royal

Table 1.1 Different categories of museums.

Museum	Collections it holds
History Museums	Museums housing artifacts and specimens related to natural history. Examples of such museums are Natural History Museum, London and Museum of Natural History, New Delhi.
Biographic Museums	Museum house artifacts belonging to famous historic person. Example of such museum includes National Gandhi Museum, New Delhi.
Art Museums	Museums exhibit permanent collections related to fine arts in the way the artist intend them to be seen. Examples of such museum is Government Museum & Art Gallery, Chandigarh; Baroda Museum & Picture Gallery, Vadodara, Gujarat.
Science Museums	Museum focuses on science and new technology. Example of such museum is Great Lake Science Centre, Ohio.
Transport Museums	Museum hold collections related to any one (aviation, railway etc.) or all of these in the same space. Examples of these museums include "The National Railway Museum" located in Chanakyapuri, New Delhi and "Railway Museum" in Mysore, which focuses on the rail heritage of India. The Transport Museum of New Delhi and Air Force Museum, Shilong are other examples.
Textile & Handicraft Museums	Museum focusing on the collection of antique textiles, costumes and craft. Examples include Calico Museum, Ahmadabad; Victoria and Albert museum, London.
University Museums	University Museums are those museums which are run by university. Examples include Ashmolean Museum at the University of Oxford in England and Ashutosh Museum of Indian Art at Calcutta.

Society in London promoted discussions, experimentation and collections, which led to the development of public museums like the University of Basil (1671), Ashmolean in Oxford (1683), British Museum in London (1759). Also, private collectors wished to ensure continuous study of their collections so they started to bequeath them to the cities they lived in. In this way private collections were moved to public domain. As these museums were for public, visitors were charged a commensurate fee for their time. In the beginning of the 19th century, the development of new ideas took place, which made museums a resource center for educating and enlightening the general public. Thus, museums were given "the role of serving" as one of the society's main political, social, and cultural institutions. During the 20th century, there was an increase in the number of museums and their operations. Also, with the development in technology and society, museums changed themselves to serve better. More use of digital media is part of museum activities starting from conservation to preservation to exhibition. Victoria and Albert museum, London have developed a process to digitize their collection to make it more approachable to the public and to preserve for a longer period. History reveals that although diverse museums were established in different parts of the world, still they have many elements in common, thus making the conception of museum international and universal. Museums have many roles to play and are formed to share facts related to past and present. With time, different museums are established at different parts of the world and can be broadly categorized as given in Table 1.1.

1.3 Textile Museums and Their History

Textile in its basic definition is a product developed by fiber & yarn using different fabrication methods such as weaving, knitting, felting, bonding, crocheting, macramé, tatting, lacing. All these are processes, which can be done through hand and machines. In earlier times, hand techniques and tools were used to create such unique textile products and involve a high level of skills. The invention of machines and tools like fly shuttle (by John Kay -1733), spinning Jenny (by Hargreaves-1733), water frame (by Richard Arkwright-1769) due to industrialization have brought a new era of textile production and consumption. Industrialization was the period of social and economic change that transformed agrarian society to industrial one. Change in production style; from manual to mechanical was also observed during this time, which provided new ways of living. The earliest and most significant transformations of the industrial revolution took place in the woollen and cotton textile industries. They gave rise to new technologies, new working methods, new trade networks, and even new fashion trends. Stories of such historical textile revolutions are part of textile museums, which exhibits history and art of textiles, putting emphasis on manufacturing process of textile industries, by locating themselves in the former factories or establish themselves in the buildings designed for the production of fiber, yarn, fabrics in ancient times. They also talks about different industries/sectors directly or indirectly connected with textiles, such as agriculture and farming of natural fibers; technologies involved in the technical products with unique characteristics, such as clothing and bedding made up of smart materials; decorative art related to fashion, carpets etc. Textile & craft museum talks about the trends of a particular timeframe related to clothing, fashion, textiles of different regions. It also shares the life led by the citizens, stories of their struggle, and how textile was an important part in survival, especially in India. The impact of industrialization worldwide and change in the ancient textile industry due to introduction to textile machinery is all presented by these museums to keep our textile history alive among the present and coming generations. The Museum of Textile Industry, Sliven, Bulgaria is one of its kind and is dedicated to the textile production from the ancient past to present day [7]. Among the three basic needs, textile in the form of clothing was always one of the most important part of human existence as part of modesty, developing relations etc. Its construction was done not only for protection or covering but also huge amount of trading of textiles is part of our glorious history, as from plain Jamdani to lustrous silk is being transported to different parts of the world for emperors. The famous silk route, exclusive embroidery of the Mughal era, and the replacement of natural fibers to synthetics are all important parts of developing basic understanding and knowledge, especially for textile artists and designers. The textile and craft museums are situated in different states and cities all around the globe having so much to share in this context. Through their display, they exhibit the traditional fabrics that are part of our societies for years. The tools used, the techniques employed, and the changes that took place with time are expressed through real objects and literature, all of which are part of museum collections and displays. Prince of Wales Museum, Mumbai provides a good example of textiles displayed in visitor friendly manner, which helps visitors to self-analyze the textile technicalities and learn about the same by going through the write-ups mounted near each textile. To understand the importance of textile museum, the mention of Sabarmati ashram would be appropriate as it links us to the khadi movement by taking us to a journey of production and process of khadi, its relevance in our life. For hand weaving, yarns are prepared using spindles. The museum of Sabarmati ashram has a unique collection of around 32 types of spindles used by textile artists/weavers in earlier centuries. Not only this, the spindles housed in the Sabarmati ashram belong to different states and thus explains the similarity and dissimilarity of the antiques. It is very fascinating to visit various textile museums as the real-time visual information received through the displayed objects clearly explains the similarities and dissimilarities of the crafts of different states and countries. For example, the relationship between the phulkari of Punjab and Haryana can be studied by visiting museums of both the states where displayed costumes and textiles embellished with such embroideries shares their relation before partition.

Textile museums worldwide are visited in nearly all parts of the world by teachers, academicians, scholars, and design enthusiasts. In a specific period, they provide an analysis of the clothing and design patterns of the region. Textile museums have information about the clothing practices and fashion objects used in ancient times, representing people's lifestyles in earlier eras. One such example is "Egyptian Textile Museum," Cairo, Egypt. The museum housed and displays fine collections of textile from the Pharaonic, the Roman, Coptic, and Islamic eras, which clearly explains that the history of Egypt is not only documented on papyrus or stones but beautifully knitted and ornamented on textiles such as bedsheets, mummy covers, cushion cover, etc. [8]. Different textile museums have remained successful in reviving ancient art and crafts to some extent, as these buildings exhibit antique woven sculpture and share the history of the artifacts of historical importance, such evidences are part of Central Museum of Textiles in Lodz, Łódź, Poland. The textile museum exhibitions, such as "Interwoven Globe" organized by The Metropolitan Museum of Art, New York in the year 2013 to 2014 [9] shares the journey of textiles which plays an important part of historical garments, fashion collections, paintings, tapestries other handloom and handicraft objects, which attract visitors with their aesthetic appeal [10]. Their preservation plays an important part in making the antique and historical textiles available for public viewing. Much of the time, conservation strategies rely on the climatic conditions of the place. The archival materials are employed to preserve and conserve textiles, which are unique, fragile, and irreplaceable. The interpretation of the museum collections has expanded toward educational exhibitions, documentary productions, and social media activities, which have increased the presence of antique textile and heirloom worldwide. Thus, the museum represents complete ensembles from various time periods and geographic cultures and the vast array of embellishment techniques. Museums are an important source of inspiration for many designers; it may be fashion, interior, product, textile. These designers are on lookout for authentic and ancient art works that could inspire them to create innovative and creative designs that are suitable in this modern world. Recently, a report shared the revival of an old art form from Laos called Laotian art, which mostly deals with animals, birds, and reptiles. This art form had become extinct after the civil war in Laos, and recently came under the notice of the American weavers who are planning to revive it. Such stories suggest that museums leave room for the revival of an ancient art form and making them a fashion statement of a particular period by preserving their remnants [11].

It is also important to note that every continent is well known for its collection stored in one or the other museum related to textiles and some examples that can be quoted in this context includes the American Textile Museum, US depicting American heritage through textile. The Textile Museum of Canada, dedicated to collection, exhibition, and textile research founded in 1975 by Max Allan and Simon. In 1985, a special textile museum "National Textile Museum, Malaysia was established which stands out for its unique textile collections. The finest woven, spun, painted and printed textiles are part of the Calico Museum founded in 1949 in Ahmedabad, Gujarat and wide range of old wooden looms, spinning wheels are part of collection of Japanese textile museum "Yukari." Textile museums also collaborates with fashion & textile designers all around the globe to showcase their talents, innovations, creativity in the premises of museums. One such example is The Fashion and Textile Museum, London, England [12]. All the above mentioned museums have their own unique collection but the basic approach is to share the knowledge of textile with society and preserve the ancient textile art.

1.4 Importance of Textile Museums

To understand the actual importance of museums and why they matter, let us discuss four values of museums for an individual and society.

1.4.1 Connective Value

Museums play an important role in providing an opportunity to visitors, staff and others connected directly or indirectly to enjoy; experience and understand the past, define the present, and get educated for the future. Museum exhibits stimulate thoughts and facilitates debate thus providing a platform for thinking rationally and expressing with freedom. They inspire, enthral, and enlighten visitors by allowing them to move toward discoveries and innovations. They initiate discussions regarding social issues, current affairs, political issues, which not only attract regular visitors but also those who otherwise do not visit museums. With its partnership with different sectors of society like private, public, educational institutions, professionals, and amateur artist, museums provide a space to amalgamate thoughts, skills, and ideas. National Handloom and Handicraft Museum, New Delhi brings Indian traditional art and craft close to the society by displaying the unique and ancient textiles, sculptures, paintings, etc. Also, it connects us to the ancient old procedure of craft making by showcasing the work of artisans and allowing them to interact with the visitors explaining to them the intricacy of the developed product.

Collections are a launching point for discussions which communities wrestle with and get inspired. In the museum space, one can hear multiple perspectives (visitors view about art collections) of the world that emerge from experiencing collection related to art and craft and the meaning that the collections spark for individuals. Government Museum and Art Gallery, Chandigarh displays unique art and artifacts as permanent collections, hold exhibitions, provide platform to interact with great artist, historians and people who are directly or indirectly connected with art. Development of skills and critical thinking through such spaces helps in improving life. Moreover, museum exploration provides the space that makes visitors understand diversified thoughts, embrace them with the silence and patience to listen to others, and create space for empathy, energized by the power of art, are some of the greatest opportunities museums can offer to audiences. In our daily work, we must go beyond the walls of our buildings to open up and respect exchange of ideas. Also, development of deeper understanding & diverse perspective can be seen in different individuals, groups and organizations having cultural differences due to social, political, geographical situations [13]. A writer & philosopher "George Santanya", said, "Those who do not learn history are doomed to repeat it." Museums drive people to evoke change at all levels of their life by sharing powerful stories of tragedy and overcoming adversity. Don Bosco Museum, Shilong is one extensively designed museum where the collections exhibited are worth watching as museum showcases artifacts related of art, history, life, culture, and craft of different tribes of all the seven sister states (Assam, Meghalaya, Tripura, Mizoram, Manipur, Arunachal Pradesh, Nagaland) of India. Looking into the life of North Eastern tribes of India, unique tradition and culture is expressed through costumes, ornaments, dance and other rituals, which have become the means of their identity. Almost each tribe has its own distinctive shawl, which is woven by the womenfolk, and in which they take great pride. Each shawl expresses their story showcased in different designs and colors. [14]. Tribal museum known as VAACHA- Museum of voice provided gamut of information about the tribes of Gujarat. It is an important part of Adivasi academy located in Tejgadh (90 km east of Vadodara). The Vaacha Museum, Gujarat is a house for various traditional artwork which involves Rogan Art of Kutch, Bamboo craft of South Gujarat, leather puppetry of Maharashtra, Jute weaving and many more [15]. The museums serve visitors with programs that resonate with their interests while maintaining quality and uniqueness. The result is an environment of trust and unity in which society can thrive. Such interaction establishes dialogue with communities, thus building relationships and generating bonds.

1.4.2 Collective Value

Museums exist due to its collections. These collections form the core elements for all the museum activities such as display, exhibition, education, research etc. The items collected by museums are essential and have value in our lives irrespective of their local, regional, national, and international significance. Collections incorporate and express our shared history & identity thus form the core value through which other values are created. The Pohon Budi Gallery of National Textile Museum, Malaysia exhibits the origin of textiles from pre-historic times along with the tools and traditional techniques and raw materials involved in textile making. It also show cases country's predominant textiles like Songket, Kainpis, Rungus, and many more [16]. Museums cannot collect and store everything therefore objects of certain category and quality are accepted into the collection. Considering the appropriateness of the artifact for a given area of interest of that museum, the artifact is selected and added to the collection. Accessioning is a process through which an object is formally included in the collection by giving a unique accession number.

Accessioning procedure is different in different museums, but in most cases it begins with:

a. Donation: A person wants to donate his or her artifact to a museum.
b. Or an object is purchased or traded on the recommendation of curator.

Museum collections were cataloged; traditionally in a card index. There is a change in the specification in the cards with time. Older cards included very few fields (description, purchased, gifted, exchanged or deposition or storage location) whereas new cards have all the above fields including an image in black and white color confirming the object's appearance. Most museums collect artifacts and heirloom according to a specific area of specialization. For e.g. textile and handicraft museum will collect and display textile of different cultures and states. One such museum is located in Ahmedabad, Gujarat in western India, known as "Calico Museum". It is an indeed one of the foremost textile museum of India, housing remarkable, distinguished and comprehensive collections of fabric spanning varied and remote to India, exemplifying handicraft textiles across five centuries. The collection includes Mughal textiles of 15th to 19th centuries, regional embroideries, tie & dye and religious textiles. This museum has become as one of the major reference areas for our surviving handicraftsman, Indian handicraft and Indian textiles machines. This popularity can be related to the fact that Ahmadabad city is one of the major trading centers of textiles since fifteenth century. The publications related to historical studies along with scientific examinations of textile processes such as dyeing, weaving and block printing are part of this museum. Other artifacts includes temple hanging, miniature paintings, Jain art, sculpture etc. [17]. Similarly collections of Victoria & Albert Museum, London spanning four centuries is a storehouse of an exhaustive collection of dresses collected from all around the world. The collection includes rare pieces of 17th century gowns, 18th century 'mantua dresses, 1930s evening wear, 1960s day wear and post war couture; along with the pieces from 21st century designers [18]. Another example is the Himachal State Museum, Shimla exhibiting the traditional costumes and jewellery of Himachal Pradesh and the textile materials used as raw material. The collection of the museum explains the changes took place in jewellery (Figure 1.1) such as silver was changed to gold, heavy jewellery and ornamentation items for ear, nose, wrist, ankle were made lightweight. The traditional costumes in Figure 1.2 also explains the cultural and environmental conditions of the area [19].

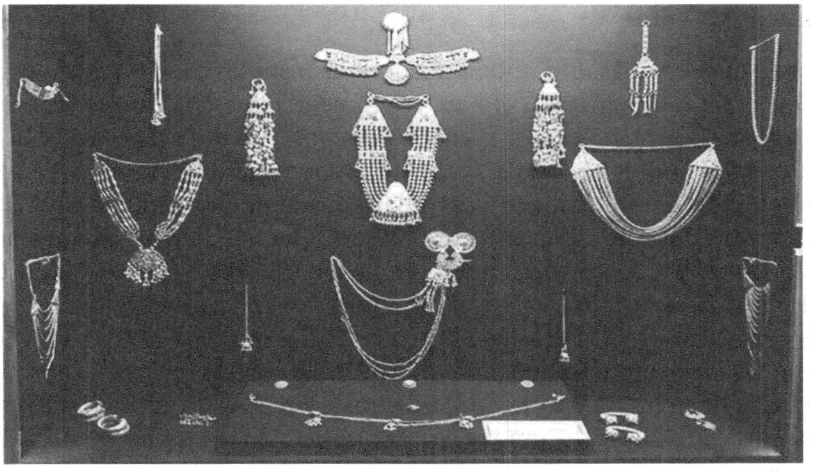

Figure 1.1 Traditional Jewellery of Himachal showcased in Himachal State Museum, Shimla. Source: https://www.himachalstatemuseums.in/jewellery-5/.

Figure 1.2 Traditional **Costume** of Himachal Pradesh displayed in Himachal State Museum, Shimla. Source: https://www.himachalstatemuseums.in/jewellery-5/.

1.4.3 Educative Value

Museums are considered as educational institution from last few decades. It can be said that whosoever visit museums either for entertainment, research, learning, experience (aesthetic, inspirational, recreational, interactive) are exposed to didactive data (videos, conversation with officials and educators). According to the American Association of Museums (AAM) "museums exist permanently essentially for aesthetic and educational purpose." Talking about museum collections, they are eclectic. One can find art and artifacts ranging from household appliances to visual art, from history to new technology. Museums, such as Albert Hall Museum, Jaipur, is a treasure house of collection related to subject matters namely art, jewellery, garments, textiles, arms and armor, miniature paintings, musical instruments, clay art, sculpture coins, etc. Along with the objects a library and a bookshop is an integral part of maximum number of museums. Here, book lovers, educators, researchers, and students can explore the unique and rare collection of otherwise unavailable books in their surrounding market.

Looking more into the role of museums in education, collections have their importance in academic and scientific research which are shared with visitors in the form of exhibitions, presentations, and talks. One such example is San Jose Museum of Quilts & Textiles, US who conducts weekly textile talks on different textile- and art-related topics. Silk Museum, Bsous, managed by Association Memory and Development was created by George and Alexandra Asseily is now considered as one of the important reference related to ecological, cultural and economic history. Finally, in 2001, the old silk factory was transformed into "Silk Museum," restoring its patrimonial identity and various functions. It acquaints visitors about silk rearing also called sericulture, shares the history of silk route and the ancient exchanges of silk which took place between the land of the Cedar and the East and the West [20].

When it comes to visitor's responses, people cite "Improvement in general knowledge and positive learning experience in such informal environment" as major reasons to visit a museum. A survey conducted by Netherland Museum Association on "The effect of museum education" revealed that museum education gives pupils a sense of pride, enjoyment along with better learning experience. The effect of educative values of museum indicated by teachers in the decreasing order of their percentage includes improved knowledge about people and culture, more imagination and originality, improved collaborative skills/ expressive ability, more positive self-image, and link to our culture [21]. While comparing the knowledge shared by museum and books, it is important to note that the information on different topics and objects is indeed available in books and other educational publications but the impact of such information is felt and lived in museums. Improvement in cognitive development and experiential learning opportunities are seen as essential contributions of museums in the educational sector. Living and Learning Design Centre of Gujarat is much more than a museum dedicated to the crafts of Kutch and working as an academic and resource center to uplift craftsmen and traditional craft by preserving, revitalizing, and promoting the glorious craft heritage of Kutch. Apart from displays, the craftsmen are practicing the craft of weaving, block printing, pottery, metal work, etc. thus providing the visitors experiential learning experience [22]. Tehran Carpet Museum of Iran was created to keep the art of carpet weaving alive and to gather the unique pieces of carpet from all over the country. This museum is a place for antique art and weaving lovers, historians, and researchers. One can learn about the Iran's iconic weaving centers, such as Kerman and Tabriz. Another important feature of the museum is that it provides experiential learning to the visitors as the weavers sitting at a loom demonstrate making of carpet and showcasing the tools, natural rug dye materials, and the other raw materials used [23].

Compared to the structural and rigid method of regular education, the learning experience of consciously/unconsciously gathering and acquiring knowledge is a valuable way to spend time in museums as it helps in the holistic development. The collections stored in museums share the stories about all those who were once part of our society, and these stories help us to understand our world as we know it today [24]. Behind the scenes, necessary research is conducted by museums on the objects collected and displayed to provide the authentic information to the visitors. Collaborative work in this context is performed by partnering with outside entities, such as universities, organizations, associations so that researchers, graduate students, and professors get the opportunity to take advantage of museum unique collections. The Grado Zero Escape Innovation of "OricalcoFabric" by the Grado Zero Escape in 2001 uncovered massive research potential of shape memory material. The innovation in the form of a first shape memory shirt made by NiTi Alloys is now part of the Chicago Museum of Science and Industry [25]. In this sense, the museum has started working as service facilitators instead of just being the collector of antique objects. Hence, modern museums are settings in which research toward the preservation of cultural heritage is undertaken on one hand and training, publications, knowledge-document centers, performances, meetings, discussions and workshops are provided on the other [26]. Education is defined as the process of experience, generally called learning, which brings in desirable changes in human behavior concerning knowledge, outstanding skill, and attitude [27]. Therefore, the role of museums in education can be summarized as environment for learning and inspiration; multifaceted approach; people from different backgrounds come at the same stage and see each other's world; visitors grow in confidence and help society to understand the world and places as well as culture within it, young people get an opportunity to learn and see things in practical and older people share their past with others, bringing out the reason, ways, and innovation that took place in their life journey. Fuller Craft Museum, USA has grown with time and has provided different learning opportunities to the visitors and people. The museum first started as Brockton Art Center-Fuller Memorial for lectures and exhibitions, with no collections and eventually changing to Fuller Craft Museum which focuses solely on contemporary craft collections of objects made with the help of wood, fiber, etc., by the makers. The museum not only educates the visitors about the objects housed in it but also inspires them to connect with the minds and methods of the makers of the craft [28].

1.4.4 Economic Value

The museum sector is a place of vibrant and exciting elements of cultural displays, busy schedules of events and meeting and engaging visitors. Creative and cultural industries are making their place in this 21st century due to more awareness about the importance of sustainability and skill. Museums and galleries deliver impressive economic benefits through increased tourism and ever-growing public appetite for culture [29].

Accumulation, management, and showcasing cultural heritage and information communication in today's digital age requires skilled people and volunteers. The scope of jobs is tremendous in museum because of their diversified category and objects stored/displayed in museum. From a craft person to an architecture and a textile designer to a restorer, all are required in museum development. The traditional methods and new technologies, such as digitization, have become important to restore the information and its dissemination to the masses in the required format. Along with the understanding of different roles of museum in context to connective, collective, educational values, museums are also considered an essential sector having great impact on the national economy by generating GDP and stimulating jobs [30]. Looking into the statistics, a report on "Museums as economic engines: A national Report 2017" by The Andrew Mellon Foundation reveals that US museum itself has been visited by US society around 850 million times and to meet such demand, museums provide each year direct & indirect employment by providing 372,100 jobs and generates income of approximately $15.9 billion [31].

Apart from the conservation of antiques and their display, museums also provide other important services taken care by the different museum staff members, such as specialized training, research-oriented collaborative activities and promote events as part of the museum promotions. The museum community consist of people who work as full time, part time, temporary or voluntarily and help in social, intellectual and personal development of museum in context of serving society. Jobs in some museums require specialized training, while others need basic education and common sense. Different areas, such as exhibitions, education, and public events require teamwork with internal, as well as external staff. Different departments—curators, information technology, finance, human resource, educators, registrar, marketing, media, conservators, and librarians form an important part of the museum and have a role to play.

a. Museum exhibition officer: Museum education officer is responsible for all types of exhibitions. It involves planning, organizing, administering and producing temporary, permanent, or touring exhibitions.
b. The job of curator is to build up museum collections, often in specialized areas keeping in mind the museum's storage and display capacity requirements.
c. Museum educators inspire people/visitors by explaining, educating, conveying information about the museum collections by adopting different means, such as tours, lectures, seminars, workshops, publications. Thus, bringing the collections closer to wider audience.
d. Conservator: A conservator's role is to care, conserve, and restore artifacts for collections, which require understanding and application of scientific methods. Conservation includes applying knowledge about physical and chemical properties related to the objects and storage materials. They work on different research projects and manage the laboratories of the museum.
e. Museum attendants: Security of the collected and exhibited artifacts, heirloom and visitors is the core responsibility of the museum attendants. They sometimes engage with the public/visitors in clarifying their doubts, providing, and demonstrating different crafts.

In many cases, museums are considered an important part of the tourism industry due to tourist engagement and holding importance in all-round development. Museums are considered as part of the tourist travelling plan, and for many, the trip is incomplete without the museum exploration. The economic impact of museums in home localities is immense as cultural destinations [32]. It also largely depends on the type of museum buildings, exhibition it hold and its presence on different digital platforms. Blue House in Mexico city is one such example which is organized and converted into a Museum "Frida Khalo Museum" by Diego Rivera as the final wish of Frida Khalo. She was a renowned Mexican artist and known for her artwork, elaborated costumes, and hairstyle. Popular art pieces and remarkable paintings by both the artist and the pre-Columbian sculptures, photographs, and beautiful garden that give access to the exhibition of Frida's dresses are all reasons for tourist attraction [33]. The economic importance of the museum is directly related to the money spent by the visitors. Studies in this relation reveal that museums generate multiple sales of its own through services, like admission fee, souvenir shops, and cafeteria. On the other side, due to museum attraction, visitors spend on accommodation, travel, food & beverages and in the local market on goods and services, and by economic activity—especially in recreation and hospitality sectors.

The presence of museums in a city attracts visitors/travelers, which plays an instrumental role in increasing the local economy by supplying a passing trade and offering local people employment. Guggenheim Museum Bilbao is one such museum, which served as a test case to attract visitors to nontourist destinations. A study was conducted to evaluate the number of visitors visiting Guggenheim Museum Bilbao, which was developed in a nontourist spot. There was an increase in the footfall of foreign visitors by 44% due to development of the museum in 2000 [31]. National Handloom and Handicraft Museum, New Delhi, is another important example of economic value. This museum is one of the well-known Handicraft Museum of India holding more than 35,000 rare and distinctive pieces related to embroidery, painting textiles, and clay craft. Apart from permanent displays, they invite number of traditional craftsman related to crafts, such as Gond Painting, Pattachitra painting, toy making through fibers (Figures 1.3 and 1.4), from all over the country to be in residence and

Figure 1.3 Display of Gond painting of Bhopal & Craft men showing Pattachitra painting of Orissa (National Handloom and Handicraft museum, New Delhi).

Figure 1.4 Display of coir toys of Orissa at National Handloom and Handicraft Museum, New Delhi.

demonstrate and display their craft. Thus, attracting and educating school children, art students, artists and designers all over the world through which crafts men also get new market opportunities and bring their traditional craft close to a diversified population. An article shared the influence of Gond art/painting on designers and how these designers have brought forth Gond art and amalgamation in wearable easy comfortable silhouettes. Designers take such inspirations during their visits to such museums where the craftsmen and artist are displaying and performing the art [34]. An article in The Indian Express, revealed that VIP's such as, First Lady Michelle Obama did some pre-Christmas shopping at craft bazaar in the National Handicrafts and Handloom Museum on November 9, 2010, Monday [35].

1.5 Objectives of Textile Museum

a. To create a place that welcomes all the visitors across the globe, race, age, ethnicity, abilities, sexual orientation, gender identity, gender expression, political affiliation, religion, geography, or culture.
b. Create a culture of providing human experience through textiles and to connect people of varied cultures and eras.
c. Organizing exhibitions and programs related to textile and textile art.
d. To ignite, inspire, and initiate the conversations about global collections of textiles and contemporary art practices.
e. In the spirit of respect, peace, and friendship, museums create space for audience and visitors to share textile practices of present and past.
f. To provide a platform to local, national, and international textile artist and craftsmen to showcase their work and engage with visitors.
g. To work as educational institution through diversified exhibitions and displays and find innovative approach to provide educational programs to the visitors both online and offline.

1.6 Function of Textile Museum

Majority of visitors visiting the museums have feeling that museums are meant to display artifacts. But as discussed about the meaning and definition of museums, they have many functions apart from just displaying and exhibiting. Some of the functions of museums are listed below are interrelated and connected to each other.

a. **Creating diversity in one roof:** The first and the most primary function of museum is to collect the objects that have historic, scientific, and artistic value. No museum can survive without collection. The collection takes place through various modes, which includes purchase, gift, loan, exchange, exploration, excavation, Treasure Trove Act, bequest, etc. Each museum has their own collection policy, which depends on the vision, mission and objectives of the museums and also the need for the kind of information a museum wants to share with the society. The collected objects are documented, which is the main administrative function of many museums and involves maintaining the record of each object with its photograph. The information documented facilitates in understanding the composition, provenance, and necessary treatments required to preserve the textiles.
b. **Conservation & Preservation of Textiles:** Second important function is to conserve and preserve the collected objects through scientific methods. The textile materials are organic in nature they need to be protected from spontaneous change in the environmental factors, like light, dust, and dirt. Observation, analysis, and use of preservation techniques and skills of the conservator plays a crucial role in conservation and preservation. Regular care is crucial both in preventive and curative conservation.
c. **Storage & labeling of Textiles:** Museums according to the collected artifacts develop a plan to store objects, keeping into consideration the construction and quality of textile material with which the artifact is prepared. This involves the use of archival materials, proper furniture and space. While storing the labeling of objects is done to identify and locate the stored artifacts with minimum handling. This function helps museums to display the heirloom/artifacts/antique textile in rotation for visitors.
d. **Research and Scientific study:** Every museum needs to conduct research to provide the authentic information to its visitors and audience. The research done by museums is done for documentation, for museum publication, to follow required conservation techniques, dissemination of correct and valuable information of the collected objects in museums and to facilitate the educators, students, research scholars through an informal environment.
e. **Exhibitions and interpretations:** Display and exhibitions are core area of museums ensuring ongoing access to the historic and cultural information, entertainment and the aesthetic pleasure to the audience/visitors. The interior design or exhibition design is an important part that needs to be taken into consideration while handling the antique objects for display. The museums plan a theme and use different display aids to make it presentable thus increasing their community reach.
f. **Collections Stewardship:** The museums are the store house of antique objects which are priceless and it is the utmost responsibility of museum to maintain a strong security system to avoid vandalism, theft, fire, etc.

New technologies, like alarms, CCTV cameras, are used by the museums and are in control of experienced security force.

g. **Public services and collaborative activities:** Various activities are organized by museums for specific target groups or to provide the knowledge about new technologies or historic object along with the static exhibitions. Showcasing documentaries, designers work, and traditional folk art is another collaborative work done by museums in the service of society.

1.7 Textile Museum and Their Future Prospects

The textile industry is growing many folds in terms of technology, innovative, and smart materials and bringing closer the textile craft and other art mediums through digital platform. With the introduction of nanotechnology, the textile materials are developed to provide numerous functionalities in terms of comfort, aesthetics, and many more. Such advancements and technologies have become part of textile museums, where the museums are using them to conserve and preserve the antique objects. The high-end instruments and tools, such as SEM microscope, archival materials, new display techniques, use of digitization for the storage of old, and fragile artifacts and use to AR, VR technology to provide a real-time experience to the visitors are shaping the museums of future. With more and more involvement of public and private sectors with museums, the future society will appreciate the presence of such museums as an essential part of their development.

1.8 Conclusion

Books will always provide required information, but museums will help us understand it and develop the required knowledge through visuals presented in displays. Museums with its different values build a better future and create a broader perspective toward life by connecting us with past and other cultures. The struggle of textile designers, craftsman, and other artist is explained as stories through the tangible heritage in museums. In the coming chapters, we will be exploring different aspects of textile museum starting from the type of unique collections stored, exhibited, and conserved. Also, discussions on advancements in the textile & craft techniques and initiatives museums have taken to reach more textile enthusiast, will be part of this book. The chapters will emphasize on textile, craft, and costume collections of the museums. Discussion on unique textile artifacts will provide an exhaustive understanding of their developmental techniques to the readers.

References

1. Kramer, J., TS eliot's concept of tradition: A revaluation. *New Ger. Crit.*, 6, 20–30, 1975.
2. Salguero, B., Defining the museum: Struggling with a new identity. *Curator: The Museum Journal*, *63*, 4, 591–596, 2020.
3. Lehmannová, M., *224 Years of defining the museum*, International Council of Museums, Czech Republic, pp. 1–4, 2020, https://icom.museum/wp-content/uploads/

2020/12/2020_ICOM-Czech-Republic_224-years-of-defining-the-museum.pdf. Retrieved on 1 January 2020.
4. Akin, M., Passionate possession: the formation of private collections, in: *Learning from Things: Method and Theory of Material Cultural Studies*, pp. 102–128, 1996.
5. Alexander, E.P., Alexander, M., Decker, J., *Museums in motion: An introduction to the history and functions of museums*, Rowman& Littlefield, 2017.
6. Salar Jung Museum, https://www.salarjungmuseum.in/History-of-SJM.html.
7. Museum of Textile Industry, Sliven, https://bulgariatravel.org/museum-of-textile-industry-sliven/. Retrieved on 11 October 2019.
8. El Saeed, Y., *A visit to textile museum*, Egypttoday, Cairo, 2017, https://www.egypttoday.com/Article/6/21638/A-visit-to-Textile-Museum. Retrieved on 22 January 2020.
9. THE MET, *New exhibition at metropolitan museum explores three centuries of worldwide textile trade*, 2013. https://www.metmuseum.org/press/exhibitions/2013/interwoven-globe. Retrieved on 13 July 2016.
10. Borkopp-Restle, B., McNeil, P., Martinetti, S., Miller, L., Riello, G., Museums and the making of textile histories: past, present, and future. A discussion with BirgittBorkopp-Restle, Peter McNeil, Sara Martinetti, and Giorgio Riello, moderated by Lesley Miller, in: *Perspective. Actualitéen Histoire de l'art*, 1, pp. 43–60, 2016, file:///C:/Users/17953/Downloads/perspective-6297.pdf.
11. Fiber2Fashion, Textile history still alive in museums, FIBER2FASHION.COM, 2014, https://www.fibre2fashion.com/industry-article/7363/textile-history-still-alive-in-museums. Retrieved on 2 April 2016.
12. Textile Museum, http://www.india-crafts.com/textile/textile-museum.html. Retrieved on 23 May 2021.
13. Bodo, S., Gibbs, K., Sani, M. (Eds.), *Museums as Places for Intercultural Dialogue: Selected Practices from Europe*, MAP for ID group, 2009, https://www.ne-mo.org/fileadmin/Dateien/public/service/Handbook_MAPforID_EN.pdf. Retrieved on 4 June 2011.
14. Don Bosco Centre for Indigenous Cultures, https://dbcic.org/traditional-technology-gallery/. Retrieved on 15 September 2020.
15. Art and Culture, Bhasha, https://bhasha.frappe.cloud/adivasi-academy/arts-and-culture. Retrieved on 5 January 2021.
16. National Textile Museum, Department of Museums Malaysia Ministry of Tourism, Arts and Culture, http://www.jmm.gov.my/en/museum/national-textile-museum. Retrieved on 18 August 2021.
17. Calico Museum, http://www.globalgujarat.com/Calico_Museum_Of_Textile.html. Retrieved on 14 January 2020.
18. Fashion & Textile Museums, http://www.fashionandtextilemuseums.com/victoria-albert-museum/. Retrieved on 21 June 2021.
19. Himachal State Museum, https://www.himachalstatemuseums.in/jewellery-5/. Retrieved on 9 March 2018.
20. Musee de le Soie, before and after, https://thesilkmuseum.com/our-story/before-and-after/. Retrieved on 17 September 2021.
21. DSP-Group, The social significance of museums, Netherland Museum Association, pp. 1–81, 2011, http://www.dsp-groep.eu/wp-content/uploads/17elmusea_The_social_significance_of_museums.pdf.
22. Living and Learning Design Centre. LLDC, https://shrujanlldc.org/. Retrieved on 2 June 2021.
23. Schwartz, O., *Interior design rugs art fashion architecture*, Nazmiyal Collections, 2020, https://nazmiyalantiquerugs.com/blog/tehran-carpet-museum-iran/. Retrieved on 7 March 2021.

24. Smith, S., Exhibits concepts, community treasures: Why museums are important, Marketing Resource Directory, 2019, https://exhibitconcepts.com/cultural-treasures-the-importance-of-museums-in-our-communities/. Retrieved on 10 July 2021.
25. Gök, M.O., Bilir, M.Z., Gürcüm, B.H., Shape-memory applications in textile design. *Procedia-Soc. Behav. Sci.*, *195*, 2160–2169, 2015.
26. Günay, B., Museum concept from past to present and importance of museums as centers of art education. *Procedia-Soc. Behav. Sci.*, *55*, 1250–1258, 2012, https://www.nationalmuseum.af.mil/Education/Introduction-to-Becoming-a-Pilot/.
27. Singh, P.K., Museum and educator. *CHRJ XL*, VII, 1, 69–82, http://magazines.odisha.gov.in/Journal/jounalvol1/pdf/orhj-10.pdf.
28. FullerCRAFTmuseum, https://fullercraft.org/museum/about-us/. Retrieved on 12 August 2021.
29. Museum Association, The economic impact of museums and galleries, Culturehive, Arts Marketing Association, https://www.culturehive.co.uk/resources/the-economic-impact-of-museums-and-galleries/. Retrieved on 13 Oct 2021.
30. Trainer, L., What is your museum's economic footprint? *J. Mus. Educ.*, *35*, 3, 237–246, 2010, https://www.jstor.org/stable/25801357.
31. American Alliance of Museums, Museums as economic engines: A national report, Oxford Economics, 2017, https://www.aam-us.org/wp-content/uploads/2018/04/American-Alliance-of-Museums-web.pdf.
32. Piekkola, H., Suojanen, O., Vainio, A., *Economic impact of museums*, p. 139a, University of Vassa. Publication, 2014, http://www.museoliitto.fi/doc/Economic_impact_of_museums.pdf.
33. Museo Frida Kahlo, https://www.museofridakahlo.org.mx/en/the-blue-house/. Retrieved on 29 September 2019.
34. Kedia, T., *Gond art lends a unique touch to contemporary wardrobe*, Singinawa Jungle Lodge, https://singinawajunglelodge.wordpress.com/2020/06/03/gond-art-lends-a-unique-touch-to-contemporary-wardrobe/.
35. Vishnoi, A., *Michelle goes 'broke' in Delhi*, The Indian Express, 2010, https://indianexpress.com/article/cities/delhi/michelle-goes-broke-in-delhi/ Retrieved on 25 October 2019.

2

Textile Intangible Cultural Heritage of the World

Ritu Pandey[1]*, Vandana Gupta[2], Pintu Pandit[3†], Kumar Rohit[3] and Suruchi Pandey[4]

[1]Chandra Shekhar Azad University of Agriculture & Technology, Kanpur, India
[2]Chitkara Design School, Chitkara University, Rajpura, Punjab, India
[3]National Institute of Fashion Technology, Department of Textile Design, Patna, India
[4]Symbiosis Institute of Management, Pune, SIU, Pune, India

Abstract

Textiles are one of the ancient crafts known to humankind and played a vital role in the lives of humans. The textile products are made in varied shapes and sizes using indigenous raw materials through various fabrication techniques. Certain traditional fabrics, archaeological findings, or fragments of cultural textiles, such as tapestries, rugs, carpets, flags, shawls, embroideries, etc., are of cultural, historical, and sentimental values as they are part of tradition and cultural legacy of communities and civilizations. They also carry the knowledge and techniques to produce the textile craft, passed on from one generation to the next. It is thus necessary to preserve this tangible wealth of humanity as they are of great cultural significance. Besides being evidence of humanity's inventiveness, heritage textiles are also a source of inspiration for textile artists and fashion designers. There should be informative and careful decisions regarding the textiles display, storage, and handling to make them available for future generations. Thus, museums play an integral role in safeguarding and preserving global textile heritage and culture. The chapter lists the textile intangible cultural heritage and the ancient costumes of the world. Important ancient textiles of the world are described in detail.

Keywords: Ancient textile, ancient costumes, ancient craft, fashion, heritage, textile intangible cultural heritage

2.1 Introduction

Textiles are among one of the valuable commodities used by human civilization, commonly as clothing, furnishing pieces, ceremonial and symbolic articles. UNESCO conceptualized the safeguarding of intangible cultural heritage in 2003 to identity ancient art and culture that need protection [1]. Textiles play an essential role in preserving cultural heritage through museum collections and inheritance of traditional practices from ancestors. Textile intangible cultural Heritage encompasses valuable traditional ancient textile skills having creative, aesthetic, manual, emotional and economic value due to uniqueness of designs,

**Corresponding author*: pintu.pandit@nift.ac.in
†*Corresponding author*: ritupandey70@gmail.com

Seiko Jose, Sabu Thomas, Pintu Pandit, Ritu Pandey and Vandana Gupta (eds.) Handbook of Museum Textiles: Volume I Conservation and Cultural Research, (19–38) © 2023 Scrivener Publishing LLC

use of local eco-resources, and functionality. Safeguarding the textile intangible cultural heritage will ensure the inheritance and sustenance of such precious hand-technique practices. Several fashion brands in Europe vouch for brand's intangible cultural heritage and history to promote their value in the market [2].

People around the world make textiles and they have been doing this for centuries. In the most basic sense, textiles can be regarded as a material made by fibers (obtained from natural and synthetic resources) converted into yarns and fabrics. Importance of textiles can also be analyzed by understanding its role and use in different cultures and daily rituals. Textiles used and selected by a certain group of people provide their preference which is certainly based on the emotional cultural bond they are having [3]. A study conducted on the weavers' attachment to their traditional method of creating textiles explains that maintaining their traditional methods with slight introduction of modernization have a positive impact on their passion of serving their tradition and provides support to their cultural identity [4]. Apart from such studies, articles like The fabric of society: recognizing the importance of textiles and their manufacture in the ancient past, clearly shares the critical role of the textiles and manufacturing team in developing an understanding about the past [5]. This section covers the world's textiles, why textile is a part of our heritage and culture, and its preservation [1, 2]. The list of textile intangible cultural properties of the world is presented in the chapter. The chapter also covers ancient textiles and costumes prevalent in the world and their historical perspective. Towards the end of the chapter marketing strategies for museum and conservation is suggested.

2.1.1 Ancient Textiles of the World

Textile design and craft have an immense importance socially and economically; along with this it also has a huge impact on individuals and culture. The cultural tangible and intangible heritage of textiles is more than just conservation and preservation in museums, as it shares the living traditional values, perspective and skills inherited by ancestors. Understanding the textile craft and the process through which these cultural connections are developed is crucial in the context of societies as well as the cultural heritage of humankind [6–9]. Table 2.1 lists some important textile intangible cultural properties (ITICP) of the world. Intangible cultural properties recognition ensures conservation of ancient folk craft techniques. Table 2.2 presents some important ancient costumes of the world [10] (indiaculture.nic.in). Common materials used for ancient costumes were mainly linen, cotton, silk, wool, specialty hair, and animal fur.

Table 2.1 Important textile intangible cultural properties.

Intangible cultural textile	Description	Country of origin
Fibers		
Bashoufu	Banana fiber extraction and weaving	Japan
Mosi	Fine ramie	Korea
Miyako-joufu	Linen fiber production	Japan
Mud silk	Silk finishing technique	China (2011)

(Continued)

Table 2.1 Important textile intangible cultural properties. (*Continued*)

Intangible cultural textile	Description	Country of origin
Karamushi	Ramie cultivation and weaving	Japan
Bark cloth	Costume from bark	Uganda (2008)
Weaving		
Yusoku-orimono	Ancient weaving technique	Japan
Rignai and Risa	Colorful handloom clothes	India
Tsumugi-ori	Silk weaving technique	Japan
Tais	Woven textile	Timor Leste
Bingata	Stencil and resist paste weaving	Japan
Hmong	Hemp clothing, embroidery, dyeing	Lao PDR
Surin Textiles	Silk weaving, natural dyeing and Khmer pattern	Cambodia, Thailand, Vietnam
Calendar Waistband	Woven belt depicting the oral tradition, culture, rituals, history, and agricultural activities.	Peru (2008)
Aubusson tapestry		France (2009)
Al Sadu	Traditional weaving	United Arab Emirates (2011)
Sa'eed	Handmade weaving	Egypt (2020)
Brocade		
Saga-nishiki	Saga brocade	Japan
Tate-nishiki	Vertical brocade of ancient origin	Japan
Baluchari, Jamdani	Heavy brocade using intricate designs woven in silk and gold thread.	India, Bangladesh (2013)
Nanjing Yunjin	Luxury silk brocade	China (2009)
Dye		
Patola	Double Ikat (yarn dyeing)	India, Indonesia, Malaysia
Yuzen	Resist dyeing technique on fabric	Japan
Kurume-Kasuri	Ikat dyeing using indigo dye for Kimono	Japan

(*Continued*)

Table 2.1 Important textile intangible cultural properties. (*Continued*)

Intangible cultural textile	Description	Country of origin
Indigo dyeing	Indigo cake is used in the dyeing process	Japan, India
Batik	Resist dyeing	Indonesia (2009)
Beni and Shikon	The red and purple dye is extracted from safflower and gromwell roots.	Japan
Printing/painting		
Kalamkari	Fabric painting using blocks	India
Edo-komon	Fine edo pattern	Japan
Blaudruck/Modrotisk/Kékfestés/Modrotlač	Resist block printing and indigo dyeing	Austria – Czechia – Germany – Hungary – Slovakia (2018)
Mokuhanzuri Sarasa	Woodblock printing	Japan
Ajarakh	Woodblock printing	India
Ise-Katagami	Paper stencil technique for printing	Japan
Embroidery		
Bunka Shishu	Traditional landscape scenery	Japan
Li	Running, applique, and beadwork embroidery with dragon patterns	China (2009)
Applique	Superimposing of fabric pieces on cloth/fabric	India
Zmijanje	Handmade naturally dyed threads are used to embroider geometrical shapes on the wedding dress and household textiles.	Bosnia and Herzegovina (2014)
Chakan	Floral embroidery	Tajikistan (2018)
Knitting		
Chullo	Knitted hat with earflaps	Peru (2008)
Lacemaking		
Lefkaritika	Handmade Laces using needlework (satin, filling stitches)	Cyprus (2009)

(*Continued*)

Table 2.1 Important textile intangible cultural properties. (*Continued*)

Intangible cultural textile	Description	Country of origin
Alencon	A rare technique of needle lace-making requires a high level of craftsmanship. Usually, it takes seven hours to finish a square centimeter to produce decorative openwork/net in textiles using needle, thread, and parchment paper.	France (2010)
Lace-making	Pag needle-point lace creating spider pattern	Croatia (2009)

Table 2.2 Prominent ancient costumes of the world.

Ancient civilization	Important costumes	Motifs
Egypt	Sheath like a tunic, short, simple skirt, apron, girdle	Scarabaeus, lotus, papyrus, and Nile pattern
Greece	Ionic chiton, Doric chiton, Peplos, himation, diploidion, chlamys, veils, hat, girdle	Frets, guilloche, anthemion
Roman	Stola, toga, palla, palliolum, girdle	Honeysuckle, acanthus, laurels
Asiatic Dacians/ Persians	Long straight tunic, mantle, girdle, vest, pantaloon, Phrygian bonnet, miter	Spots, jigjags, checks, stripes
France	Long tunic, girdle, palla, stola, corset, mantle, houppelande, décolleté, gorget, surcoat, cotehardie, Phrygian cap, camlet, layers of tunic	Byzantine symbols, cross, circle, dove, Gothic designs
India	Uttariya, antariya, kayabandh, saree, lehenga, dhoti, kurta, Jama, headgear	Swan, God, Goddess, flowers, scroll, animal, birds
Japan	Kimono, obi, furisode, poncho, jerkin haori, happi, aino, atushi, hakama, kamishino, hitagi, nuihaku	Cranes, chrysanthemums, Scenery, cherry blossom leaves, flower buds
China	Pien-fu, chang-pao, shen-i	Dragon, fire, floral, landscape

2.2 Ancient Textiles of the World: Techniques and Historical Perspective

2.2.1 Ancient Heritage Textiles—Asia and Australia

2.2.1.1 Chinese Textiles

Cotton and silk brocade weaving is an essential part of Chinese cultural history dating back to 770 - 476 BC. Li brocades were recognized as an intangible cultural property by UNESCO in the year 2009. The ancient Chinese used silk to create various materials, from brocades to satins to silk fabrics. These techniques were used to make clothes, rugs, wall decorations, and almost anything else needed. Chinese textiles changed many times over the millennia. Gold, yellow, and red were standard colors, particularly amongst the nobility, indicating wealth, status, and divine favor. Dragons and other mythical creatures were embroidered as symbols of good fortune, bravery, or longevity. Images, such as clouds, trees, fish, turtles, birds, flowers, and mountains, could be frequently found on these textiles (Figure 2.1a–c). Each design was meant to communicate something in both practical and symbolic senses. Taking one example of bird, Phoenix was considered auspicious as it symbolizes good peace and a wise monarch [11].

2.2.1.2 Japanese Textiles

Japanese fabrics can be decorated with many images, including landscape scenes, abstract patterns, or pictures of traditional symbols like cranes, which convey good luck and good health, and chrysanthemums, the character of Japanese royalty. Japanese-made textiles from plant fibers, predominantly cotton, ramie, and hemp, were spun and woven into wearable garments, bags, rugs, and blankets (Figure 2.2). The first distinct Japanese styles of silk weaving emerged in the Yamato period (300-710 CE). It was in this era that the kimono first appeared.

Sakiori is a weaving technique involving torn old pieces of cloth into strips and re-weaving them together into a new product. Sashiko is a form of sewing in which interlocking stitching pattern is used to hold pieces of fabric together. Sashiko stitching makes a textile longer-lasting and stronger. It provided designers with an artistic opportunity to add

Figure 2.1 Textiles play a significant role in Chinese history. (a) Imperial costume, (b) 12th-century image of women making silk, painting onto silk fabric, (c) Auspicious dragon symbols in Chinese costume.

Figure 2.2 Japanese textile craft (a) Japanese kimono (b) Peasant jacket made with a variety of techniques (c) Ainu garment (d) Katazome fabric in traditional indigo blue.

a little flair through alternating colors of threads used to make the interlocking stitch pattern. Japanese rural garments, or storage, could be made using various techniques, making folk textiles unique, functional, and highly individualized art pieces. Japanese textile art is a rich tradition that evolved over thousands of years. Among the methods of decoration are embroidery and other decorative stitches, as well as painting and dyeing. Japanese textiles are the subject of research and appreciation in museums and among artists around the world.

Many Japanese textile designs involve resist dyeing. A resist is a substance that repels dye. Their highly developed dyeing techniques often use a process called paste-resist. The artist makes a thick paste of rice flour and applies it to a piece of fabric. When it dries, the material is dyed. In the areas where the paste has been applied, the color will not stick. After the fabric is removed from the dye and rinsed, those areas show through with the pattern.

Katazome is a method of dyeing historically done on pieces of cotton using a paste-resist. You can recognize these fabrics by their deep blues and intricate (often white) designs. For this process, the artist starts by making a paper stencil called a katagami (the word means "paper template"). Developing the designs and completing the stencils is an art itself. Using a special sharp knife, the artist cuts designs into a stencil made of mulberry bark paper with gum and sometimes natural fibers like silk or human hair. Then the artist places the stencil on the fabric, pushes the paste which resists it, and shows the pattern on the fabric (Figure 2.3). The artist repeats the process to cover the material, then lets it dry and dyes the fabric. For katazome, traditional dyes include a deep blue made from natural indigo and a reddish-brown hue made from fermenting the juice of unripe persimmons (Figure 2.2d). When finished, the artist rinses the fabric to remove the paste-resist and reveal the final finished product. The katazome method can produce stunningly beautiful and highly sophisticated designs.

Tsutsugaki is one such art who creates designs in free-hand paste-resist, applying the paste using a large paper cone with an applicator tip. This method can create bold designs visible from a distance, and it has been used for centuries, often for military and religious banners.

Yuzen is the method invented around 1700 CE involving a combination of freehand and stenciled paste-resist applications. It has often been used to decorate very fine kimonos. The cloth is stretched on a frame, and then a pattern is drawn on it using a nonpermanent dye.

Figure 2.3 Example of Katagami, or paper stencil.

In this case, the design is a guide for the artist to follow when applying the resist. When the drawing dries, the artist carefully traces the lines with a resist paste. It is a time-consuming process sometimes done with fine brushes. Then the fabric is dyed. This process is repeated for each layer of color and detail until the final design is completed. This method can also include hand-painting of the dye [12].

2.2.1.3 Indian Textile

Indian materials are some of the oldest trade items globally, contributing a soft splash of color to the history of global economies and interactions. Three indispensable costumes

Figure 2.4 Indian textile embellishments (a) A 17th-century pashmina rug (b) A variety of embroidery techniques (Sindhi, Kathiwari, mirror, appliqué, and beadwork) used to create the vibrant and complex textiles of India.

adorned in ancient India were Uttariya (upper costume), antariya (lower garment) and kayabandh (girdle) to hold the lower garment in place. A few more notable classics are saree, lehenga for women and dhoti-kurta-headgear for men. Indian textiles are defined mainly by their bright colors, which do not tend to fade over time (Figures 2.4a, b). Part of this is the quality of the dyes, but another part is the technique. Indian textile producers use a mordant chemical additive, which helps the dye bond with the fabric. Regions, villages, and communities throughout India have unique weaving traditions with distinct customs and patterns. In the batik process, a wax or mud-brick is used to stamp designs onto the material, which the dye cannot penetrate. Batik dyeing is mainly done on sarees and apparel. Patola originated in Gujarat, India, where a few master craftsmen are continuing the method. In this process, warp and weft both yarns are resist-dyed, preferably with natural dye, before weaving. Dyed yarns are woven on handlooms to create geometric patterns.

2.2.1.4 Turkish Textiles

Turkey was the Ottoman Empire and textile was an essential part of their economy and if any type of textile was sold it was part of the empire's treasury. The Turkish state classified cotton, wool, and silk textiles. Cotton was obtained from India, whereas wool from Europe was mainly used for soldier uniforms. Turkish brocades called kemha could have embroidered gold and silver flowers and celestial bodies adorned furniture cushions, carpets, mirrors from the 16th and 17th centuries (Figures 2.5a–c).

2.2.1.5 Persian Rugs

Weaving was probably introduced into Persia around 529 B.C. The skill was necessary for migrating people who wove coverings for tent floors. Today, many Persian rug styles reflect this history. Diverse practices developed among tribes and villages and many remain in use today. Scholars have also mentioned that the woven rugs was from the Persian Sassanid Dynasty (224 - 641 C.E.). The most delicate rugs were made of silk, wool, gold and silver threads primarily used two kinds of knots such as Persian or senneh knot which was also

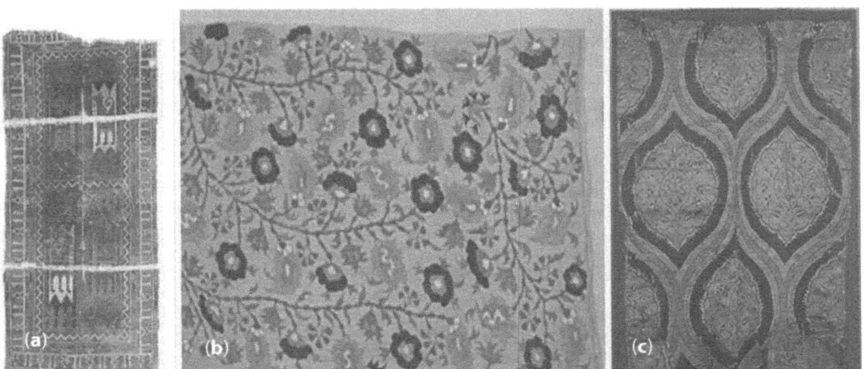

Figure 2.5 Turkish textiles (a) Carpet 11th to 13th century carpet (b) Turkish mirror cover 18th century (c) Ottoman brocade 16th century.

Figure 2.6 Persian rugs in Iran (a) Carpet weaver at work (b) Ardabil rug from the Safavid dynasty, ca. 1539 - 1540 (c) A Persian carpet with an intricate design (d) Kilim carpet with geometrical patterns.

known as "asymmetrical knot," famous for its luxuriously thick fluffiness (Figures 2.6a–d). It was also found that "Jufti knot" is used in a specific area of Iran in which a yarn is wrapped around four warp threads.

2.2.2 African Textile

African people have a long history of producing intricate textiles, which we know from evidence from the fabrics themselves to renderings on ancient tombs and pyramids (Figure 2.7a) [13]. Details of African textiles are presented in Chapter 12.

Two popular fabric forms originating in Africa are Bogolan and Kente cloth. Bogolan or mud cloth was a handwoven material. Bogolan means to be made of mud. Throughout Africa, skilled craftsmen make various textiles using available raw materials like cotton, wool, silk, and raffia, the fibers of long palm leaves. African fabrics are used for many purposes, both valuable and symbolic. They communicate social and political messages in their colors and the specific designs or patterns that adorn them. They are also used for ceremonies like weddings, funerals, initiation, rites, and of course they keep people warm and express personal style (Figure 2.7d).

Kente cloth was Ghana's national fabric (Figure 2.7c). Kente cloth was made via weaving methods. The word kente is derived from opening up the warp yarns and beating in the weft yarn during weaving. Kente cloth is a woven cloth often made from silk or cotton. This striking cloth is decorated during the weaving process with brilliantly colored geometric stripes and bands. It is made by several tribes related to the Akan ethnic group that lives around Ghana, including the Asante and Ewe people. Award-winning Nigerian textile artist Nike Davies Okundaye grew up in a culture rich in traditional textile weaving and dyeing practices.

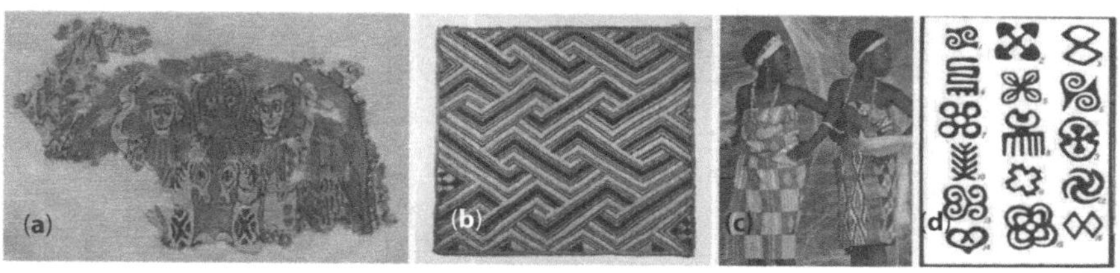

Figure 2.7 (a) Tapestry fragment from an Egyptian tunic (c. 6-7th century) [13] (b) Kuba cloth (c) Young girls wearing costumes made of Kente cloth (d) Symbols stamped onto Adinkra cloth.

Figure 2.8 (a) Moroccan sash (b) Intricate Moroccon embroidery (c) Embroidered footwear (d) Moccasins with floral beadwork.

Adire is an indigo-dyed fabric hand-painted with resist paste. Resist paste is a substance that results in dye not adhering to it when applied to a surface. The result is a hand-drawn design on a deep blue background. It is a type of cloth specific to Yoruba culture.

Africa is a large continent with a rich history and a diverse population. Some fabrics are made from materials like tree bark and tree leaves [14]. An example of this kind of fabric is Kuba cloth woven raffia which are strands of fibers collected from specific type of palm tree. Fabric preparation is a tedious process which involves different steps starting from picking the palm leaves and stripping them with special combs to obtain fibers. The fibers are dyed with natural colors and hand rubbed to provide them with softness properties. These fibers are converted into yarns which are then used to develop a fabric known as raffia cloth. It can have patterns like lines, zigzags, and triangles (Figure 2.7b). Finished pieces of material may be further decorated with embroidered and dyed designs or with tufts of fiber in geometric designs. Within Kuba cultures, eighteen sub-groups have their patterns and symbols that can be woven into the cloth.

The northwest African nation of Morocco has been called a tapestry of cultures. Morocco's colors, sounds, and tastes are composites of many different influences, and nowhere is this more visible than in their arts [15].

Morocco's textile heritage dates back hundreds of years. Once married, women would produce textiles for their homes and engage with other women to share techniques, styles, and ideas. Today, textiles are primarily a women's art form in Morocco and are still highly respected. There is a whole process of sewing or weaving required to create the base; what defines Moroccan textiles is the embroidery, needles, and thread to decorate fabrics (Figure 2.8a–d).

Azemmour embroideries are primarily created by Jewish women of European ancestry, mainly Spanish or Portuguese, and embrace European stitching patterns. Communities with ancestral ties to Algeria employed a Turkish style to create a cloth called a tension covering mirrors, warding off the evil eye. Moroccan textiles draw from various influences but are interpreted through a distinct taste for intricacy and virtuosic technique.

2.2.3 Scandinavian Textile Design

Scandinavian textiles had patterns with repeats or small identical designs in rows or grids to cover the whole cloth. Because of significant natural materials, substances like wool, linen, and cotton are used for various goods. Ekelund weavers is a Swedish company that has been around since 1692. For more than 400 years, this family-owned business has produced woven fabrics of cotton and linen with traditional and contemporary patterns.

Figure 2.9 (a) Woven rug designed by Marta Maas-Fjetterstrom. (b) Rolls of woven Swedish fabrics with geometric and nature-inspired patterns [16].

Nature-inspired designs feature patterns with plants and animals that are clear and crisp, colorful forms against clean backgrounds that are not overly fussy or cluttered. All the textile items produced are practical things like towels, dishcloths, and table runners.

In the early 20th century, several Scandinavian women designers began to gain notice for their work. Marta Maas-Fjetterström (1873–1941) was a Swedish weaver who became known for her rugs. She created designs that featured geometric and nature-inspired shapes with colors that reflected the light and hues of summer (Figure 2.9a). Mass-Fjetterström started weaving workshops in 1919 and became highly collectible.

After World War II, Scandinavian textile design and the large area of interior design became popular throughout Europe and the United States. Designs became bolder and more contemporary, with brighter colors (Figure 2.9b). But they stayed true to the Scandinavian elements of nature and clean, stylized geometry with textiles made mostly of natural materials.

2.2.4 American Textiles

Mayan textiles are one of the primary Maya art forms and are equally complex. Maya textiles characterized by vivid colors of every shade imaginable and covered in geometric and animal patterns (Figure 2.10a, b). The art of weaving is seen as a spiritual process, and the various motifs and designs within Maya textiles generally have religious and symbolic importance. Today, weaving maintains a similar solemnity within Catholic contexts [17].

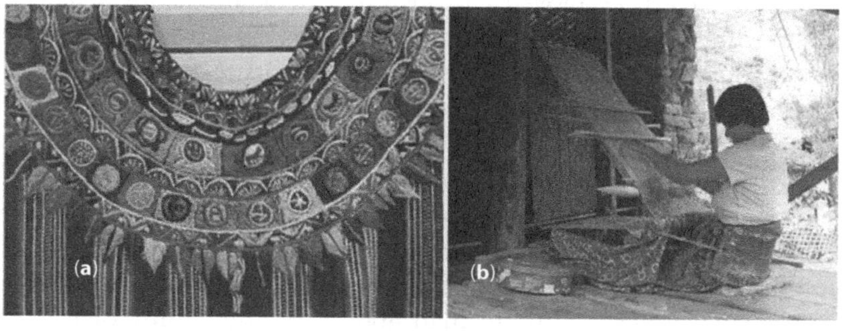

Figure 2.10 (a) Colorful Mayan textiles. (b) Backstrap loom.

Peruvian The Chavin period in Peru extends from approximately 1000 BC to 100 BC. The next period overlaps between the Moche (about 50 AD to 800 AD) and Wari (about 600 AD to 900 AD). From about 1100 to 1450 AD, the Chimú people held sway in the northern Peruvian region. Finally, the Inca (about 1450 to 1532) continued the Moche and Chimú in their design.

Backstrap looms were and still are almost exclusively used for all types of Peruvian textile construction (Figure 2.11c). In addition, intricate feather work was a hallmark of Wari culture. Examples have been found in burial shrouds in the dry Andean landscape, which acted as a perfect preserving environment for the ancient fibers. This old tapestry from the Chimu period is faded but shows the intricate designs characteristic of early Peruvian textiles (Figure 2.11a).

Realistic depictions of humans and animals were common in Chavin weavings. Also, patterns incorporating geometric patterns, chevrons, and lines are common (Figure 2.11b). The Moche textile designs were more realistic, the Wari tended to lean more toward finely woven abstract designs. Wari abstract designs were responsible for influencing artists and architects in Germany centuries late, in the 1920s. This period saw a return to textile designs characterized by more realistic designs, and there was also a sculptural aspect to their tapestries.

Navajo As the second-largest federally recognized tribal nation in the United States, the Navajo held a large amount of cultural influence over what is now the four-corners area for centuries. Their most significant art influence is their weaving. Navajo weaving does not follow an exact pattern but instead reflects a personal spiritual journey between the weaver and the blanket (Figure 2.12a–c). Still, there is a basic aesthetic that can be found across Navajo textiles. Colors tend to reflect earth tones of red, brown, grey, and black, which is unsurprising given the mostly mineral-based pigments of this dry region. While Navajo blankets have been made of wool since the Spanish introduced sheep in the 17th century, tradition suggests that plant fibers like indigenous hemp may have been used for textiles before this. Most of our current interpretations came from the late 19th century when outside cultural pressures forced the Navajo to standardize and codify certain practices.

One ubiquitous symbol in Navajo weaving is the cross (Figure 2.12b). However, it does not represent Christianity. It is the symbol of Spider-Woman and it represents her spiritual energy, teachings, and Navajo culture. The wool for Navajo weaving came from the Iberian Churra, a sheep acquired from the Spanish explorers. The sheep evolved into the Navajo-Churra, a

Figure 2.11 (a) Peruvian textiles fabric. (b) Chavin textiles. (c) Backstrap loom.

Figure 2.12 Navajo blanket geometrical design. (a) Lightening motifs. (b) Cross design on blanket. (c) Drawing of Navajo weavers.

well-suited species to the hot dry climate of the southwestern United States. This kind of sheep produced long-staple wool that was well suited to the weaving process.

2.3 Role of Museum: Textile as a Part of Heritage and Culture

The Textile Museum UK, Acquired by George Hewitt Myers in 1950, displays around 30,000 pieces of fabric and 16,000 costumes. The evolution of our attire mirrors the journey of our civilization. A large body of evidence has since been unearthed, pointing to the Indian subcontinent as cotton agriculture and weaving seat in ancient times. The fame of India's cotton. Leeds textiles was widespread and has been extensively chronicled by travelers, such as Marco Polo and others in vivid detail [18].

Museums help the masses get acquainted with their history, which eventually helps them make decisions in the future. Textile museums play an essential role in determining the textile and fashion trends of the past years. The Fashion and Textile Museum in Paris has around The Victoria and Albeit Museum, UK, houses a fine collection of dresses from the 17th century, the Ardabil Carpet, and many other artifacts which are no less than wonders. It is one of the favorite destinations for tourists in the UK. Textile museums worldwide have something unique to display, which would be very hard to find in the world outside (Figure 2.13). That is how they keep the history of the time alive (Figure 2.14a–c). Textile museums help shift the attention of the masses from high-profile designers and manufacturing units to small-time artisans of the past and present.

Figure 2.13 Textile history still alive in museums.

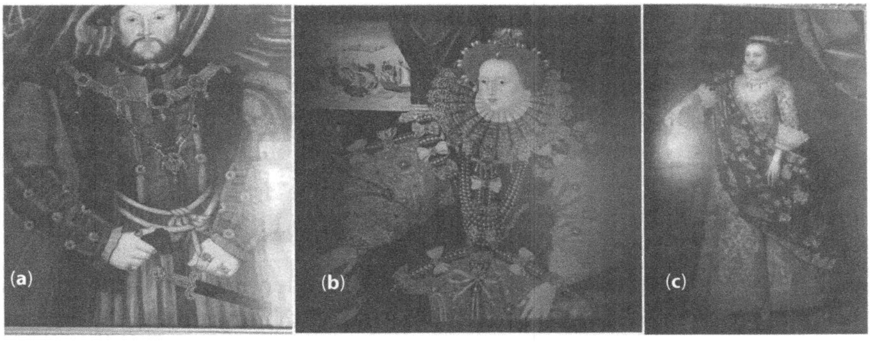

Figure 2.14 British Royal armours and costumes (a) King Henry VIII portrait depicting British Royal armour (b) Elizabeth I 1588 portrait (Royal Armouries, Leeds, UK) (c) 15th CE random portrait depicting richly embellished costume at Temple Newsam textile museum.

They contribute to the overall economy of the country, especially in the case of emerging economies. Textile museums help in keeping records alive. There are several textile museums worldwide, and they contribute to the global textile industry in their way. They contribute to the global textile and fashion industry and have a role in studying history, which is ultimately helpful in all the fields.

2.4 Marketing Strategies for Museums and Conservation

Today's era is of marketing; what gets seen is known and valued in this world. This intangible heritage of the world is to be projected as tangible in protecting and sustaining cultural heritage. Marketing museum artifacts is imperative to increase awareness of the existence and scope of the historical curation and conservation industry. Increasing footfall in museums will ensure a resulting increase in the action that includes displaying objects and artifacts in the museums. These efforts contribute to increasing awareness regarding the scope of conservation and curation as careers and the scope of museums, exhibitions and artifact preservation as a business. They were deliberating on the fact that marketing involves developing a deeper understanding of the audience, tailoring the message and the service to suit the needs and tastes of the audience, consequently meeting their needs and converting the entire service into an experience and most importantly communicating the message to the audience in a manner that attracts their attention [19].

The focal part of the communication campaign constitutes information regarding the museum, exhibition programs, opening and closing hours, entry fee (if applicable), what to expect when visiting, theme of the exhibition, career opportunities for future reference. The key audience for this communication campaign includes tourists, culturalists, people interested in local history, schools and educational institutions, researchers, historians, conservators, and curators. Given the diverse nature of the audience, the language and the idea must be curated and customized to suit the audience's expectations [20].

While the traditional marketing strategies for museums include,

- Distribution of Pamphlets and brochures at schools, educational institutions and other fairs frequented by the selective audience of these museums
- Display of posters on social media as well as on offline channels to increase visibility
- Print advertisements on newspapers, magazines, postcards, stickers, badges etc.
- Collaborating with relevant individuals with credentials that can boost footfall in the museums, essentially collaborating with brand ambassadors for the cause/exhibitions/museum activities/camps, etc.
- Organizing summer camps and other short-term events to ensure regular footfall and attention.
- Organizing and participating in charity events, auctions and other gatherings to improve visibility.
- Sponsoring and conducting short courses on the cause/theme of the museum along with career enhancement workshops for curators and conservators.
- Community exhibitions, moving exhibitions and other means of bringing the museum to the people.
- Developing theme-based establishments like Dakshin Chitra in Chennai, India which conducts folk music and dance recitals, puppet show and other means of expression to ensure the message about the theme reaches the audience and consequently acts as both an excursion choice for children and a paradise for lovers of art and history, and
- Creating selfie-spots near key exhibits and on the establishment to increase traction.

The most creative strategies for marketing include [21]:

- Increasing website traffic with Search Engine Optimization (SEO) through Google and other Search Engines by optimizing for right keywords, improving page speed and ensuring curated quality content about the museum and its causes.
- Driving local traffic by claiming and updating MyBusiness listings on Google to ensure the museum is visible and appears when users perform near me searches on Google.
- Improve website design to increase the time spent by leads on the website. Museums can also include videos of virtual tours and testimonials of employees and the audience visiting the museums.
- Crediting the museum in blogs and allowing users to create a blog on the website of the museum to share valuable information is a way to ensure readers take notice of the museum and its activities
- Infographics is one of the most definite forms of marketing any idea which helps the advertiser communicate a multitude of information in a concise and visually appealing form. Museums can choose to explain complex topics

or detailed chronological data through infographics and charts, on the website of the museum

- Investing on Pay per Click Advertising to increase digital visibility
- Encouraging social media presence and marketing events through social media announcements and interactions. Creating hashtags to increase brand awareness.
- Creating content in the form of e-books, podcasts, animated videos etc. to capture the interest and attention of the audience
- Creating a user and device-friendly mobile application that can improve traffic and investing in programmatic display advertising, which essentially appears as an image along a sidebar of a web page or app that edirects users to the website when they click on the image [22].
- Opting for other digital marketing techniques like Display retargeting, Geofencing etc.
- Monitoring ad campaigns and consumer behavior trends and re-working them based on the quantitative and qualitative feedback.

In "today's digital age", it is imperative that museums embrace the smart phone culture, encourage social media interactions, unique host events throughout the year, leverage local influencers and publishers, build lasting value partnerships with local attractions/restaurants and other establishments with a similar audience and above all, opt for creative digital marketing techniques that is synchronized with the digital footprint of the target audience to ensure museums attract and retain patrons and visitors alike [23, 24]. Post COVID-19, technologies have been part of the new normal like any other sustainable product marketing strategies. The administrator must reach out to target the population with technology as a base for conventional products as well.

2.5 Conclusion

Textile is made from combining fibers to make the fabric and used for items like clothing, blankets and rugs. In many cultures, textiles play various essential roles. Historically, some were reserved for ceremonial purposes, while others were reserved for and worn only by specific people. Today's era is of marketing; what gets seen is known and valued in this world. This intangible heritage of the world is to be projected as tangible in protecting and sustaining cultural heritage. Marketing museums is imperative to increase awareness of the existence and scope of the historical curation and conservation industry. These efforts contribute to raising awareness regarding the scope of conservation and curation as careers and the scope of museums, exhibitions and artifact preservation as a business. Thus, museums play an integral role in safeguarding and preserving global textile heritage and culture.

References

1. Petronela, T., The importance of the intangible cultural Heritage in the economy. *Proc. Econ. Financ.*, 39, 731–736, 2016.

2. Radclyffe-Thomas, N., Intangible cultural heritage in fashion marketing: From number 1 Savile Row to the world, in: *IFFTI 2016 Conference Proceedings*, pp. 126–135, China Textile & Apparel Press, 2016, March.
3. Deepshikha, P.Y. and Nath, N., Textiles as communicating links for cultural traditions, in: *Proceedings of the 7th International Conference on Kansei Engineering and Emotion Research 2018*, p. 168, 2018, March.
4. Dias, R.M., Ogle, J.P., Diddi, S., Constructing cultural identity through weaving among Ri-Bhoi women weavers: A symbolic interactionist approach. *Fashion Text.*, 7, 1, 1–21, 2020.
5. Foxhall, L., The fabric of society: Recognising the importance of textiles and their manufacture in the ancient past. *Antiquity*, 91, 357, 808–811, 2017.
6. Strand, E.A., Lindgren, S., Larsson, C., Capturing our cultural intangible textile heritage, MoCap and craft technology, in: *Euro-Mediterranean Conference*, pp. 10–15, Springer, Cham, 2016a October.
7. Strand, E.A., Lindgren, S., Larsson, C., Capturing our cultural intangible textile heritage, mocap and craft technology, in: *Digital Heritage. Progress in Cultural Heritage: Documentation, Preservation, and Protection. EuroMed 2016.* Lecture Notes in Computer Science, vol. 10059, M. Ioannides, *et al.* (Eds.), Springer, Cham, 2016b, https://doi.org/10.1007/978-3-319-48974-2_2.
8. Lefferts Jr., H.L., Textiles for the preservation of the Lao intangible cultural heritage, in: *Laos and Ethnic Minority Cultures: Promoting Heritage*, pp. 89–110, 2003.
9. Sae-Wang, R., Spirit of the loom: The conservation and commodification of ' 'Surin's textile cultural 'heritage'. *Int. J. Intang. Herit.*, 10, 85–100, 2015.
10. *National List for Intangible Cultural Heritage (ICH)*, National list for Intangible Cultural Heritage (ICH) | Ministry of Culture, Government of India (indiaculture.nic.in), 2022.
11. Zapolska, K., Confucian values presented in depictions of birds on chinese textiles from the qing dynasty period (1644–1912). *Art Orient*, 9, 115–133, 2020.
12. Reese, H., Understanding the differences between AI, machine learning, and deep learning, 2017. https://www.techrepublic.com/article/understanding-the-differences-between-ai-machine-learning-and-deep-learning/.
13. Oke, J., The Fabric From Africa, MyNativeFashion. Nigeria, 2020. https://www.mynativefashion.com/fabric-from-africa/.
14. Textiles: Materials for living. Textile Exhibit, Museum LA. Maine, USA, https://www.historyweaver.net/textile-exhibit. Accessed in 2022.
15. Muscato, C. About Moroccan textiles - Moroccan cushions. UK, 2022. https://www.moroccancushion.co.uk /about-moroccan-textiles/
16. Rovine, V.L., African fashion, global style: Histories, innovations, and ideas you can wear. Bloomington, Indiana University Press, 2015.
17. Muscato, C., Mayan textile history. 2022. study.com, https://great-home-decorations.com/mayan-textiles-history/
18. Bhandari, V., Jewelled textiles: Gold and silver embellished cloth of India. New Delhi, Om Books International, 2015.
19. Kawashima, N., Knowing the public. A review of museum marketing literature and research. *Mus. Manag. Curatorship*, 17, 1, 21–39, 2022.
20. Komarac, T., Ozretic-Dosen, D. and Skare, V. Understanding competition and service offer in museum marketing. *Acad. Rev. Latinoam. De Adm.*, Emerald Publishing, 30, 2, 215–230, 2017. https://doi.org/10.1108/ARLA-07-2015-0159.
21. Kirezli, O., Museum marketing: Shift from traditional to experiential marketing. *Int. J. Manag. Cases,* Access press, UK, 13, 4, 173–184, 2011.
22. Gurevich, A. and Haynie, S., Museum marketing strategy: How to attract more visitors. Hearst Bay Area, 2019. https://marketing.sfgate.com.

23. Sullivan, M.A., Bruegel and the creative process. pp. 1559–1563. Routledge, London, 2017. https://doi.org/10.4324/9781351162289.
24. Pandey, S., Pandit, P. Pandey, R., Pandey S. Marketing strategies for upcycling and recycling of textile and fashion, in: *Recycling from Waste in Fashion and Textiles: A Sustainable and Circular Economic Approach*, pp. 253–275, 2020. http://doi.org/10.1002/9781119620532.ch12.

3

Important Woven Textile Specimens in World Museums

Karthika Audinet

Montgomery College, Applied Technologies Department, Gudelsky Institute for Technical Education, Rockville, Maryland, USA

Abstract

Surviving examples of textiles of varied complexity and geographic origins from prehistory to the pre-Jacquard era across the globe are grouped in museums based on their belonging to specific cultural periods and geographic areas. As the range of textiles is vast, many attempts to facilitate the comprehension of textiles produced in different parts of the world have been made by isolating textile structures, standardizing terminology, and classifying fabrics. Cross-cultural influences between geographical regions of the world have also been studied. Over the abovementioned long period, weavers across the world mastered the creation of specific types of textiles built upon three foundational weaves—tabby, twill, and satin—while developing complex woven structures. Examining this progression provides a finer appreciation of their evolution. Based upon degrees of sophistication in textile structures of specimens found in museums across the world, this chapter presents 46 examples of woven textiles representative of technical mastery.

Keywords: Woven structures, complex woven structures, tabby, twill, satin

3.1 Introduction

This chapter presents 46 examples that are representative of the mastery of specific woven textiles from the prehistoric period to the pre-Jacquard era (before 1804) across the globe. During this long period, weave structures, weaving devices, materials, and patterned textiles developed into a plethora of variations, including complex and luxurious textiles. From this vast diversity, certain textiles became reputed for their mastery of combining material, weave structure and/or patterning. Some rare examples are exemplary in terms of innovation and yet others are representative of the major types of variations in fabric development. This chapter attempts to highlight these salient features in eight groups that correspond to the common categorizations used in museums. The examples chosen range from simple structures with two sets of elements (one warp and one weft) to complex structures (two or more sets of warps and wefts).

Email: karthika@audinet.net

Seiko Jose, Sabu Thomas, Pintu Pandit, Ritu Pandey and Vandana Gupta (eds.) Handbook of Museum Textiles: Volume I Conservation and Cultural Research, (39–86) © 2023 Scrivener Publishing LLC

Studying textiles that are outstanding for their degree of innovation, sophisticated means of production or combination of materials, structure, and pattern cannot be complete without considering the reasons that led to their design. Many of these developments were due to multiple regional cultural exchanges and technological advancements. By correlating technological characteristics and advances of woven textiles with their geographic and historic contexts, we can see similar structures in different cultures, compare, and comprehend differences and perceive the gradual buildup of complex textiles till the 1800s.

3.2 Methodology

Across the globe archaeologists have found proof of textile making in the form of fragments of dyed plant and animal fibers, cords, mats, baskets, nets, twined, and woven cloth dated from 160,000 years to 6000 years BC. Of all the varied textile structures created, woven structures have seen the maximum innovation to date. As civilizations invented, adapted, and eventually adopted techniques on a large systematic scale as studied by Gille Bertrand [1], textiles and techniques played an important role in the development of sophisticated cultures across the globe.

Many of the most exquisite textiles in world museums were commissioned by powerful patrons and royalty and not by the common man. This allowed for specific cultural identities to be developed as these textiles with their associated aesthetics, techniques, and symbolism trickled down, became more accessible and were adopted by people of the geographical area to become what is now perceived as separate cultures that form our collective global textile heritage.

Museums around the world are repositories of this global heritage. Small local museums located in situ may hold rare textiles, while others like the Victoria & Albert Museum in London and The Metropolitan Museum of Art in New York hold encyclopedic collections of more than 75,000 and 36,000 textiles respectively. These collections represent a broad range of techniques, periods, and cultures. As a medium of pictorial and decorative design, textiles acquired a place in the history of art; and textiles of the finest quality of design and craftsmanship have been collected and gathered into museums of the decorative arts [2]. Other criteria for a textile to be accessioned by museums are age and archaeological context, material and technical importance, rarity, anthropological, cultural, and religious symbolism, significance, textile design, aesthetics, and innovation.[1]

Literature on historic world textiles consists of either narratives that cover all types of textiles from prehistory to the present categorizing them under themes that explore materials, technology, social influences, and significance [3] or illustrated guides that treat woven textiles as one among seven categories of textiles[2] [4]. Books that analyze weaving techniques across time and space may leave out important areas such as the Americas [5]. Technical manuals and scientific papers referred to by scholars of woven structures or weavers remain inaccessible with fragmented subject matter. Museum publications devoted to the subject of

[1]The history of collections and the predilections of curators and collectors have added a strong serendipitous element to the extant material.

[2]The 7 textile categories treated by Gillow are Non-Loom, Loom-Woven, Painted and Printed, Dyed, Sewn, Embroidered and Embellished with other elements.

woven textiles by subject experts mostly focus on a particular civilization, culture or period or the collections of the specific museum of affiliation. A book based on an exhibition, its associated conference and workshops held at the China National Silk Museum in 2018 comes closest to the aim of this chapter, as it highlights looms, the evolution of mechanical and patterning technologies and the transfer of some of these technologies around the world [6].

The vast scope of the topic and the inevitable need for editing leaves us with difficult choices. Of all the multiple aspects for the study of textiles, the specimens mentioned in this chapter have been selected and grouped with 3 criteria in mind—historical dating, geographical or cultural provenance, and mastery of woven techniques and structures. Historically, certain geographical areas saw a lot of changes and movements of peoples. This is especially true when it comes to the Roman, Byzantine and Ottoman empires. Persia and Central Asia also saw a lot of disruption with resettlements of skilled weavers as well as nomadic tribes. The author has grouped some of these cultures according to geographic location and not according to empires in flux.

Apart from prehistoric textiles, the choice of seven categories[3] was made by researching prominent museums, their collections of historic woven textiles, and their method of classifying them. This method was felt to be far less subjective than relying on literature where omissions or personal bias can take precedence or relying on institutions that protect cultural heritage such as the UNESCO.[4] The Metropolitan Museum of Art "... possesses one of the finest and most comprehensive collections of textiles anywhere in the world." Their textile collections are categorized by civilizations and cultures, namely Islamic Art; Greek and Roman Art; European Sculpture & Decorative Arts; Egyptian Art; Asian Art; Arts of Africa, Oceania & the Americas; Ancient Near Eastern Art. This categorization allowed the author to further confirm the above-mentioned groups and their importance by analyzing museum collections that stood out for their volume of woven textile specimens, number of research efforts made to understand them, and exhibitions that highlighted various aspects of each of these groups and their textiles' deep histories. As the chapter is meant to illustrate important specimens in multiple world museums, the author widened the selection of specimens by including items from more specialized museums such as the Abegg-Stiftung (Riggisberg, Switzerland) or The George Washington University Museum and The Textile Museum (Washington D.C.). The choice of objects from smaller museums was guided by well-developed online collections and availability of photographs and information in English or French.

All weave structures are derivatives of the three basic weaves: tabby, twill, and satin.[5] By studying woven textile specimens in important museum collections, the author has

[3]These seven geographic regions are Peru; China; the Indian Subcontinent; Egypt, Sudan, Mali, and Ghana; Japan; Iran and Central Asia; Italy and France.

[4]While defining importance, the author studied some textiles that have been identified by the UNESCO which evaluates nominations for textiles in need of protection. Textiles listed by the UNESCO posed two problems for inclusion of specimens in this chapter. The first problem is that the UNESCO criteria for importance are not focused on how a specific woven structure contributes to the global knowledge of woven textiles. The second problem is that only some museum specimens have the distinction of being protected by the UNESCO. Nevertheless, as the UNESCO protection remains an important vetting process, it is mentioned in the text when pertinent [7].

[5]Some understanding of these basic weaves is a pre-requisite. Readers can find detailed explanations of primary textile names, weave structures and illustrations in the seminal work of Irene Emery [8].

chosen examples from pre-history and each of the seven geographical regions enumerated above that demonstrate increasing proficiency in blending materials, weave structure, and technology. Starting with simple twining and tabby weave the groups of textiles have been ordered to gradually introduce more complex structures.[6] The woven structures listed by the Centre International d'Étude des Textiles Anciens (CIETA, Lyon, France) in their technical analysis courses allowed the author to validate the choices of the main structures chosen in this chapter.

In addition to the references cited, the author has relied on the descriptions provided by the various museums listed, other available data and the author's personal research to complete or compliment information regarding specimens. The description of each textile specimen in the chapter includes technical and construction analysis correlated with aspects such as loom technology, design, historical development, and use. A short glossary of terminology has been included at the end of the chapter. For some complex structures, the author has preferred to use French terms as no English equivalents do justice to them.[7]

The textiles and museums enumerated in this chapter are by no means an exhaustive list, but it is hoped that this descriptive classification will give the reader an idea of the main range of fabric structures that evolved in different parts of the world to form our knowledge base.

3.3 Prehistoric Textiles

The history of textiles goes back to prehistoric times when early human beings began to use plant and animal fibers for clothing, hunting, containers, furnishings, architecture, and burials (shrouds, grave liners). Soon after human beings began to twist fibers into cords (along with the mathematical comprehension of pairs, sets, numbers) they were starting to see shapes, represent ideas in terms of sculptures and cave paintings and depict abstract ideas. They saw color, and they began to manipulate materials even further. They were beginning to not just use textiles for practical purposes—they were beginning to design them.

Early textile design involved complex processes such as:

1. Material processing, manipulation of fibers, and yarn making.
2. Dyeing using natural resources (plants and minerals and probably insects).
3. Structures such as felting (which occurs naturally and was probably the first textiles used after furs and skins), netting, knotting, looping, twining, and weaving.
4. Pattern—combining materials, stripes, textures, structures and woven or painted motifs.
5. Auxiliary crafts such as development of tools and devices.

[6]Detailed information on complex textiles can be obtained in the book by Dorothy K. Burnham based on the vocabulary of technical terms created by the Centre International d'Étude des Textiles Anciens, in 1964 [9].

[7]This aspect has been validated by the Centre International d'Étude des Textiles Anciens, CIETA (Lyon, France). The CIETA coordinates terminology and methods of analysis used by specialists in ancient and historical textiles.

Apart from textile pseudomorphs on metal, carbonized textiles, imprints on pottery, metal or ground surfaces, most textiles disintegrate before bone, shells, or stone. Freezing cold, anaerobic waterlogged contexts, peat bogs, salt mines or dry desert conditions [10] allowed remnants of textiles from certain sites around the world to survive. To name a few examples of actual specimens found by archaeologists: microscopic fibers of dyed yarns in Georgia (32,000 years old), twined sandals in Oregon (10,000 years old), twined cloth patterned in stripes in Peru (6200 years old), a twill weave mat in Armenia (5100 years old), a tabby weave linen tunic with pleating in Egypt (5000 years old), bespoke trousers with multiple patterns in China (3000 years old), and an almost intact pile carpet in Siberia (2600 years old). These are important testimonies to early human ingenuity and textile design in terms of structures, dyes, patterning, as well as attention to details. They illustrate production of furnishings and clothing to facilitate physical well-being and mobility of the users, and the progression from utilitarian to valuable commodities. Two-dimensional fabrics were manipulated to form three-dimensional baskets or clothing often made to size as seen in the following examples (1.3.1 to 1.3.6).

Four textile structures have been examined—twining, tabby weave, twill weave, and pile. Although twining and basketry differ from weaving in that the warps and wefts are manipulated by hand with no system to open a shed these two techniques may have preceded or developed alongside tabby weave textiles. They could have catalyzed a natural urge to develop quicker, more efficient methods when interlacing pliable cords and yarns. This probably prompted prehistoric weavers to find ways of keeping flexible cords or yarns

Table 3.1 Examples of prehistoric textiles.

Sl. no.	Museum	Accession number	Date	Brief description
1.3.1	The Museum of Natural and Cultural History	Catalogue no: 1-3563, Accession number: 56	9235 Cal BP	Close and spaced weft twining
1.3.2	Museo Cao	2009.052.01.B	6200 Cal BP	Spaced weft twining
1.3.3	Institute of Archaeology and Ethnography NAS RA	001.2010.001	Chalcolithic, approximately 4040-3800 Cal BC	Twill weave
1.3.4	Petrie Museum of Egyptian Archaeology	UC 28614bI	3000 BC	Tabby weave
1.3.5	Turpan Museum (in 2014)	Find number: 2003SYIM21:19	1074-935 Cal BC	Twill weave with repetitive patterning
1.3.6	Hermitage Museum	1687-93	500–400 Cal BC	Pile with symmetrical knot

under tension either with simple frame looms as seen in Peru and Egypt[8] or warp weighted looms as evidenced by loom weights found in Armenia and Austria.[9] Trans-material influences abound in archaeological finds. If a basket could be woven in twill weave, the structure could very well have been translated into a woven textile. Warp weighted looms were especially suited to create twill structures [12] as seen in fragments from Hallstatt, Austria. Animal skins were surely inspirational for the creation of pile structures. Archaeological evidence proves that Chinese weavers were the earliest to think of both tensioning and repetitive patterning devices, which in turn led to weave structures that were derivatives of tabby and twill. The six examples that have been chosen below illustrate the above-mentioned structures (Table 3.1).

3.3.1 Textile Specimen 1.3.1

The Museum of Natural and Cultural History, at the University of Oregon, USA, holds nearly 100 examples of sandals in sizes that belonged to men, women, and children. Recovered from cave sites in Oregon, the earliest sandals, known as Fort Rock sandals, were woven using the weft twining technique in sagebrush bark (*Artemisia Tridentata*) and other fibers carbon dated to 12,000 to 9000 years old [13]. Sandals like #1-3563, Accession number: 56 were weft-twined in one piece, i.e., thick warp elements were twined with close spacing to make the sole, split into finer warp elements, folded back and twined again with wide spacing between the wefts to cover the front of the foot.[10] Cords attached to the twined body of the sandal secured the sandals around the ankles. This is an excellent example of comprehension of inherent properties of a material, its manipulation and harnessing its qualities. An inherently stiff material was transformed into pliable textile material to create protective sandals. The thick bark for the sole was split to form a more pliable textile for the top of the sandal.

3.3.2 Textile Specimen 1.3.2

Thousands of textile fragments were buried during sacred rituals practiced over 3000 years at the preceramic mound of Huaca Prieta in Peru. #2009.052.01.B is an approximately 6200 years old striped fabric woven using the weft twining technique now housed in the Museo Cao, Peru. It has a striped warp in cotton (*Gossypium Barbadense*) dyed with an indigoid dye; and a 2-ply yarn of naturally colored beige cotton speckled with shiny white milkweed fiber (*Asclepias* sp.) [14]. The fine warp threads measuring less than 1 millimeter in diameter may have been stretched on a rudimentary frame loom. Other examples from the site show the use of circular warps, a sophisticated range of twining techniques, patterning with supplementary warps and wefts (continuous and discontinuous), as well as balanced and

---

[8] A painted facsimile # 33.8.16 of a scene from a pharaonic tomb dated to 1897–1878 B.C. depicts weavers at a horizontal loom at The Metropolitan Museum of Art (New York).

[9] Traces of warp weighted looms have been found from the Eastern Mediterranean through central Europe and up to Scandinavia and Iceland covering a period of 9000 years. A Greek oil flask #31.11.10 dated to 550-530 BC at the Metropolitan Museum of Art shows women weaving on a warp weighted loom [11].

[10] This specimen is a Fort Rock style sandal, but it is from Catlow Cave, not Fort Rock Cave. It is about 9325 years old.

warp faced tabby weave, knotted netting, and looped textiles. Many fragments also show red, yellow, and black ochre staining.

3.3.3 Textile Specimen 1.3.3

Nineteen fragments of woven reeds have been found in the Areni-1 cave in Armenia, now housed at the Institute of Archaeology and Ethnography, National Academy of Sciences of the Republic of Armenia (Yerevan). One of the fragments found in the same trench as #001.2010.001 was dated to 4040 to 3800 cal. BC. Specimen #001.2010.001 is woven using twill weaves embellished with contrasting matt and glossy surfaces of inner and outer reed (Calamagrostis). The sequence of lifting the reed elements in twill weave has been changed in a corner to create a three-dimensional basket shape. Although most of the fragments found were clearly related to funerary rituals, they were also used as floor coverings as seen in stacks of fragments that were fastened to the ground with a wooden nail [15] (Figure 3.1).

3.3.4 Textile Specimen 1.3.4

Examples of linen in tabby weave at the Petrie Museum of Egyptian Archaeology (London) give us an idea of Egyptian mastery of this fiber. A Neolithic fragment, #UC 72770 dated to about 5000 BC is rather coarse with 12 by 9 threads per centimeter. By the bronze age, when the first kings ruled over unified Egypt, the Egyptians had mastered linen, splicing the stiff fiber, and spinning it into fine thread. An almost intact tunic, #UC 28614bI found at the Tarkhan burial is made of linen woven in tabby weave. Dated to 3482-3102 BC, the tunic has a thread count of 22 to 23 by 13 to 14 threads per centimeter, showing a definite progress in terms of fineness of cloth. The construction of the garment shows little cutting. A rectangular fabric folded over and whip stitched on one side forms the body; two pleated pieces placed at an angle form a V-neck bodice and sleeves. The natural stiffness of linen may have lent itself to pleating as a form of decoration [16]. Pale grey stripes are also present indicating efforts to color the fiber.

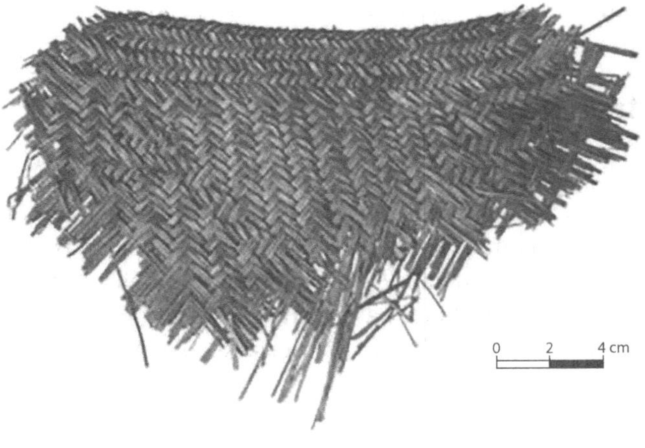

Figure 3.1 Twill weave reed mat #001.2010.001. Reproduced with permission from Institute of Archaeology and Ethnography, National Academy of Sciences of the Republic of Armenia.

3.3.5 Textile Specimen 1.3.5

Two trousers with woven patterning were excavated at Yanghai cemetery (Xinjiang, China) and radiocarbon dated to 13th to 10th century BC. The trousers finds number #2003SYIM21:19 and #2003SYIM157:14, were housed at the Turpan Museum (Xinjiang, China), in 2014 [17]. The less damaged #2003SYIM21:19 is made of three independently woven pieces of fabric each, involving no cutting in the construction. Two long rectangular pieces formed the two sides from waistband to ankle and a cross-shaped crotch-piece bridged the gap between the legs. All three parts were shaped on the loom, in the correct size to fit a specific person. Great attention to detail is seen—the yarns of the three fabrics and threads for final sewing match in color and quality, sewn edges are hidden under braids and patterns are carefully aligned. Most studies have been done on the construction and design of the bespoke trousers, which seem to be the predecessors of modern riding trousers, improving comfort and mobility[11] [18]. The contrasting colors of natural wool on a twill ground (as well as a turquoise weft in #2003SYIM157:14) is especially remarkable. The stepped patterning is probably made using tapestry weave. The fine lines of the neat repetitive horizontal zig zag motifs and sudden cropping of the lattice pattern is suggestive of early patterning tools. These specimens would merit further study of the woven techniques used.

3.3.6 Textile Specimen 1.3.6

Dated to the 5th to 4th century BC, the world's oldest and practically intact pile carpet #1687-93 was discovered in a burial site in Pazyryk, Siberia and is now at the Hermitage Museum (Saint-Petersburg, Russia).[12] Measuring 183 * 200 cm, it is made with a dense wool pile of 277 knots per square inch with symmetrical knots [20]. Especially striking is the combination of technical skill and sophisticated patterning with multicolored yarn. Composed of a central field surrounded by five borders, the motifs range from abstract crosses, floral buds, and mythical griffins to a realistic herbivore motif and men on horses. Another multicolored pile carpet fragment with 300 asymmetrical knots per square inch was found in a burial mound 100 miles west of Pazyryk. These and other textile fragments give us insights into the lives of nomadic horsemen and the cultural exchange of textile techniques and/or products between China, Central Eurasia (Scythia and Persia), and Armenia [21].

3.4 Textiles from Peru

Pre-Columbian textiles especially from Peru are one of the worlds most sophisticated in terms of mastery of structures and textile techniques, brilliant dyes and range of colors, the use of a variety of fibers (namely cotton and camelid fibers such as Alpaca), iconography and patterning, and attention to detail. As mentioned previously, 6200 years ago Peruvians at Huaca Prieta were already proficient at creating a variety of structures. Successive cultures across the Andes, such as the Chavin, Paracas, Recuay, Wari, Nazca, Chimu, and Inca

---

[11]Their age corresponds to the spread of mobile pastoralism in eastern Central Asia.

[12]The carpet is under consideration to be declared world heritage by the UNESCO [19].

Table 3.2 Examples of Pre-Columbian textiles from Peru.

Sl. no.	Museum	Accession number	Date	Description
1.4.1	The George Washington University Museum and The Textile Museum	91.466A, B and C	300-100 BC	Triple cloth
1.4.2	Brooklyn Museum	38.121	100-300 CE	Warp wrapping
1.4.3	The Museum of World Culture. Now in The National Museum of Archaeology, Anthropology, and History	1935.32.0118	100-300 CE	Discontinuous warp and weft
1.4.4	The Metropolitan Museum of Art	1987.394.706	700-900 CE	Tapestry weave
1.4.5	The George Washington University Museum and The Textile Museum	91.341	750-950 CE	Discontinuous warp and weft
1.4.6	Dallas Museum of Art	1986.39	1200-1400 CE	Gauze weave
1.4.7	The George Washington University Museum and The Textile Museum	91.504	1750 CE	Tapestry weave

cultures dating from 900 BC to 1800 CE explored unique methods to weave on rudimentary looms, such as warp wrapping, twill and tapestry weaves, double and triple cloths, warp and weft substitution, supplementary warp and weft, brocading with supplementary wefts, complementary warp, crossed warp patterning, discontinuous warp and weft[13], looped pile, twining and gauze weaves, and on rare occasions warp ikat. They also practiced braiding, sprang, embroidery, tie dye, appliqué, and patchwork. A great amount of attention to details was given, such as borders with fringes, tassels, and embroidered figures. Their iconography ranged from geometric, zoomorphic, and anthropomorphic motifs, arranged either diagonally or in vertical lines with a lot of play of permutations and combinations of colors.

[13]Weaving with both discontinuous warps and wefts is a technique that is unique to Peru.

Camelid hairs absorb dyes very well, and together with dyes from natural sources, Pre-Columbian cultures derived an astonishingly beautiful and wide range of colors.[14]

The National Museum of Archaeology, Anthropology, and History (Lima, Peru) contains the biggest collection of textiles from the Paracas period (800 to 100 BC)[15] and fine examples with complex weaving techniques exist in many museums around the world. This list of seven specimens is meant to illustrate some of the woven techniques mentioned above. (Table 3.2).

3.4.1 Textile Specimen 1.4.1

The George Washington University Museum and The Textile Museum holds several fragments in tabby weave double cloth from Peru. Most are in two contrasting colors like #TM 91.697 and present both sides of the fabric with the same patterns but with reversed colors as is typical of double cloths [23]. Three fragments of the Recuay culture #91.466A, B and C are made of triple cloth in camelid hair allowing three distinct sets of contrasting colors of warp and weft—red, yellow, and gray—to interlace in tabby weave. We see that these fragments have been woven systematically in three layers to create motifs in multiple color variations on both the face and back of the fabric. This is an important example of triple cloth unlike many other textiles that have multiple sets of color, with what is better qualified as warp and weft substitution [24] as the weaver chooses to either weave sets of warps and wefts or leaves them floating within the structure of the complex fabric. Peruvians also produced fancy versions like The George Washington University Museum and The Textile Museum's #91.852A from the Chimu culture made in cotton and camelid fiber, which combines tabby double cloth using two sets of warp and weft along with supplementary weft patterning.

3.4.2 Textile Specimen 1.4.2

The "Paracas Textile" #38.121 at the Brooklyn Museum, New York, is one of the most renowned Andean textiles in the world, highly valued for its complexity and mysterious imagery. Mantles with similar borders or fragments of similar borders can be found in several other museums and are usually the focus of attention for the embroidery technique (cross looped needlework) used to create elaborate figures representing flora, fauna, cultivated plants, and humans impersonating gods [25]. The field with motifs made of wrapped warp in varied colors is a rare example of a method that comes close to the concept of ikat. The warp made of cotton is carefully wrapped with dyed camelid hair around individual warp ends to create motifs. The fabric is then woven with a balanced tabby weave using a cotton weft creating hazy motifs due to the interaction of the colored wrapped areas and the natural white of the cotton weft. Both faces of the fabric are thus identical [26]. This inner field was probably created earlier than the surrounding border and a consensus of its provenance and age has not yet been reached. The mantle was most likely used as a ceremonial object.

---

[14]One of the most important dyestuffs they discovered was cochineal [22].

[15]The online collection is described in Spanish and thus presents some difficulties.

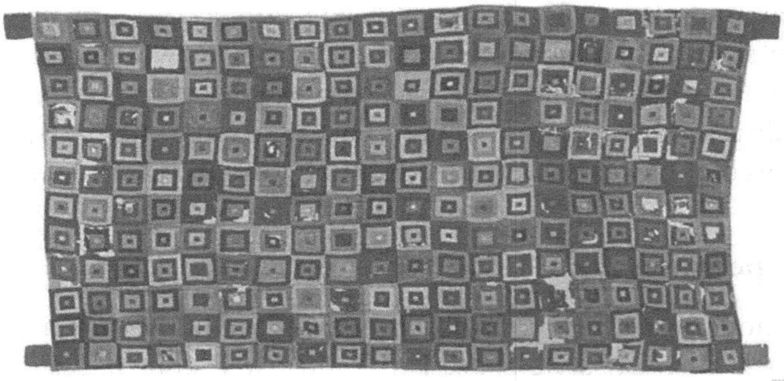

Figure 3.2 Mantle in discontinuous-warp-and-weft technique, #1935.32.0118. Reproduced with permission from The Museum of World Culture, Gothenburg.

3.4.3 Textile Specimen 1.4.3

The Museum of World Culture, Gothenburg, Sweden, held #1935.32.0118,[16] a mantle with 240 squares in 22 different colors in camelid hair. The unusual pattern is made of concentric squares containing four smaller squares of varying sizes and colors. Only three squares have the same combination of colors [27]. The mantle illustrates a common device used by weavers of the Paracas culture when creating motifs—that of mixing up colors in such a way that the viewer is dazzled by the varying color combinations. The squares of the mantle have been woven in a balanced tabby weave with multiple discontinuous-warps-and-wefts joined together with double interlocking technique. Some form of scaffolding yarns may have been used to construct the discontinuous warps. Since the color areas are too small for lifting groups of warp threads at a time, a needle may have been used to do the interlacing of warp and weft [28]. Measuring 180 * 89 centimeters, the mantle was probably shortened, and the original may have had even more squares along with a tubular border in the cross looped needlework mentioned in 3.4.2 above (Figure 3.2).

3.4.4 Textile Specimen 1.4.4

Moche-Wari Pre-Columbian weavers mastered tapestry weave structures turning the inherent constraints into design features. They either stitched-up slits caused by color blocks employed in the tapestry technique or exaggerated them to make perforated patterned borders. They also devised a clever way to close the slits by dovetailing two adjacent warp ends with an outline color weft as seen in The Metropolitan Museum of Art's #1987.394.706 in cotton and camelid hair. The Dallas Museum of Art's #1991.377 is another example of clever use of tapestry weave to create interesting fringes and repetitive slits to create an open work fabric.

3.4.5 Textile Specimen 1.4.5

Spectacular Wari tunics or their fragments with vivid tie-dyed colors, such as the well-preserved specimen #91.341 at the George Washington University Museum and The Textile

[16]This object was repatriated to The National Museum of Archaeology, Anthropology, and History in Peru in 2017.

Museum are common in Pre-Columbian textile collections. They are especially fascinating due to their complex method of manufacture, which involved weaving separate units of discontinuous warps and wefts stabilized on scaffolding made of cords stretched across frames in tabby weave, disassembly into individual parts, tie-dyeing the parts, followed by reassembly to produce color blocks with sharp edges not normally possible with tie dyeing [29].

3.4.6 Textile Specimen 1.4.6

A variety of methods were used to skillfully weave lacelike textiles in cotton. Of these unusual textiles, the Chancay style gauze is the most well known. Spaced warp and weft mesh fabrics were made with fine tightly spun cotton warp ends in alternating S and Z spun directions that are crossed over each other and kept in place with wefts. The wefts could be either wrapped or knotted around the crossed warp ends [30]. Once this mesh was made, the fabric was further embellished with embroidered motifs as seen in #1986.39 at the Dallas Museum of Art. Several lesser-known variations exist in terms of structure, as well as color, the most common being in natural undyed cotton, followed by some examples that are either all brown, or dyed in one color such as indigo as seen in #DTM 12809 at Deutsches Textile Museum (Krefeld, Germany) [31]. The Michael C. Carlos Museum of Emory University, Atlanta, has an unusual polychrome example #2002.1.85 of Chancay gauze with discontinuous warps. It is probable that, for some of these textiles, two weavers worked together pushing beyond technical limitations to bring together laborious techniques of gauze weaving and discontinuous warping in a single textile [32].

3.4.7 Textile Specimen 1.4.7

After the arrival of the Spanish in the Americas in the 1500s, silk textiles from China and Japan reached South America via ships like the *Manilla Galleon*. From this point onwards Pre-Columbian textiles were no longer insular. Fine Inca tapestry cloth called *cumbi*, usually reserved for royalty was as valued for its silken feel and brilliant colors, as silver and cochineal by the conquistadors. A fine example of artistic cross-fertilization at the George Washington University Museum and The Textile Museum #91.504 is made of cotton, camelid hair and silk in tapestry weave perfected by Incas but with iconography inspired by Chinese Ming silks and Christian didactic themes promoted by missionaries [33].

3.5 Chinese Textiles

As we have seen in the previously mentioned trousers from China, woven patterning with regularly repeating, identical motifs began as early as the 13th to 10th century BC in China. Prolific in their textile techniques, early examples show twining, plain and twill weaves and their derivatives, supplementary warp, tapestry weave, warp faced compound weave, and plain and patterned gauze weave. Looped structures, painted patterns, dyed and resist patterning and felts have also been excavated at multiple sites. From the 8th to the 1st century BC, the Chinese knew the use of wool but their most important discovery, silk (*Bombyx mori*), was known some 8500 years ago [34].

The period between the 5th century BC to the 2nd century CE (Warring States and Han Dynasty) saw the evolution of treadle and multi-heddle looms that attest to early understanding of mechanisms that facilitated the lifting of groups of warp threads to weave patterned textiles with relative ease, rapidity, and repetition. This is also the time when weaving technology reached a historical peak with the renowned warp faced *jin* polychrome patterned silks that were highly valued and widely distributed woven on pattern looms. The China National Silk Museum (Beijing) has an excellent collection of ancient Chinese looms and their associated textiles [35]. It is also probably in China that large pattern repeats began to be woven on drawlooms [36]. It is believed that the design of these early looms influenced the invention of later looms and the spread of technology and woven structures throughout Eurasia, the Middle East, and Europe.

Chinese textiles starting from the Han dynasty until the Qing dynasty are notable for woven structures such as the above-mentioned compound warp faced *jin* silks, tapestry, brocade, gauze, twills and twill damask, satin, satin damask, satin grounds with velvet and cut and uncut pile velvets. The satin weave structure and fabric were invented in China [37]. Silk gauze weave textiles were usually embellished with silk embroidery as seen in two remarkable examples at the Metropolitan Museum of Art, #32.30.10 and #1981.410. Examples of complex gauze fabrics with small lozenge patterns or floral patterns, and clamp resist dyed gauze are held at the China National Silk Museum. The earliest Chinese velvet made of low uncut warp pile is dated to 2nd century BC. This technique may have died out in China and was probably re-introduced in the 13th century from the Middle East and/or

Table 3.3 Examples of Chinese textiles.

Sl. no.	Museum	Accession number	Date	Description
1.5.1	Abegg-Stiftung Museum	5302/ 5304	400–300 BC	Striped warp faced compound tabby weave
1.5.2	Cotsen Textile Traces Study Collection, % The George Washington University Museum.	T-1215a, b, c and d	127 BC-72 CE	Polychrome warp faced compound tabby weave
1.5.3	Hunan Provincial Museum	N/A	960 to 1127 CE	Twill damask
1.5.4	The Metropolitan Museum of Art	1989.282	12th to 13th C	Brocade
1.5.5	The Metropolitan Museum of Art	2001.595	14th C	Lampas
1.5.6	The Metropolitan Museum of Art	40.27.2	Second half 16th C	Satin Damask
1.5.7	The Metropolitan Museum of Art	60.1	Late 16th to early 17th C	Tapestry weave

Europe [38]. Lampas and proto-lampas structures influenced by Iranian weavers were also absorbed and adapted to the Chinese repertoire.

Almost no flora was depicted as motifs in ancient China before the Qin dynasty (221 to 206 BC) [39]. Daoist, and later Buddhist motifs (1st c. AD), as well as influences through trade via the silk routes were absorbed into Chinese iconography. Motifs that have become ubiquitous are mountains, rocks, clouds, flames, animals, such as tigers, dragons, phoenix, rabbits, various birds and fish, stylized flowers such as the lotus and peonies and abstract versions of lichen as well as intricate geometric motifs. Proficient in dyeing silks with natural dyes, luxurious textiles blended shades of polychrome silk, gold, and silver metallic threads. Both flat strips of gold leaf and metal wrapped around a silk core were used. Gold leaf on animal substrates (influenced by Islamic textiles) and paper (Asian provenance) were used [40].

UNESCO cultural heritage lists from ancient China include both the tangible and the intangible. Among these are the excavated pattern loom models from the Han Dynasty, two *jin* textiles[17], and the sericulture and silk craftsmanship of China, which include the weaving and looms of both the Han and minority groups [41]. The following selection of 7 woven specimens illustrates some of the technically advanced mastery of woven structures in China (Table 3.3).

3.5.1 Textile Specimen 1.5.1

The Abegg-Stiftung Museum silk specimens, #5302 and 5304 belonging to the same fabric, are remarkable for the warp stripes colored in vermillion threads dyed with cinnabar as well as the regular and intricate patterning with a repeat height of 4 centimeters. This warp faced compound tabby weave fabric has two series[18] of warps with approximately 58 to 64 (warp) ends per series and 36 to 42 (weft) picks per centimeter. The textile is made of several lengths sewn together and remains of padding and lining suggest that it may have been used as a blanket [42]. Patterning with geometric and stylized animal motifs that are mirrored symmetrically along the warp in three colors shows a clever mastery of loom technology. By lifting specific groups of (warp) ends to achieve one half of the motif and reversing the lifting order of (warp) ends to obtain the remaining half, loom set up and lifting of (warp) ends would have been eased. The Metropolitan Museum of Art's oldest Chinese textile fragments #2002.558 and #2002.559 are similar fragments. The Cotsen Textile Traces Study Collection, in the care of the George Washington University Museum also has a comparable fragment #T-0327c 18770 that has been carbon dated to 381-207 BC. These are useful samples to consider the development of this weave structure from two sets of warps to more elaborate fabrics that used up to five sets of warps simultaneously as in the polychrome patterned textiles mentioned below (3.5.2) (Figure 3.3).

3.5.2 Textile Specimen 1.5.2

Within the family of weave structures that probably inspired the creation of the taqueté weave structure, are warp faced compound tabby weave polychrome patterned textiles

---

[17] 2 *jin* textiles have been listed under the UNESCO Silk Roads Programme.

[18] In this context, the word 'series' can be used to denote a distinct set.

Figure 3.3 Striped warp faced compound tabby weave #T-0327c-d. Reproduced with permission from Cotsen Textile Traces Study Collection, courtesy of The George Washington University Museum and The Textile Museum.

collectively known as *jin* textiles. Four fragments of one such textile, #T-1215a, b, c and d, in the collection of the Cotsen Textile Traces Study Collection in the care of the George Washington University Museum are identical to another specimen that was replicated at the China National Silk Museum. The replica was woven on a loom that was created based on a pattern loom called the Laoguanshan loom with 84 pattern heddles.[19] The original *jin* silk fragment used as an arm protector has 220 (warp) ends and 24 (weft) picks per centimeter, and 5 sets of warps that appear on the face simultaneously to create the pattern. It is now in the collection of the Xinjiang Institute of Cultural Relics and Archaeology (Xinjiang) [43]. Similar fragments #5301/5303/5336 exist at the Abegg-Stiftung (Figure 3.4).

3.5.3 Textile Specimen 1.5.3

Silk twill damask or *Ling* was most popular during the Tang Dynasty from 618 AD. The most common twill damask is woven with wefts and warps in the same color. The patterned areas could be altered by changing one or more of three factors: warp and weft threads, twill direction and float sizes. In an example at the Hunan Provincial Museum (Changsha) dated from 960-1127 CE, a warp faced twill is juxtaposed with a weft faced twill to create a contrast in the way light bounces off the floats so that one area looks glossy while the other looks matt. A similar twill damask woven with motifs of boys tucked among floral scrolls and lotus pods, #52.8, at the Metropolitan Museum of Art is dated to the 11th to the 12th century. In addition to the Buddhist belief in rebirth from the lotus, multi seeded lotus pods with boys were thought to symbolize plentiful offspring [44].

[19]The pattern repeat recorded by the Abegg-Stiftung is 8.5 centimeters high.

Figure 3.4 Polychrome warp faced compound tabby weave #T-1215a. Reproduced with permission from Cotsen Textile Traces Study Collection, courtesy of The George Washington University Museum and The Textile Museum.

3.5.4 Textile Specimen 1.5.4

Brocades produced from the 12th to the 13th centuries constitute an important group of textiles, falling into two categories: those that clearly belong to the Chinese Jin Dynasty traditions and those that were influenced by weavers from eastern Iran who were relocated to China. Gold brocading wefts composed of flat strips of animal substrate gilded on one side are found in many examples. Although slight technical variations are found in terms of foundation wefts or warps, the continuous gold brocaded wefts are mostly bound by warp ends in a twill. Stock patterns with both symmetrical as well as asymmetrical motifs may have been used to weave multiple fabrics differing in color and technical details. Brocades of this nature made in the imperial workshop had finer strips gilded on both sides and floated across the back between the repeats of the motifs. Available evidence suggests that gold brocade represented the most luxurious textile of the Jin dynasty [45].

3.5.5 Textile Specimen 1.5.5

The lampas weave in China emerged during the Yuan Mongol dynasty (1271 to 1368). This hybrid weave structure combined Chinese and Iranian elements to produce *nasij*, or cloth of gold lampas. Introduced by Iranian weavers (mentioned in 3.5.4), a main warp and weft were used to outline designs on a ground made of flat gold continuous supplementary wefts bound by a separate binding warp. The *nasij* was one of the most luxurious materials in the Central Asian Yuan court, woven in imperial workshops by weavers with high technical expertise. The Metropolitan Museum of Art's #2001.595 is a *nasij*, with red silk outlines of the pattern on a gold metallic thread ground [46]. Variations with simpler structures were also woven concurrently with the *nasij* lampas, using gold thread. One such textile known as *Jin Duan zi*, was woven with only one (main) warp, a ground weft and continuous or discontinuous supplementary weft gold thread for patterning. The supplementary gold threads were bound by the main warp. *Jin Duan zi*, with dark grounds sometimes use a proto-lampas weave structure incorporating binding warps that were yellow to blend better with the gold patterning and to achieve strong color contrasts [47].[20]

3.5.6 Textile Specimen 1.5.6

As mentioned earlier, the satin weave structure was invented in China and derived its current name from Tsia-Toun [48]. During the Ming period (1368 to 1644), the terms *duan* and *zhusi* were used for "plain damask" and a type of damask with yarns spun from a mix of silk and ramie [49]. Written descriptions of these textiles that are either bi-color or with patterns that reflect light differently suggest that they were probably woven with the long floats of warp and weft satins juxtaposed next to each other creating typical damask fabrics. Silk satin damasks like the Metropolitan Museum of Art's #40.27.2 with its combination of highly symbolic Buddhist motifs (elephants and stylized lotus flowers) and Hapsburg double-headed crowned eagle were woven in China for the Iberian market. Its five-end satin has a blue warp dyed with an indigoid dye while the yellow of the weft was made from the Asian pagoda tree (*Sophora japonica*). A white twill selvage may indicate that this damask belongs to a group of silks that were exported via the port of Macau.

3.5.7 Textile Specimen 1.5.7

Tapestry weave textiles in wool from the Xinjiang region of northwest China during the Han dynasty may have been the precursors of this technique that privileged weft patterning [50]. Regular tapestry weave can be heavy, and relatively stiff due to the high number of (weft) picks that are packed down, hiding the (warp) ends. By the 11th century, during the Tang Dynasty, the technique was adapted to silk and gained popularity in the Song Dynasty (960 to 1279) [51] when the discontinuous silk wefts were combined with metallic thread. As in painting, this technique has no limitations to the number of colors that can be used and the freedom it afforded, is clearly seen in the lustrous light weight silk tapestry examples going all the way up to the Qing Dynasty (1644 to 1912). These fine silk tapestry weave textiles known as *kesi* were used in lieu of paintings, as wrappers, wall hangings, furnishings, and clothing as well as rank badges[21]. The Metropolitan Museum

[20] Lampas in silk and metallic threads with geometric patterns, supposedly derived from designs on ancient textiles, began to appear in the later Ming period (late 16th century). It continued to be woven during the Qing dynasty as seen in two similar pieces: The Metropolitan Museum of Art's # 34.98 and #34.66 at the Seattle Art Museum.

[21] Textile patterns or symbols could indicate rank or identity. Large badges sewn onto coats indicated the rank of the official wearing it in China, Korea and Vietnam.

of Art's #60.1 is a massive wall hanging with spectacular colors and motifs, probably belonging to a powerful person.

3.6 Textiles from the Indian Subcontinent

From the Indian subcontinent arose an extraordinary breadth of textile techniques and materials—block printed; painted; resist dyed; embroidered[22]; warp, weft and double ikats; felts; split ply braids; flat and pile carpets; and woven textiles. Weave structures range from the three basic weaves—tabby, twill, and satin, derivatives of these three basic weaves, supplementary weft and brocading, double cloth, samite, lampas, velvet, and symmetrical knotted carpets. Climatic conditions in the subcontinent are such that scant examples of early woven textiles remain. A wool fragment in tabby with stripes and tapestry weave patterning #1907, 1111.105 at the British Museum, found in China dated to the 1st century CE may be one of the earliest examples of woven wool from India [52]. Six cotton pile fragments found at Berenike in Egypt, dating to the same period, have also been attributed to India [53]. Written records dating to as early as the 1st century BC speak about fine trade textiles; we see people wearing patterned fabrics in the cave paintings of Ajanta dated to the 5th century; and enough evidence has been found to prove that the people of the Indus Valley civilization knew how to cultivate, grow, dye, weave, and trade cotton textiles. Cotton woven in different weights in tabby weave was the most popular both for internal consumption as well as trade as early as the 1st century CE. Dhaka muslins with their gossamer thin warps and wefts in single ply cotton (*Gossypium Arboreum var Neglectum* [54]) and tabby weave were the most coveted.

Tussar silk (*Antharea paphia*) was widely used in India. Two areas in North India developed woven silks using samite and lampas weave structures—Gujarat in northwest India and Assam in northeast India. Other silks including Mulberry (*Bombyx Mori*), Muga (*Antharea Assamensis*) and Eri (*Samia Cynthia*) are found in the repertoire of Indian saris. A large range of woven saris with their individual stock patterns of checks, stripes, brocaded motifs, patterned borders, and decorative ends (*Pallus*) exist. Some efforts have been made to record saris from different parts of India [55]. Notable woven saris with figured motifs are *Banarasi, Baluchari, Paithani,* and *Kancheepuram* in silk and *Jamdani* in cotton. Silk yarns wrapped with gold and silver were often used for patterning, and textiles combining silk and cotton were also popular, especially with the Mughals from the 16th century onward.

Drawloom technology was already present in India when the Mughal rulers arrived in the 16th century. In Kashmir, sashes, shoulder mantels and shawls in soft lightweight Pashmina (*Changthangi goat*) and Shahtoosh (*Pantholops hodgsonii*) were woven using twill tapestry weave. From simple border patterns during the time of the first Mughal emperors, these shawls developed into elaborate patterns that were highly sought after in Europe in the 17th and 18th centuries.

Mughal patronage encouraged the development of Indo-Persian taquetés, samites, lampas and velvets. Sumptuous lampas and velvets in silk and gold-wrapped thread as well as

[22]Indians mastered natural dyes, resists and mordants and traces of resist dyed patterned cottons have been found in large numbers in Egypt dated to as early as the 5th century. The earliest carbon-dated block printed fragment exported to Egypt from India dates to 8th century CE.

Table 3.4 Examples of textiles from the Indian Subcontinent.

Sl. no.	Museum	Accession number	Date	Description
1.6.1	The Metropolitan Museum of Art	1993.2a–m	15th -16th C	Samite weave
1.6.2	Calico Museum of Textiles	864, 1425	late 1500s or early 1600s	Lampas weave
1.6.3	The George Washington University Museum and The Textile Museum	6.29	1700s	Double cloth in twill and tabby and twill damask
1.6.4	Calico Museum of Textiles	35	Mid 1700s	Lampas weave
1.6.5	Victoria and Albert Museum	8653(IS)	1855	Tabby weave
1.6.6	Victoria and Albert Museum	IS.80-1957	1870	Twill tapestry weave
1.6.7	Victoria and Albert Museum	IS.5-1989	1900s	Tabby weave, ikat

elaborate hand-painted or embroidered cotton textiles were used as wall panels, door covers, cushions and floor coverings. A floral tent panel #1981.321 at the Metropolitan Museum of Art dated to ca. 1635 is in cut warp pile silk velvet with gold leaf motifs. Delicate velvets also seem to have been used as floor coverings. The Calico Museum (Ahmedabad) has a patterned velvet, #303 dated ca 1650 used as a floor covering [56]. Pashmina was also used for floor coverings in plush pile weave with silk warp, of which only fragments remain[23].

Mention must be made of the famed Patolas in tabby weave with resist dyed patterning in both warp and weft (double ikat). Prized in India as well as in Indonesia, written accounts of these double ikats appear in Indian sources in the 14th century and appear as traded goods to Southeast Asia in European accounts of the 16th century. It is not clear whether the initial double ikat textiles were made in cotton, silk, or both [57]. Ikat textiles are also found in southern and eastern India. The following 7 examples of textiles have been chosen for their level of complexity and prestige amidst importing countries (Table 3.4).

3.6.1 Textile Specimen 1.6.1

A textile made up of a collection of 15 fragments of which two are in a private collection and the rest #1993.2a–m are at the Metropolitan Museum of Art is a rare example of its type. Woven using a weft faced compound samite weave it has two sets of warp and weft in deep blue and yellow. It is among the earliest in a small group of medieval silks from India, the production of which has been attributed to the important textile centers of Gujarat, the Deccan, and the Assam–North Bengal region where drawloom technology and weaving of

[23]As Mughal velvets were highly influenced by Persian textiles, we have preferred to detail velvets in the Persian textile section (3.9).

patterned silks had been going on before the arrival of the Mughals in 1526 [58]. Several complex patterned silks were exported to Tibet and this one was possibly retrieved from a Buddhist Monastery in Tibet.

3.6.2 Textile Specimen 1.6.2

The Calico Museum holds two rare lampas weave textiles woven probably in pre-Mughal Gujarat, in northwest India, #864, and #1425. Being an important hub of international trade, and a production center of exquisite textiles as early as the 5th century, there might have been special workshops with early drawloom technology. The more complete of the two identical pieces, #864, has unique tree motifs and is an important precursor to the later lampas textiles created for Mughal royalty. The main warp is dyed in lac red with pink or red binding warps, red ground wefts and pattern wefts in yellow, white, blue, green, and turquoise. The silk yarns used for the different warp and weft elements are varied- gummed, degummed, twisted and untwisted [59]. They were probably used as a tent screen or floor spreads. Silk lampas were also made in Assam in northeast India in mulberry and tussar silk, with woven Hindu motifs and inscriptions [60].

3.6.3 Textile Specimen 1.6.3

The George Washington University Museum and The Textile Museum's #6.29 is a sash or *patka* considered to be a technically sophisticated example of draw loom weaving with important differences compared to similar sashes made in Persia. Probably invented as part of a courtly costume by the Mughal emperor Jehangir these sashes are woven with two contrasting weave structures: one for the horizontal and side borders, and another for the field [61]. The borders, with floral motifs are in double cloth, with a ⅓ weft faced twill face in metallic thread and a tabby weave back in silk, while the field is a twill damask with a warp faced 3/1 twill in silk and 1/3 weft faced brocaded motifs of 2 wavy lines or clouds in metallic thread [62] (Figure 3.5).

3.6.4 Textile Specimen 1.6.4

The Calico Museum's #35 is a tent panel woven in a lampas weave with a silk 5-end satin weave foundation patterned with silk supplementary wefts bound in a 1/2 twill weave. As is typical of lampas with satin grounds, the proportion of warp is much higher than weft in the ground (5 (warp) ends for 1 (weft) pick or 75: 15/centimeter). A binding warp of about 16 (warp) ends per centimeter binds the floating wefts on the back when they are not being used for patterning on the face. Both ground and patterning wefts are untwisted silk for maximum sheen. The ground is red, and the patterned floral motifs are in white, blue-green and salmon

Figure 3.5 Double cloth with weft faced twill and tabby, and twill damask. #6.29. Reproduced with permission from The Textile Museum, Washington, D.C., Acquired by George Hewitt Myers in 1928.

pink [63]. This opulent tent panel is typical of the interiors of tents when Mughal rulers travelled in India.

3.6.5 Textile Specimen 1.6.5

Although the art of weaving Dhaka muslin is lost, many efforts to revive the same have been made [64]. These efforts have been supported by the study of museum specimens and swatches that were compiled into 18 volumes in 1866, by John Forbes Watson, the British Reporter for the Products of India at the India Office in London. Watson's systematically recorded samples cut from textiles that were sent for the 1855 Paris International Exhibition, and previous museum acquisitions allow us to see different qualities of renowned muslins [65]. The Victoria and Albert Museum (V&A, London) #8653(IS) named *Abrawan* or running water is an example of a specific grade of fine muslin.

3.6.6 Textile Specimen 1.6.6

The Kashmir shawl was woven on drawlooms with multiple discontinuous wefts wrapped around slim quill like "*Kanis*" or individual shuttles. The twill tapestry weave shawls developed over three hundred years under four different periods of political rule: Mughal, Afghan, Sikh, and Dogra. During the reign of Jehangir (1605-1627), the flowering plant or the beginnings of the boteh motifs first appeared. Later, the boteh was defined by its shape filled by floral mosaics, gradually filling up entire shawls with large sweeping curves and patterns. Exports to Europe and especially France was important and spurred intense creativity both in India and France. Beautiful examples of these shawls exist in the Victoria and Albert Museum. For example, the #IS.80-1957 with elegant, elongated paisley motifs is hand woven in Pashmina in Kashmir, dated to 1870, while #T.72-2010 in Pashmina is woven on a Jacquard loom in France and dated to 1850. The Bharat Kala Bhavan, at the Banaras Hindu University in India and the Calico Museum have multiple examples of Mughal mantles, waistbands and shawls dating from the mid 18th century [66].

3.6.7 Textile Specimen 1.6.7

Patolas are plain weave silks with double ikat, a process where both warp and weft are tie dyed and the design is meticulously woven to avoid "bleeding" or feathering of edges when the warp and weft threads are woven together. Based on the degree of precision achieved, the technical quality and consequent value of a *patolu* (singular) is determined. Traditionally, a set system of motifs and colors were assigned to each of the fifteen or so *patolu* sari types used by specific communities or purposes [67]. Spectacular *patolas* with elephant patterns like #IS.5-1989 at the Victoria and Albert Museum produced as ceremonial cloth for the Indonesian royalty were of the highest quality due to the density of weave, and complexity and accuracy of dyeing [68].[24]

---

[24] A fragment of warp ikat in plain weave cotton dated to 7th to 9th century CE excavated at Nahal Omer, Israel, is attributed to India due to the direction of the spun yarn [69]. The ikat technique may have begun as single ikats in cotton as seen in this fragment and Ajanta paintings or in silk. Ikat fragments found in Deir 'Ain 'Abata, a Byzantine monastery in Jordan, are attributed to India [70].

3.7 Textiles from Sudan, Egypt, Mali, and Ghana (North and West African Textiles)

Cotton (*Gossypium herbaceum*) was cultivated in southern Egypt and northern Sudan as early as the 2nd to 3rd century CE [71]. Examples from the cemetery of Karanog in Nubia show the use of natural cotton and cotton dyed in shades of a blue indigoid dye. An example of cotton embroidered with blue yarn requires further study to ascertain if the blue yarn is wool. Three fragments, #E7511A.1.2, #E7511A.2, and #E7511A.3 at The University of Pennsylvania Museum of Archaeology and Anthropology (Philadelphia) are described as having short twisted and closely woven loops and merit further analysis of weave structure and dating.

Linen processing reached its pinnacle in terms of fine yarn counts in tabby weave in Egypt by 3000 BC to 1300 BC. Egyptian linen (*Linum usitatissimum L.*) processing was a fine art, using pale green linen from young plants, bleaching it in the sun to obtain shades of white or, in rare cases, stained with red ochres or dyed in indigo (The Metropolitan Museum of Art, #20.3.203a and #09.184.217 respectively). Linen was widely used in woven textiles throughout the eastern Mediterranean region, in Greece, Syria, Turkey, Iran and Iraq. Wool and silk became important fibers starting from the Roman period in Egypt from 30 BC to 395 CE. As these fibers were easy to dye in natural dyes, they were especially used as colored wefts in tapestry weaves on natural linen warps. Purples, dyed with the precious Murex[25] or mixtures of Madder and indigotin, or polychrome shades of red, blue, green, and yellow are seen in decorative textiles loosely called "Coptic textiles" dated from the 4th to the 13th centuries CE. Wool was also used for both warp and weft. However, linen continued to dominate as the fiber was so inextricably linked to Egyptian culture, history, and geography.

Some of the best examples of Coptic textiles in terms of preserved size and color are in museums such as the George Washington University and the Textile Museum, the Dumbarton Oaks Research Library and Collection (Washington DC), The Metropolitan Museum of Art, The Brooklyn Museum, and The Cleveland Museum of Art in the USA, as well as museums around the world, such as the Musées Royaux d'Art et d'Histoire (Brussels), the State Hermitage Museum (St. Petersburg), the Abegg-Stiftung (Riggisberg), and the Musée d'Art et d'Histoire (Geneva) [72].

Linen tabby weave in a range of qualities was predominant until the Roman period, after which tabby derivatives became typical: basket weave with paired threads, and tapestry weave with colored wefts. Along with these weave structures, uncut loop pile appeared as seen in examples dated to 400 CE [73, 74]. See section 3.9 for taqueté textiles in Egypt which was by then well connected to Iran and Central Asia.

Textile finds from burials in the Bandiagara cliffs of Mali carbon dated to as early as the 10th century to 16th century are important because this is a period of which scarcely anything is known about weaving and clothing in Sub-Saharan Africa. West Africa is in many respects the heartland of textile production in Africa with Central Mali probably being the original locus from which weaving was diffused across West Africa [75–77]. Regarded as the earliest textiles known from the region, these textiles are preserved in museums such

---

[25]Murex dyes were reserved for royalty.

Table 3.5 Examples of textiles from North and West Africa.

Sl. no.	Museum	Accession number	Date	Description
1.7.1	The University of Pennsylvania Museum of Archaeology and Anthropology	E7511C.1, E7511C.2.1, E7511C.2.2, E7511C.2.3	2nd-3rd CE	Weft faced tabby weave
1.7.2	The George Washington University Museum and The Textile Museum	71.78	4th to 7th CE	Tapestry weave, tabby weave, supplementary weft loop pile.
1.7.3	Musée National du Mali	92.05.173	11th–13th CE	Tabby weave
1.7.4	British Museum	Af1947,06.1	Acquisition date: 1947	Tabby weave with supplementary weft

as the National Museum van Wereldculturen (Leiden) and the Metropolitan Museum of Art on long-term loans from the Musée National du Mali (Bamako). The structures and patterns of the fragments resemble textiles woven by the Fulani and Mande weavers in more recent eras, suggesting connections with populations who lived closer to the centers of trans-Saharan trade. Cotton was widely cultivated in Sudan; indigo dye was imported from north of the Sahara, while sheep's wool came from the region of Mauritania to the northwest [78].

Although developed much later in the 17th century, the *Kente* strip weave cloths from Ghana with their abstract motifs ought to be mentioned due to their dynamism and symbolism. Originally reserved for royalty, *Kente* is believed to have its roots in a long tradition of weaving dating back to 3000 BC [79]. The Democratic Republic of Congo is reputed for *Kuba* cloths—cut pile embroidery on Raphia (*R. taedigera*) tabby weave fabrics. Wild silk (Anaphe and Epanaphe moths) is used for weaving and embroidery in Nigeria and fine gauze cotton weaves are found in Ethiopia.[26] Unfortunately, as in the Indian subcontinent, few ancient textiles have survived in the regions with tropical climatic conditions in the African continent. Four examples of woven textiles have been listed below to illustrate the proficiency of weaving as described above (Table 3.5).

3.7.1 Textile Specimen 1.7.1

The cemetery of Karanog, ancient Nubia, (southern Egypt and northern Sudan), yielded woven cotton textile fragments with border patterns in two shades of blue and natural cotton dated to the 2nd to 3rd century CE [80] as seen in four related fragments

[26]Other noteworthy textiles from West and Central African countries are indigo-dyed resist fabrics using stitch, folding and tie dye as well as surface application of cassava paste as in the *Adire* textiles of Nigeria. *Bogolanfini* are mordanted and painted with iron rich clays from Mali. Tribal weaves, especially floor coverings from Algeria, Morocco and Tunisia are also outstanding.

Figure 3.6 Weft faced tabby weave with patterned border, #77.3. Reproduced with permission from The Textile Museum, Washington, D.C., Acquired by George Hewitt Myers in 1934.

numbered E7511C.1, E7511C.2.1, #E7511C.2.2, and #E7511C.2.3 at The University of Pennsylvania Museum of Archaeology and Anthropology [81]. Three fragments #77.1, #77.2, and #77.3, at The George Washington University Museum and the Textile Museum appear to be similar. The fabric is woven with a weft faced tabby with a tan colored warp and blue (indigoid dye) weft. The patterning is done in 2 different shades of blue and tan in tapestry weave. We can distinguish expertise in the regularity and intricacy of the motifs (Figure 3.6).

3.7.2 Textile Specimen 1.7.2

By 300 CE, decorative purple tapestry weave motifs began to be used on the plain tabby weave linen. This required either a shift in weave structure which was done by allowing some warp ends to float freely at the back or special pieces were woven separately in tapestry weave and then attached to tabby linen. Some structures show the use of 3 wefts – a ground weft in linen, supplementary weft in wool and discontinuous wefts in fine linen wrapped around warp ends to provide delicate outlines or strokes within motifs. Uncut loop pile appeared as seen in examples dated to 400 CE. [82] One such example is a Coptic textile, #71.78 at The George Washington University and The Textile Museum in undyed linen and polychrome wool with plain weave, tapestry weave, and supplementary weft-loop pile [83]. Weft-looped pile in linen or silk was woven in the Mediterranean, especially Egypt in the 3rd to 8th centuries [84] (Figure 3.7).

Figure 3.7 Tapestry weave, tabby weave, supplementary weft loop pile, #71.78. Reproduced with permission from The Textile Museum, Washington, D.C., Acquired by George Hewitt Myers in 1941.

3.7.3 Textile Specimen 1.7.3

Textiles belonging to the Tellem civilization excavated in the burial caves in the Dogon region of Mali are regarded as the earliest textiles known from West Africa. #92.05.173 from the Musée National du Mali is a blanket fragment from the Bandiagara burial cave. Cotton and wool blankets were used to wrap the dead [85]. They are patterned in a variety of stripes, checks, and geometric motifs and exhibit finely woven cotton, indigo dyes and both plain weave and weft patterning, as well as some woolen textiles. The textiles are composed of narrow strips sewn together edge to edge to make a larger cloth. This tradition of strip-weaving 1-to-18-inch widths and then stitching them selvage to selvage to make them into large rectangle wraps is common in West Africa [86].

3.7.4 Textile Specimen 1.7.4

Alternating warp faced and weft faced blocks are woven in tabby weave with supplementary weft motifs in narrow strips of cloth as seen in #Af1947,06.1 at the British Museum [87]. For the creation of Kente cloth, a long, narrow woven fabric is cut at fixed intervals to produce up to twenty-four bands, sewn together selvage to selvage. The Metropolitan Museum of Art holds several of these prestige cloths made of cotton and silk. One such example, #2010.557 is made using a rare technique which uses the principle of supplementary warp threads that are used to create meandering lines that go across the fabric in vertical lines, transition to supplementary wefts to make horizontal lines and go back to being supplementary warp again [88].

3.8 Textiles from Japan

Japan is reputed for a range of highly sophisticated textile techniques and traditions. A large range of silk textiles such as gauze, crepe, satin, and twill damasks destined for the upper classes were woven and piece dyed, or embellished with a variety of applied, clamp, bound, stitch and rice paste resist methods, as well as free hand painting and embroidery in polychrome silk and gold foil. Apart from tabby weaves, five unique variations of textured twill weaves were employed including intermittent twills on tabby ground and 2 twills combined on a single ground differing either in direction or size [89]. They also wove three types of silk gauze fabrics using different leno weave structures called *sha* (two (warp) end repeat), *ra* (four (warp) end repeat with two fixed and two movable (warp) ends), and *ro* (tabby weave rows alternated with rows of leno weave) [90]. Most noteworthy are the woven silk textiles manufactured during the 7th and 8th centuries. Among the woven silk fabrics, warp and weft brocaded textiles in polychrome silk and gold leaf on strips of paper loosely grouped under the term *nishiki* woven on draw looms are the most significant. Motifs can look painterly, large-scale, asymmetric or placed strategically on certain parts of the garment. The seasons, their flora or foliage and colors are important in clothing choices during the year.

Japanese clothing became inextricably entwined with social standing especially during the Momoyama and Edo periods from the late 1500s till the 19th century. During the Edo period (1603-1868), the shogunate issued 200 decrees ordering the merchant class to restrain themselves from wearing luxurious clothes. A clever and sophisticated fashion aesthetic called *soko itari* developed to get around the royal decrees: On the exterior, kimonos were discrete, but the linings were rich and ornate [91]. Stripes—vertical, horizontal or checks—became popular among Kabuki actors, merchants, and peasants. Woven vertical

Table 3.6 Examples of Japanese textiles.

Sl. no.	Museum	Accession number	Date	Description
1.8.1	Shōsō-in	N/A	Mid 8th century	Brocade
1.8.2	Shōsō-in	Listed as "Shokusei Kesa"	756 CE	Tapestry weave
1.8.3	Kyoto National Museum	46-996-1	14th century	Brocade
1.8.4	The Metropolitan Museum of Art	19.93.10	1615-1868	Twill ikat and brocade
1.8.5	Kyoto National Museum	165-3767-1	17th Century	Satin weave
1.8.6	The Fowler Museum	X2008.6.42	19th century	Single and Double ikat

stripes, or *Kissho gara* (lucky stripes), are considered auspicious, a symbol of eternal happiness. The *Mansuji*, or 10,000 stripes, looks like a flat color, but on closer inspection, it is made up of fine pinstripes. *Kameda jima* originally woven by poor farmers are in shades of blue, navy, black, grey, brown, and russet in cotton [92]. Folk weavers used bast fibers such as banana fiber (*Musa liukiuensis*), ramie *(Boehmeria nivea)*, hemp (*Cannabis Sativa L.*), recycled cotton fabrics and indigo resist patterning. Cotton was not an important fiber in Japan until after the 15th century. On the island of Okinawa, two unique fibers, *Bashofu* (banana) and *TonByan* (exact origin unknown) developed along with a rich tradition of ikats. *Kijōka-bashōfu* is recognized as one of the Important Intangible Cultural Properties of Japan.

The Kyoto National Museum (Kyoto, Japan) houses fragile textiles while many temples including the famous Shōsō-in, the treasure house of the Tōdai-ji Temple (Nara, Japan) have remarkable specimens of woven textiles that are deemed important from both historical and technical viewpoints. The list below illustrates some of the unique features of Japanese woven textiles (Table 3.6).

3.8.1 Textile Specimen 1.8.1

Nishiki encompasses a wide range of both warp and weft faced patterned textiles. The Shōsō-in at the Tōdai-ji Temple holds several examples of patterned polychrome silk warp or weft faced *nishiki* woven on draw looms. Warp faced *nishiki* seldom exceeded three or four colors with a few exceptions going up to 6 colors in the warp, and a monochrome weft. In the 8th century, inspired by Chinese textiles, weft-faced *nishiki* became very popular in Japan. This was the inverse of warp faced *nishiki* with a monochromatic warp and brocaded colored wefts including gold leaf paper, with the possibility of achieving many more color variations as wefts could easily be changed. An armrest cover with weft patterned floral tendril roundels enclosing phoenixes on a purple ground is said to have been inspired by a similar Chinese fragment unearthed at Turpan and now preserved at Musée Guimet (Paris) [93].

3.8.2 Textile Specimen 1.8.2

Originally, the tapestry weave technique in silk reached Japan from China. The Shōsō-in temple holds a silk priest's robe *(Shokusei Kesa)* in tapestry weave patterned with random shapes that resemble the irregular bark of a tree. The silk warp and weft threads are woven in groups of two (doubled). Almost invisible to the eye, fine ground (weft) picks in a contrasting blue color runs throughout the fabric enriching the overall color and strengthening the structure. The Kyoto National Museum (Kyoto) preserves a similar priest's robe (*Kenda Kokushi no Kesa*)[27] which was brought back by a Japanese priest who had travelled to China in the 8th century. It has a similar yet more prominent ground weft that appears at regular intervals like stitching [94]. Outside styles and techniques especially from China were further refined and adapted to Japanese taste, and sense of quality until the late eighth century AD.

---

[27] Accession number unavailable.

Figure 3.8 Twill ikat with brocade, #19.93.10. The Metropolitan Museum of Art.

3.8.3 Textile Specimen 1.8.3

As we have seen in the previous two examples, textiles from China were always held in admiration and used as models in Japan. While the textiles imported in the 7th and 8th centuries can be viewed collectively among the temple treasures of Hōryū-ji Temple (Nara, Japan) and the Shōsō-in, those from the medieval and early modern times are spread out in various repositories, being transmitted either as religious textiles like priest's robes, *Kesa* or as *meibutsugire* ("famed fabrics") used in the practice of the tea ceremony. Known as *Keitō Kinran* reddish-brown twill silk with supplementary gold continuous weft patterning is seen in the brocade #46-996-1 at the Kyoto National Museum. The gold brocade patterning wefts are bound in twill weave by the main warps (rather than supplementary warps). The ground wefts on the reverse side float under the gold wefts [95]. This simple method of binding the brocading wefts and allowing the ground wefts to float on the reverse indicates that the technique is an older weave structure.

3.8.4 Textile Specimen 1.8.4

Japanese Buddhist monks' robes went from humble robes of poverty to splendid patchworks or *Kesa* as seen in 1.8.2. The Metropolitan Museum of Art's #19.93.10 is a brocaded ikat fabric that was initially a Noh costume dated to the Edo period between 1615 and 1868. The ground is woven in a twill weave ikat in two colors, and the patterns are brocaded with silk and metallic thread (Figure 3.8).

3.8.5 Textile Specimen 1.8.5

The #165-3767-1 at the Kyoto National Museum is a patterned satin-weave silk dyed with stitched tie dye in horizontal bands of black, red, black, and white. Subsequently, each band is densely filled with small motifs done in embroidery and gold leaf. This is typical of the *Keichō Kosode*[28] garments popular in the early years of the Edo period. Documents

---

[28]Kosode are predecessors of kimono.

recording early Edo-period customs note that on ceremonial occasions the wives of the military elite wore *kosode* "without background", where the entire garment was covered with gold leaf [96].

3.8.6 Textile Specimen 1.8.6

E-gasuri, or pictures made with ikat, dating from the early 19th to mid 20th centuries can be seen in weft ikat or double ikat (warp and weft) textiles. Ikat was popularized during the second half of the eighteenth century, and became an important commodity produced in many rural areas of the country. Always woven in narrow widths of about 30 to 35 centimeters, the cloth was sold in long rolls. The X2008.6.42 at the Fowler Museum at UCLA (Los Angeles) has both weft and double ikat with indigo dye [97].

3.9 Textiles from Iran and Central Asia

Woven textiles belonging to salt miners of Chehrabad in Iran, dated to the Achaemenid (550-330 BC) and Sassanid periods (224-651 CE), give us a rare glimpse into common people's attire in Persia.[29] Dye analysis of some of these wool fragments show that they were dyed in red (*Rubia tinctorum L*), blue (probably *Isatis tinctoria L.*), yellow (probably *Tamarix* sp.) and green (mix of blue and yellow) [98]. Apart from wool, linen and silk were widely used throughout the Eastern Mediterranean and Western Asia, with silk being reserved for luxury textiles. The wool textiles from Chehrabad may have been inspired by the rich silk textiles that were being imported from Sogdia (a province of the Achaemenid empire in present day Uzbekistan, Tajikistan, Kazakhstan, and Kyrgyzstan) and China [99].

The Pazyryk carpet (see 3.3.6), dating from 5th to 4th century BC bears testimony to the prowess of weavers from the area between the Caspian Sea in the east, to Mongolia and China in the west, Persia in the south, to Russia in the north. Compound tabby weave *jin* silks (see 3.5) were traded all the way to present day Syria, which in turn became an important center of textile production. Silk fabric with repetitive curvilinear motifs woven with a draw-cord device for raising a succession of sheds was found in the coffin of St. Paulinus at Trier, dated to approximately 395 CE, and was probably woven in a Syrian workshop [100]. By the 3rd century CE, weavers in Central Asia had developed the weft faced version of the Chinese warp faced compound tabby weave structure—taqueté, which spread all over Syria, Egypt and Persia and was transferred back to China [101]. Taqueté fabrics while offering slightly less drapeability than warp faced structures are easier to weave. Harnesses with a lesser number of warp ends are lighter to move, and colors can be changed with multiple wefts. Taquetés evolved in Egypt as early as the 4th to 5th century as weavers strove to get pure color. Early examples like the Metropolitan Museum of Art's #2008.322.8 from Egypt are in wool.

---

[29] Photographs show one fragment with a tapestry weave patterned edge. Another colorful specimen has two different strips tied together: one strip is patterned with a checkerboard pattern and stripes and the other has vertical and horizontal stripes. No analyses of the textile structures were available at the time of publishing this chapter. These fragments are now at the Archaeological Museum of Zanjan, Zanjan province, Iran.

As the Sasanian empire stretched from Central Asia to current day Syria, it played a major role in cultural exchanges with Byzantine Egypt and the early Islamic eastern Mediterranean from 224- 651 CE. Along with trade and movement of people, rich tapestries and silk fabrics evolved with mixed Mediterranean, Roman, Greek, and Iranian elements [102]. Weft-faced polychrome samites continued to be made in the Islamic era that followed. The Victoria and Albert Museum's #1314-1888 is typical of this genre of woven samites with a main warp and a binding warp in a 1:1 proportion in red silk and 4 colors (red, green yellow and cream) for the wefts that are bound in a 1/2 twill weave [103]. The fabric is similar in color and weave structure to several Byzantine silk fabrics in the V&A dated from the 7th to the 9th century CE. The Sogdian Empire at the crossroads of an extensive trade network from the 6th to the 11th century CE also developed similar woven silks. During the 13th and 14th centuries various lampas with exotic patterns combining Iranian, Chinese, and Central Asian elements as mentioned in 3.5.5 were traded to Central Asia (Ilkhanid empire) Iran (Mamluk empire) and Italy where they were copied and/or inspired new creations [104].

Under royal Safavid patronage, the Persian silk industry reached its pinnacle with well-developed urban workshops for silk textiles, provincial centers of rug weaving and small farms cultivating silk in the Caspian region from 1501 to 1722. Silk Taffetas in tabby weave, ikats with supplementary wefts, patterned silk velvets (continuous cut pile, or mixed cut, uncut and voided velvets), velvets with discontinuous warps and lampas with gold and silver metal thread brocading were produced for both internal consumption and external trade. Designs for figured textiles relied heavily on illustrations for compositions and subject matter such as popular characters from Persian literature and poetry, flora, and fauna, hunting scenes and other popular pastimes. Skilled designers (*Naqshbands*)calculated the lifting of warp threads on drawlooms ensuring fluid lines to depict motifs in all over patterns using straight, step and large repeats [105].

Meanwhile, starting in the 14th century the Ottoman empire began developing motifs and textiles with a definite identity of their own. Due to Islamic influence in Uzbekistan, figured textiles gave way to flamboyant large-scale abstract patterns of colorful warp ikats in silk with cotton wefts as well as silk ikat velvets like the Art Institute of Chicago's #2005.606. Some examples of remarkable structures have been illustrated below (Table 3.7).

Table 3.7 Examples of textiles from Iran and Central Asia.

Sl. no.	Museum	Accession number	Date	Description
1.9.1	Abegg-Stiftung	5682	Mid 7th-8th CE	Samite weave
1.9.2	Cleveland Museum of Art	1996.2a	ca 700	Samite weave
1.9.3	The George Washington University Museum and The Textile Museum	3.219	Mid 16th century	Cut and voided velvet
1.9.4	The Metropolitan Museum of Art	52.20.15	Mid 16th century	Taqueté weave
1.9.5	Victoria and Albert Museum		1539-1540	Ardabil carpet

3.9.1 Textile Specimen 1.9.1

The Abegg-Stiftung possesses important examples of a group of woven silks in the weft faced samite weave structure, mostly with a red ground and polychrome weft faced motifs of animals in roundels which were found in Tibet, but which have a strong Sasanian flavor. The #5682 at the Abegg-Stiftung with deer in pearl bordered roundel motifs on a red ground is a large Sasanian silk that has been dated to mid 7th to 8th century CE. Woven in a 1/2 twill samite weave structure, it has a 3:1 warp proportion (3 (warp) ends for the face and 1 (warp) end for binding) and polychrome wefts. Similar silks were exported to Europe in early medieval times and were used to wrap relics. This Sasanian silk with a Central Asian (Sogdian) provenance, has a large-scale design that was favored in Central Asia and Tibet. The Metropolitan Museum of Art's #2004.255 is dated to 500-700 CE, with similar pearl borders and a typically Sasanian winged horse motif. The influence of these textiles' structures and/or motifs can be seen across Eurasia as far as Japan as they were widely exported via the Silk Road.

3.9.2 Textile Specimen 1.9.2

Sogdian patterned silks are recognizable for their symmetric arrangement of motifs such as pairs of deer, birds or hunters surrounded by roundels of pearls, the latter being strongly influenced by the above-mentioned Sasanian culture (ca 224 AD to ca 640 AD). Other influences for design elements came from Egypt and the Byzantine Empire. These symmetric motifs clearly allow the weaver to use a specific tie up of warp ends for half of the motif and inverse the tie up to complete the motif, thus easing the set-up of the patterning device or drawloom. One of the most complete and well-preserved Sogdian silks is the outer fabric of a child's coat, #1996.2a at the Cleveland Museum of Art. Typical to these types of weft-faced samite structures from Sogdiana, there are five weft colors and a proportion of three warp ends for the face and one end for binding wefts in a 1/2 twill [106]. Improving upon the taqueté weave, samites with twill weave weft floats allowed weavers to achieve unadulterated colors.

3.9.3 Textile Specimen 1.9.3

In the history of textiles, silk velvets from Safavid Persia represent the zenith of structural technique and decorative complexity. The #3.219 at the George Washington University Museum and The Textile Museum is one such spectacular example of a cut and voided velvet with a densely woven 5 end satin foundation and a plush pile made with supplementary warps in multiple colors. The textile is embellished with supplementary wefts of metallic strips bound in a twill weave [107]. The Metropolitan Museum of Art's #52.20.11 is a similar fragment that had a metallic thread background (now mostly worn off) and shows finely executed figures with permutations and combinations of colors for each repeat. These thick, heavy textiles were used for furnishings, cushions, interior hangings, tent panels, and ceremonial robes (Figure 3.9).

Figure 3.9 Cut and voided velvet with supplementary weft. #3.219. Reproduced with permission from The Textile Museum, Washington, D.C., Acquired by George Hewitt Myers in 1938.

3.9.4 Textile Specimen 1.9.4

Two important examples of early woolen taquetés produced in Sasanian Iran (similar in structure to #2008.322.8 from Egypt at the Metropolitan Museum of Art mentioned above) are #2000.624.2 and #48.43 dated to 6th to 7th century and 7th-9th century respectively. The latter has a double pearl border and stylized horses with neck ribbons drawing from early Sasanian/Sogdian motifs (mentioned in 3.9.1 and 3.9.2). The Cleveland Museum of Art holds multiple fragments of #1975.45, a taqueté carbon-dated to 899 to 1160 in silk from Iran. By the early 16th century, this woven structure was being used for luxury textiles in silk with silver or gold metal-wrapped thread in the Ottoman Empire. Original small-scale designs gave way to eccentric large-scale designs by the mid- to late 16th century, as in The Metropolitan Museum of Art's, #52.20.15, a fragment of an Ottoman Kaftan in silk and metal-wrapped thread. These examples allow us to speculate about how silk may have replaced wool in taquetés, and how the samite structure may have replaced the taqueté structure due to the purer colors the structure offered.

3.9.5 Textile Specimen 1.9.5

Persian and Ottoman carpets, generally known to be of the highest quality in terms of fineness, materials (silk and wool) and intricacy of pattern as well as organized production are world renowned. The Ardabil Carpet, #272-1893, dated to 1539-1540 at the Victoria and Albert Museum is one of the largest, most beautiful (in terms of complexity of pattern), and historically important examples. Made in the town of Ardabil, north-west Iran, a pair of carpets were commissioned by the court for the shrine of a Sufi leader, which, by the 16th century, had become a place of pilgrimage. The pair of carpets have a white silk warp and ground weft. The knotted wool pile is in ten colors, with an average of 350 asymmetrical knots per square inch. The second carpet, which has suffered more damage, is now in the Los Angeles County Museum of Art (Los Angeles, USA).[30]

3.10 Textiles from Italy and France

Prehistoric textiles such as those found in the salt mines of Hallstatt, Austria show the use of S and Z plied yarns to create subtle textural variations in tabby weave fabrics. Twill weaves combined with colored stripes in warp and weft were used for patterns such as the houndstooth and clever combinations of weave structure, color, warp drafting, and treadling are important European traditions. These initial innovations, with wool and linen being the two most common materials led to highly innovative textiles when the Roman and Byzantine empires began to import oriental textiles, and Europe began exploring the world subsequently.

Trade textiles, especially from Central Asia and Persia, and the spread of pattern loom technology catalyzed the Spanish and Italian textile industry during the Middle Ages and the Renaissance from the 14th to the 15th centuries. Silk damasks, brocades, lampas and velvets in polychrome silks and yarns in precious metals were used for clothing for the rich and powerful, religious vestments for the church, as well as furnishings. The weavers from Spain and the Italian peninsula produced most of the European luxury silks during this time and continued to dominate the sector well into the seventeenth century [108]. Patterned silk velvet was the most expensive and prestigious of all woven textiles.

Although the idea of pile structures originated in different places, Oriental velvets reaching Europe from the mid-13th century onward surely had an impact. Early European velvets had simple checkered patterns requiring no pattern harnesses and just 2-colored warps. Pattern velvets, on the other hand, require one set of harnesses for velvet figures and another for the ground. By the fifteenth century, Italian silk weaving had reached its zenith competing with luxury textiles manufactured in the Near East, Persia, and China. The new patterns and techniques achieved are among the greatest achievements of European ornamental design, breaking away from the motif repertory of antiquity and Renaissance. Inspiration for patterns went back and forth with lampas patterns being adopted for velvets and subsequently, monumental scale velvet designs influencing lampas and damasks. Ottoman and Spanish velvets imitated Italian velvets until they found their own artistic inspiration [109].

Beginning in the 1600s the French King, Louis XIV and his superintendent of finance, Colbert took a conscious decision to encourage the French textile industry. In 1536, the first Italian drawlooms were imported to Lyon in France. After the end of Louis XIV's reign (1715), and especially during the ensuing two decades of the 18th century, France and Italy remained the main centers of creation and dissemination of *Bizarre* silks noted for their motifs of oriental inspiration. Woven in a variety of structures, they could vary from simple monochrome or bicolor damasks to rich lampas brocaded with metallic threads. These textiles became so popular that centers for manufacturing them sprouted in several countries especially as Huguenot weavers from France fled religious persecution and settled in other countries of Europe taking the technique with them [110]. During this period, French

[30]Justice cannot be done to the vast sector of carpets within the scope of this chapter. Readers are invited to explore books that are focused on the subject.

Table 3.8 Examples of textiles from Italy and France.

Sl. no.	Museum	Accession number	Date	Description
1.10.1	Museo del Tessuto, Prato	N/A	Last quarter 15th century	Silk cut voided velvet
1.10.2	The Metropolitan Museum of Art	38.182.2	17th century	Genoa jardinière velvet
1.10.3	Musée des Tissus et Des Arts Décoratifs	MT 28996	Second half of 17th century	Brocatelle
1.10.4	Victoria and Albert Museum	T.187-1931	ca. 1770	Lampas

textiles went through an intense period of creativity with designs, colors, weave structures and combinations of materials, all the while developing a distinct French style [111]. France gained supremacy in the production of innovative textiles with royal and noble patronage, setting new trends for the rest of Europe. They combined textures derived from tabby, twill, and satin weaves for the grounds with traditional weave structures of brocades, damasks, and lampas, introduced linen wefts in traditionally pure silk textiles (*Brocatelle, Lampasette*), interpreted ikats as either warp painted (*Chiné a la Branche*) taffetas or velvets, and improved drawloom devices to weave bands of colored supplementary weft to create small repeats of motifs (*Droguets liserée*) and shawls inspired by Kashmiri shawls.

Until the invention of the Jacquard loom, patterned fabrics were woven using laborious tie up systems on three types of looms in Europe. Multi harness looms with up to 80 harnesses and button-drawlooms (operated with the help of buttons that allowed for the lifting of specific warp ends) were initially used by lone weavers to weave less complex figured fabrics. Drawlooms such as the ones invented by Philippe de Lasalle in 1775 employed two persons—the weaver and a person who would pull on individual pattern cords, which lifted groups of warp ends. This removable device was called a "*Semple*". Physically demanding, it was difficult to find people willing to take a draw person's job in 18th century France [112]. Searching for ways to expedite laborious weaving techniques, several weavers and inventors came up with ideas for the mechanized lifting of warp threads to weave patterned textiles. Among the people who contributed to the Jacquard loom mechanism are Basile Bouchon in 1725, Falcon in 1742 and Vaucanson in 1745. The Jacquard mechanism, patented in 1804 greatly expedited weaving.[31]

Leading up to this invention were a series of technological improvements made to looms as well as textiles. Remarkable woven textile specimens from Italy and France are illustrated below (Table 3.8).

[31]The Jacquard mechanism did away with draw persons and cords and instead relied on cards with holes punched in them. Needles would pass through the punched holes to lift warp threads based on a binary code of 0 and 1, zero signifying no change in warp and 1 signifying that the warp is lifted. This method of programming patterns was borrowed by Charles Babbage in 1839-40 for the very first computers [113].

3.10.1 Textile Specimen 1.10.1

During the 15th and 16th centuries, Venice and Florence were renowned for their sumptuous velvets incorporating gold and silver threads with large floral patterns. The example from the Museo del Tessuto, Prato is a silk cut voided velvet with three pile warps and gold brocading wefts with the heraldic device of the Medici family, manufactured in Florence during the last quarter of the 15th century [114]. As an expression of power, wealth, and taste, specially woven fabrics incorporating a family coat-of-arms or other motifs associated with the family's reputation were particularly valuable [115]. The Abegg-Stifftung's #1764 Italian fragment of a chasuble has a wavy vine design in gold velvet with cut and uncut pile, brocaded with silver gilt yarn (gold-coated silver wound around a silk core) dated to 1450 to 1460. The cut pile has three different heights, and the silver yarns used in the weft for brocading themselves have small loops (bouclé), and shimmering gold and silver speckled effects (*allucciolato*).

3.10.2 Textile Specimen 1.10.2

Jardinière velvets with multicolored floral motifs usually on a white ground first appeared during the 17th century and remained popular until the 19th century due to their rich details and finesse. Initially woven on draw-looms requiring a minimum of two people, one to do the actual weaving and another to pull up warp threads, floral motifs increased in size as loom technologies and the Jacquard loom were invented. Genoa, in Italy was most reputed for these velvets [116]. The Metropolitan Museum of Art's silk *Jardinière* velvet #38.182.2 from Genoa dates from the 1700s and has the typical cut and uncut velvet on a smooth satin ground. Another example of *Jardinière* velvet at the Royal Ontario Museum, #961.202.2, is an early 18th century example.

3.10.3 Textile Specimen 1.10.3

17th century textile design in France was dominated by Italian taste marked by stylized symmetrical arabesques in flat bicolor color application. The #MT 28996A was a popular fabric called *Brocatelle* woven with an 8-end satin ground with 2 silk warps (ground and binding warp) with a ground weft in linen and motifs in a continuous supplementary weft of silk bound in a 1/2 twill by the binding warp. This combination of structure and material based on the damask fabric structure, gave the appearance of bulk and slight relief of the motifs against the smooth warp satin ground [117].

3.10.4 Textile Specimen 1.10.4

"The Partridges" are one of the most well-known lampas fabrics made by designer, engineer, and maker Philippe de Lasalle. The Victoria and Albert Museum's #T.187-1931 has a ground made of a dense red silk warp in a tabby derivative, "*Cannellé*" and motifs that are brocaded and bound by a binding warp. This elaborate furnishing was commissioned in the early 1770s as part of the decoration for a salon in the Palais Bourbon in Paris. Ten years later, in the early 1780s, it was remade with a different background color for the Russian empress Catherine the Great who used it for walls, furniture covers, and curtains.

Figure 3.10 Lampas #50.8a, The Metropolitan Museum of Art.

In addition to the Metropolitan Museum of Art where it has been executed with a pale blue background (#50.8a) it is seen with a yellow background at the Abegg-Stiftung (#5064). These fabrics were woven using Lasalle's invention of a "*Semple*" which allowed for large vertical repeats of detailed pictorial designs that run across the width. In the case of the partridges, the width of the fabric is only about 55 centimeters wide while the full vertical repeat is about 192 centimeters high [118] (Figure 3.10).

3.11 Conclusion: Toward an Understanding of the Historic Foundations of Woven Structures

Ten thousand-years ago twined textiles were being woven in North America; 6200 years ago twining and tabby weave have been recorded in South America. A twined fragment dated to Japan's Jomon period—5000 BCE to 3500 BCE—has been found, [119], and iron age twill textiles were found in Austria. Textile making is closely associated with human evolution into mostly sedentarized civilizations. Textile techniques, tools and structures evolved across the globe at differing speeds. In the case of the sandals from Oregon, three clear phases of design can be discerned over a period of 700 years starting with a specific twined technique used for Fort Rock sandals, changing to two other types of sandals with structural variations. This gradual development of techniques and products may have led to the development of loom-like devices. A back tensioned loom with cords made of twisted rice straw at the Buyeo National Museum, (Chungcheongnam-do, South Korea) was probably used to weave sandals with a similar design and structure [120]. Continuity and change in weave structures ran parallel to each other as can be seen over large periods of time. With time, laborious weft twined structures dwindled to mere additions in woven cloth as edges and decorative finishes. A version of warp twining persisted in the form of gauze weaves while weavers embraced weaving tabby and twills on frame looms for their ease of weaving, capacity to interlace finer threads and expedite production.

From this foundation, arose two early groups of woven textiles in Peru, and China. Both demonstrate a creative range of woven structures, and mastery over specific fibers and natural dyes that have not been found elsewhere. Pre-Columbian textiles from approximately 7000 BC to 1450 CE show prowess in weaving eight categories of loosely grouped structures that consistently broke boundaries [121]. Innumerable variations in yarn counts, thread counts, colors, supplementary elements, layers, warping methods and embellishment techniques were mastered using what appears to be rudimentary tools. Insular to the rest of the world until the Spanish discovered them in the 1500s, they skillfully wove fine camelid hair, cultivated cotton, and obtained brilliant natural dyes from plants and insects. In contrast to the tedious labor-intensive techniques that they developed and valued, they seem to have invented few devices and tools unlike the technological advances in China.

Diametrically opposite on the globe, Chinese textiles developed an array of woven structures, embellishment techniques and natural dyes initially using wool and soon excelling in the use of silk.[32] They too were relatively insular to the rest of the world till about 100 BC when trade began with the West while silk production and weaving had reached a high degree of technical sophistication. By the 3rd century CE, they had devised looms with treadles and frames holding multiple heddles that allowed for increased productivity of sophisticated woven structures with repetitive patterning. Trade and cultural exchanges along the silk route gave the impetus for further developing patterning devices when Chinese weavers tried to imitate imported silks with polychrome weft patterning. The result was the draw loom, the ultimate patterning loom of pre-Jacquard era which seems to have achieved its final form by the 7th to 10th centuries [122].

This difference in the mentality toward finding technological solutions coupled with the opening of trade is an important factor to consider in China's influence on the rest of the world, the momentum given to woven textiles, patterning with precious materials and dyes, and their ensuing value. Trans cultural exchanges due to trade, wars, disruptions and displacements, migration, and employment of skilled artisans from other geographical regions spread textile techniques, their related structures, and patterns to other geographical regions: the Indian Subcontinent; a vast area encompassing the Roman and Ottoman empires, Persia, and Central Asia; and Japan.

By the early 18th century, India was the largest exporter of textiles in the world. Trade, exchange, and adaptability to new markets stimulated eclectic pattern development. Weave structures and their associated textiles are often closely linked to the fibers used. In the Indian context, this seems to be an important factor that led to deep specialization within certain types of textiles made by weaver communities as seen in the case of Dhaka muslins

[32]Scientific analysis confirmed that carbonized remains of a fabric from a Neolithic site in China are made of silk dating to 5300 to 5500 years ago.

and Kashmir shawls which were woven using laborious hand processes. In most of the Subcontinent simple pit looms with treadles and two harnesses were preferred for the comfort they offered in the hot and humid climate. Small drawloom mechanisms to weave supplementary weft patterns on borders of the unstitched Indian garments were adapted to pit looms by the 13th century. However, in important hubs of trade and manufacture, complex draw looms for weaving samites, taquetés and lampas were operational in the Pre-Mughal era. While textiles had already reached a high level of sophistication, the arrival of the Mughals in the 16th century coupled with Persian influence gave rise to novel patterns, improvements to drawloom technology and the establishment of organized workshops. Indian textiles both absorbed external influences and influenced world textiles.[33]

Archaeological, iconographic and papyrological evidence shows that at least 6 types of loom structures[34] were used in ancient Egypt starting from the Pharaonic era to the Roman and Byzantine periods that lasted from 1st century CE till the 7th century CE [123]. This range of loom types shows that there was an effort to address issues related to the weavers' comfort as well as warp tension and ease of weaving specific materials and structures. In their quest for pure undiluted color, tabby weave fabrics evolved to incorporate polychrome tapestry weave, along with the use of wool.[35] Warp-faced compound tabby weaves from China inspired weavers in the Roman empire and Persia to develop weft faced compound structures like taqueté and samite weaves. These structures allowed weavers to obtain pure color while using heddle looms which speeded up the process and allowed for repetitive patterning. By the 13th century, fine silks were being woven on horizontal drawlooms in Egypt. These compound structures gave rise to even more complex weaves like the proto-lampas and lampas.

Central Asian samites with motifs within pearl bordered roundels were imitated all the way up to Japan returned from China. Drawlooms became the dominant system for patterning silks when envoys from Japan and skilled immigrants from China and Korea settled in Japan from the 7th century onward. The Japanese elite clothed themselves in all kinds of silk textiles, including patterned polychrome silks that combined external influences with aesthetics anchored in principles of Zen Buddhism taking many arts including that of woven textiles to further heights.

The Sasanian Empire had a long-lasting influence beyond the borders of Iran even after the decline of the dynasty in the 7th century. This is associated with their role in international trade with the Mediterranean world, Central Asia, China, India, and the Arabian Peninsula. Silk samites, damasks, lampas and velvets were all woven across this vast geographical expanse [124]. By the 14th century, warp pile velvets in silk embellished with gold and silver became the most sought-after textiles across Europe, Persia, and India. By the 15th century drawlooms were being used across Eurasia. Continually searching for novelty, French manufacturers looked to cultures from the east and the colonies. Multiple

---

[33] Apart from influencing patterning in textiles, profits from the Indian textile industry instigated colonial practices of cotton and indigo plantations, inspired natural dye usage and the development of roller printing in France.

[34] Horizontal looms; vertical looms with one or both beams that could be moved or rotated; vertical looms tensioned with the foot; warp weighted looms used mostly by neighboring Greek or Nubian weavers, and tablet looms which were small and easily transportable.

[35] Wool or wool and cotton, or wool and linen were combined to create large tapestry weaves or taquetés that were used as furnishings in Coptic Egypt, the Roman empire and Persia.

derivatives of the three basic weaves (tabby, twill and satin, and supplementary warp and weft were used to pattern silks. Lampas underwent huge change when Jean Revel began to design florals with shaded supplementary wefts using graph paper. Phillippe de Lasalle devised improvements to the drawloom and large repeats with realistic shading were now possible. In 18th century France the cumulative effects of technical advances to expedite the weaving of complex brocades resulted in the invention of the Jacquard loom which was finally patented in 1804.

By comparing regional textiles such as the insular Pre-Columbian textiles with Eurasian textiles that saw competition, exchange, and change, we see how cultural exchanges went back and forth enriching textile technology, structures, and iconography. Woven structures and their evolution during the period studied in this chapter show that certain features were the focus of makers and their patrons. Beginning with comfort and safety for the user, textiles increased in fineness, pure color in patterning, ease of production along with improved quality and rapidity, curvilinear designs, and large repetitive patterning and textiles as luxury items seem to have been the main preoccupations.

Questions, such as drapability, weight, maintenance, and hygiene, that came with the industrial revolution changed the way we perceive textiles. Combined with mechanization and computerization, textile production has seen a complete change in manufacture and impacts on society and environment. Yet, many of the structures enumerated above have been adapted especially for high end woven furnishing fabrics or form the basis for modern complex textiles. However, the savoir faire acquired by weavers in Italy and France are threatened as factories close because the global economy is moving toward cheaper, and simpler textile structures for clothing and furnishings. The history of woven textiles, far from being globally fragmented, is an accumulation of cultural exchanges, techniques, knowledge and talents, and deserve to be valued as our common heritage. "Heritage is our legacy from the past, what we live with today, and what we pass on to future generations. Our cultural and natural heritage are both irreplaceable sources of life and inspiration." [125].

References

1. Bertrand, G., *Histoire générale des techniques*, vol. I, 1962, Vol.II, 1965; re-edited as Histoire des Techniques (Encyclopédie de la Pleïade), Éditions Gallimard, Gallimard, Paris, France, 1978.
2. Buck, A., *Standards in the museum care of costume and textile collections*, Museums & Galleries Commission, London, 1998.
3. Schoeser, M., *World textiles, a concise history (World of Art)*, Thames & Hudson, London, 2003.
4. Gillow, J. and Sentance, B., *World textiles, a visual guide to traditional techniques*, Thames & Hudson, London, 1999.
5. Geijer, A., *A history of textile art*, Pasold Research Fund in association with Sotheby Parke Bernet Publications, Totowa, N.J, 1979.
6. Zhao, F., Sardjono, S., Buckley, C., (Chief Eds.), *A World of Looms: Weaving Technology and Textile Arts*, Zhejiang University Press, Hangzhou, China, 2019.
7. World heritage, http://whc.unesco.org/en/about. Website consulted on 25 February 2021.
8. Emery, I., *The primary structure of fabrics: an illustrated classification, 1966, re-edited*, Thames & Hudson, London, 2009.
9. Burnham, D.K., *Warp and weft: a textile terminology*, Royal Ontario Museum, Ontario, 1980.

10. Grömer, K., *The art of prehistoric textile making: the development of craft traditions and clothing in Central Europe*, Naturhistorisches Museum Wien, Vienna, 2016.
11. Schoeser, M., op. cit. [3].
12. Zhao, F., op. cit. [6].
13. Connolly, T.J. and Cressman, L.S., *The fort rock sandals: 10,000 years of shoes*, J.M. Erlandson, and S.B. McClure, (Eds.), pp. 1–5, Museum of Natural and Cultural History, University of Oregon, Eugene, 2011, and personal conversation with Endswig P. on 29 September 2021.
14. Splitstoser, J.C., Dillehay, T.D., Wouters, J., Claro, A., Early pre-hispanic use of indigo blue in Peru. *Sci. Adv.*, 2, 9, 02 September, 2016.
15. Stapleton, L., Margaryan, L., Areshyan, G.E., Pinhahi, R., Gasparyan, B., Weaving the ancient past: Chalcolithic basket and textile technology at the Areni-1 Cave, Armenia, in: *Stone Age of Armenia. A Guide-Book to the Stone Age Archaeology in the Republic of Armenia*, B. Gasparyan, and M. Arimura, (Eds.), pp. 219–232, Kanazawa University, Kanazawa, Japan, 2014.
16. Petrie Museum of Egyptian Archeology, https://www.ucl.ac.uk/museums-static/digitalegypt/textil/wovenlinen.htm Website consulted on 2 October 2021.
17. Kramell, A., Li, X., Csuk, R., Wagner, M., Goslar, T., Tarasov, P.E., Kreusel, N., Kluge, R., Wunderlich, C.-H., Dyes of late Bronze Age textile clothes and accessories from the Yanghai archaeological site, Turfan, China: Determination of the fibers, color analysis and dating. *Quat. Int.*, 348, 214–223, 2014, and personal communication with Kramell, A., 23 February 2021.
18. Beck, U., Wagner, M., Li, X., Durkin-Meisterernst, D., Tarasov, P.E., The invention of trousers and its likely affiliation with horseback riding and mobility: A case study of late 2nd millennium BC finds from Turfan in eastern Central Asia. *Quat. Int.*, 348, 224–235, 20 October 2014.
19. *Treasures of the Pazyryk Culture*, UNESCO by Permanent Delegation of the Russian Federation to UNESCO, 16 January 2018, https://whc.unesco.org/en/tentativelists/6283/ Website consulted on 20 February 2021.
20. Hermitage Museum, https://www.hermitagemuseum.org/wps/portal/hermitage/digital-collection/25.+Archaeological+Artifacts/879870/. Website consulted on 21 Feb 2021.
21. Eiland Jr., M. and Eiland III, M., *Oriental carpets, a complete guide*, pp. 14–15, Bullfinch Press, Hachette Group, New York, 1998.
22. Phipps, E., Cochineal red: the art history of a color. *Metrop. Mus. Art Bull.*, 67, 3, 2–21, Winter, 2010.
23. Cahlander, A. and Baizerman, S., *Double woven treasures from old Peru*, pp. 11–121, Dos Tejedoras, Minneapolis, MN, 1985.
24. Cahlander, A., op. cit. [23].
25. Mantle, ("The Paracas Textile"), Brooklyn Museum, John Thomas Underwood Memorial Fund. https://www.brooklynmuseum.org/opencollection/objects/48296. Website consulted on 22 February 2021.
26. D'Harcourt, R., *Textiles of ancient Peru and their techniques*, University of Washington Press, Seattle, WA, 1974.
27. Mantle with squares 1935.32.0118, 2021, https://collections.smvk.se/carlotta-vkm/web/object/96656. 2021. Website consulted on 2 February.
28. Rowe, A.P., Interlocking warp and weft in the Nasca 2 style. *Text. Mus. J.*, III, 3, 67–78, 1973, and personal communication on October 28, 2021.
29. Rowe, A.P., Tie-dyed tunics, in: *Catalogue of Wari: Lords of the Ancient Andes*, pp. 193–206, The Cleveland Museum of Art, Cleveland OH, 2012.
30. Rowe, A.P., The Art of Peruvian textiles, in: *Andean Art at Dumbarton Oaks*, vol. 1, E.H. Boone, (Ed.), pp. 336–337, Dumbarton Oaks, Washington, DC, 1996.
31. Nagy, K., The pre-Columbian textile collection of the German Textile Museum Krefeld. *Pre-Columbian Textile Conference VIII/Jornadas de Textiles Pre-Colombianos VIII*, June 2020,

Center for Textile Research, University of Nebraska, Lincoln, NE, DigitalCommons@University of Nebraska, Lincoln, 2019.
32. Evans, A., Representations of duality in a Chancay knotted weft wrapping headcloth in the collection of the Michael C. Carlos Museum, Master's thesis, Emory University, Atlanta, GA, 1995.
33. Phipps, E., Hecht, J., Martín, C.E. (Eds.), *The colonial Andes: tapestries and silverwork, 1530–1830*, The Metropolitan Museum of Art, New York, and the Yale University Press, New Haven and London, 2004.
34. Gong, Y., Li, L., Gong, D., Yin, H., Zhang, J., Biomolecular evidence of silk from 8,500 years ago. *PLoS One*, 11, 12, e0168042, 2016, https://doi.org/10.1371/journal.pone.0168042.
35. Zhao, F., Wang, Y., Luo, Q., Long, B., Zhang, B., Xia, Y., Xie, T., Wu, S., Xiao, L., The earliest evidence of pattern looms: Han Dynasty tomb models from Chengdu, China. *Antiquity*, 91, 356, 360–374, 2017.
36. Exhibition review, a world of looms: Weaving technology and textile arts in China and beyond, 1.2 treadle looms, 2018, http://www.chinasilkmuseum.com/yz/info_98.aspx?itemid=26752 Website consulted on 1 March 2021.
37. Hardouin-Fugier, E., Berthod, B., Chavent-Fusaro, M., *Les etoffes: dictionnaire historique*, pp. 342–348, les éditions de l'amateur, Paris, 1994.
38. Okumura, S., Velvet and patronage: the origin and historical background of ottoman and italian velvets. *Textile Society of America Symposium Proceedings*, p. 1008, 2016, https://digitalcommons.unl.edu/tsaconf/1008.
39. Rawson, J., *Chinese ornament, the lotus and the dragon*, British Museum Publications Ltd, London, 1984.
40. Mao, H., Comparative study of Nasij and Jin Duan zi of mongol period. *Asian Soc. Sci.*, Published by Canadian Center of science and Education, 16, 5, 1–13, 2020.
41. UNESCO, Silk road program, https://en.unesco.org/silkroad/Website consulted on 21 February 2021.
42. Personal communication with Dr. Regula Schorta, Abegg-Stiftung regarding thread count and proof of fragments probably belonging to a blanket, October 25, 2021.
43. Zhao, F., op. cit. [6] pp. 202.
44. Watt, J.C.Y. and Wardwell, A.E., When silk was gold: central asian and chinese textiles, pp. 36–37 and 58–59, The Metropolitan Museum of Art, New York, NY, 1997.
45. Mao, H., op. cit. [40].
46. Watt, J.C.Y. and Wardwell, A.E., When silk was gold: Central asian and chinese textiles, pp. 137–141, The Metropolitan Museum of Art, New York, NY, 1997.
47. Mao, H., op. cit. [40].
48. Hardouin-Fugier, E., op. cit. [37].
49. Watt, J.C.Y. and Wardwell, A.E., When silk was gold: Central asian and chinese textiles, pp. 137–141, The Metropolitan Museum of Art, New York, NY, 1997.
50. Tunstall, A., Woven paintings, woven writing: Intermediality in *Kesi* silk tapestries in the Ming (1368-1644) and Qing (1644-1912) dynasties, Dissertation towards PhD, Columbia University, New York, NY, 2015.
51. Kesi, http://www.chinasilkmuseum.com/yz/info_98.aspx?itemid=26752. Website consulted on 1 March 2021.
52. Cohen, S., The earliest indian textile. *Text. Hist.*, 30, 1, 5–15, 1999.
53. Wild, J.-P., First indian carpets, in: *Drawing the Threads Together: Textiles and Footwear of the 1st Millennium AD from Egypt*, A. De Moor, C. Fluck, P. Linscheid, (Eds.), pp. 74–85, Lannoo Publishers, Tielt, Belgium, 2014.
54. Islam, S., *Muslin! our story*, Drik Picture Library Ltd, Dhaka, 2016.

55. Chisti, R.K. and Singh, M., *Saris of India: Tradition and beyond*, Roli Books, New Delhi, India, 2013.
56. Jain, R., Mughal velvets in the collection of the Calico Museum of Textiles, in: *Technical Studies Monograph*, vol. 2, pp. 2–3, Sarabhai Foundation, Ahmedabad, Gujarat, 2011.
57. Guy, J., *Woven cargoes: Indian textiles in the east*, pp. 88–89, Thames and Hudson, London, 1998.
58. Cohen, S., A group of early silks: the tree motifs, in: *Woven Splendours Indian Silks*, vol. XLVI, No. 3, J. Dhamija (Ed.), pp. 19–36, Marg Publications, Bombay, 1995.
59. Cohen, S., op. cit. [58].
60. Crill, R., Vaishnavite silks: the figured textiles of assam, in: *Woven Splendours : Indian Silks*, vol. XLVI No. 3, J. Dhamija (Ed.), pp. 37–48, Marg Publications, Bombay, 1995.
61. Sonday, M. and Kajitani, N., A type of Mughal sash, in: *Handwoven Fabrics of India*, J. Dhamija, and J. Jain, (Eds.), pp. 40–46, Mapin Publishing Pvt. Ltd, Ahmedabad, 1989.
62. Jain, R., The Indian drawloom and its products. *Text. Mus. J.*, 32 and 33, 50–81, Washington DC, 1993-1994.
63. Jain, R., Indian lampas -weave silks in the collection of the calico museum of textiles, technical studies monograph No.3, pp. 38–39, Sarabhai Foundation, Ahmedabad, Gujarat, 2013.
64. Islam, S., op. cit. [54].
65. Watson, J.F., *The textile manufactures and the costumes of the people of India*, Printed for the India office, by G. E. Eyre and W. Spottiswoode, London, 1866.
66. Ames, F., *The kashmir shawl and its Indo-French influence*, 3rd edition, Antique Collectors' Club, Woodbridge, Suffolk, 1997.
67. Desai, C., *Ikat textiles of India*, Chronicle Books LLC, San Francisco, CA, 1988.
68. Guy, J., op. cit. [57].
69. Shamir, O. and Baginski, A., The earliest cotton ikat textiles from Nahal 'Omer Israel 650–810 CE, in: *Global Textile Encounters*, M.-L. Nosch, Z. Feng, L. Varadarajan, (Eds.), pp. 65–74, Oxbow books, Oxford, October 2014.
70. Politis, K.D., The monastery of aghios lot at deir 'ain 'abata in Jordan, http://www.hsnes.org/pdf/Lot/monastery/article.pdf, consulted in 2021.
71. Yvanez, E. and Wozniak, M.M., Cotton in ancient Sudan and Nubia, archaeological sources and historical implications, in: *Revue d'ethnoécologie*, No. 15, CNRS, MNHN, Université Paris 7, Paris, 2019.
72. Bühl, G., Krody, S.B., Williams, E.D., *Woven interiors: furnishing early medieval egypt*, the George Washington University Museum and The Textile Museum, Washington, DC, 2020.
73. Petrie museum of archeology, op. cit. [16].
74. Bühl, G. *et al.*, op. cit. [72].
75. LaGamma, A., *Sahel: art and empires on the shores of the Sahara*, The Metropolitan Museum of Art, New York, NY, 2020.
76. Gillow, J., *African textiles*, Chronicle Books, San Francisco, 2003.
77. Bolland, R., *Tellem textiles: archaeological finds from burial caves in Mali's Bandiagara Cliff*, Tropenmuseum/Royal Tropical Institute, Leiden, the Netherlands, 1991.
78. Exhibition, caravans of gold, fragments in time, case study: some of africa's oldest textiles, 2021, https://africa.si.edu/exhibitions/current-exhibitions/caravans-of-gold-fragments-in-time-art-culture-and-exchange-across-medieval-saharan-africa/fragments/.
79. Fening, K.O., History of kente cloth and its value addition through design integration with African wild silk for export market in Ghana. *ARPPIS Scholar*, ICIPE, Nairobi, Kenya, 2006.
80. Yvanez, E. *et al.*, op. cit. [71].
81. *Personal communication with Allison McLaughlin, assistant registrar*, The University of Pennsylvania Museum of Archaeology and Anthropology, Philadelphia, PA, 2021.

82. Petrie museum of archeology, op. cit. [16].
83. Bühl, G. *et al.*, op. cit. [72].
84. Okumura, S., op. cit. [38].
85. LaGamma, A., op cit. [75].
86. Gillow, J., op. cit. [76].
87. Gillow, J., op. cit. [76].
88. Man's prestige cloth, https://www.metmuseum.org/art/collection/search/320679, Website consulted on 11 October 2021.
89. Matsumoto, K., *Jodai Gire 7th and 8th century textiles in Japan from the Shōsō-in and Hōryū-ji*, Shikosha, plates 131 and 129, pp. 225–226 and 241, Alan Kennedy, New Haven, CT, 1985.
90. Tatsumura, H., Takashimaya., N., Asashi, S., *Creation and re-creation, the fabrics of Heizō Tatsumura I*, p. 28, Asashi Shinbunsha, Japan, 1996.
91. https://edostripe.com/edo-stripe, Website consulted in 2020.
92. https://edostripe.com, op. cit. [91]
93. Matsumoto, K., op. cit. [89].
94. Matsumoto, K., op. cit. [89].
95. https://artsandculture.google.com/asset/_/2gElSnCSsrVGmA Website consulted on March 10, 2021.
96. Kimono and textiles, Kyoto National Museum, https://artsandculture.google.com/exhibit/kimono-and-textiles-kyoto-national-museum/NAICu6LOm0ubIQ?hl=en, Website consulted on 9 March 2021.
97. https://www.fowler.ucla.edu/wp-content/uploads/2016/07/Krauss-release.pdf, 2021, Website consulted on 11March 2021.
98. Mouri, C., Aali, A., Zhang, X. *et al.*, Analysis of dyes in textiles from the Chehrabad salt mine in Iran. *Herit. Sci.*, 2, 20, 2014.
99. Aali, A., Abar, A., Boenke, N., Pollard, M., Rühli, F., Stöllner, T., Ancient salt mining and salt men: the interdisciplinary Chehrabad Douzlakh project in north-western Iran. *Antiquity Project Gallery*, 86, 333, 2012. http://antiquity.ac.uk/projgall/aali333/ 2021.
100. Wild, J.P., *Textile manufacture in the Northern Roman Provinces*, Cambridge University Press, Cambridge, UK, 1970.
101. https://trc-leiden.nl/trc-digital-exhibition/index.php/out-of-asia/item/193-2-the-taquete-family-of-weaves, consulted in 2021.
102. Kitzinger, E., The horse and lion tapestry at dumbarton oaks. a study in coptic and sassanian textile design, in: *Dumbarton Oaks Papers*, vol. 3, pp. 1+3–59+61-72 (107 pages), Dumbarton Oaks, Trustees for Harvard University, Washington DC, 1946.
103. Williams, B., The Sasanians, https://www.metmuseum.org/exhibitions/listings/2012/byzantium-and-islam/blog/topical-essays/posts/sasanians, Website consulted on 31 October 2021.
104. Watt, J.C.Y., op. cit. [44].
105. Allan, J.W., Thompson, J., Canby, S.R. (Eds.), *Hunt for paradise : court arts of Safavid Iran, 1501-1576*, pp. 275–276, Exhibition Catalogue with contributions of Asia Society Museum, Museo Poldi-Pezzoli, Palazzo reale di Milano, Skira, Lausanne, Switzerland, 2003.
106. Watt, J.C.Y. and Wardwell, A.E., When silk was gold: Central asian and chinese textiles, pp. 34–37. The Metropolitan Museum of Art, New York, NY, 1997.
107. Bier, C. (Ed.), Textile arts of Safavid and Qajar Iran 16th–19th centuries, in: *Woven from the Soul, Spun from the Heart*, pp. 57–84, Textile Museum, Washington, D.C., 1987.
108. Watt, M., Textile production in Europe: silk, 1600–1800, in: *Heilbrunn Timeline of Art History-essays*, Department of European Sculpture and Decorative Arts, The Metropolitan Museum of Art, 2003.

109. Peter, M. and Hohmann, H.B. (Eds.), *Velvets of the fifteenth century*, Abegg-Stiftung, Riggisberg, 2020.
110. Vingtain, D., Leonelli, M.-C., Dutocq, E., Valansot, O., *Merveilles d'or et de soie: trésors extiles de notre-dame des doms du XVIe au XIXe siècle*, pp. 20–21, Exhibition Catalogue at the Palais des Papes, Avignon, 2000-2001.
111. Berthommier, C., *Le génie de la fabrique: excellence, création, innovation des industries textiles de Lyon et sa region*, vol. 1 and 2, Musée des Tissus, Lyon, France, 2016.
112. Zhao, F., Sardjono, S., Buckley, C., (Chief Eds.), *A World of Looms: Weaving Technology and Textile Arts*, pp. 181–182, Zhejiang University Press, Hangzhou, China, 2019.
113. Essinger, J., *Jacquard's web, how a hand loom led to the birth of the information age*, Oxford University Press, Oxford, 2004.
114. De Marinis, F., *Velvet: history, techniques, fashions*, p. 202, Idea Books, New York, 1994.
115. Watt, M., Renaissance velvet textiles, in: *Heilbrunn Timeline of Art History*, The Metropolitan Museum of Art, New York, NY, 2000, http://www.metmuseum.org/toah/hd/velv/hd_velv.htm (posted August 2011).
116. https://www.luigi-bevilacqua.com/en/jardiniere-velvets-history, consulted in 2021.
117. Berthommier, C., op cit. [111].
118. Miller, L.E., Departing from the pheasant and the peacock: the role of furnishing textiles in the Career of Philippe de Lasalle (1723-1804), furnishing textiles. *Studies on Seventeenth- and Eighteenth-Century Interior Decoration*, Riggisberger Berichte, no. 17, Abegg-Stiftung, pp. 79–90, 2009.
119. Zhao, F. *et al.*, op. cit. [6].
120. Zhao, F. *et al.*, op. cit. [6].
121. D'Harcourt, op. cit. [26].
122. Zhao, F. *et al.*, op. cit. [6].
123. Mossakowska-Gaubert, M., A new kind of loom in early Roman Egypt? How iconography could explain (or not) papyrological evidence, in: *Egyptian Textiles and their Production: 'Word' and 'Object'*, 2020.
124. Von Fircks., J. and Schorta, R. (Eds.), *Oriental silks in Medieval Europe*, p. 35 and 47, Abegg-Stiftung, Riggisberg, 2016.
125. World Heritage: http://whc.unesco.org/en/about. Op. cit. [7]

List of Museums

Abegg-Stiftung Museum, Riggisberg, Switzerland
Bharat Kala Bhavan, Banaras Hindu University, India
British Museum, London, UK
Brooklyn Museum, New York, USA
Calico Museum, Ahmedabad, India
Dallas Museum of Art, Texas, USA
Deutsches Textilmuseum, Krefeld, Germany
Dumbarton Oaks Research Library and Collection, Washington, DC, USA
Hunan Provincial Museum, Hunan, China
Institute of Archaeology and Ethnography, National Academy of Sciences of the Republic of Armenia, Yerevan, Armenia
Museo Cao, Trujillo, Peru
Musée d'Art et d'Histoire, Geneva, Switzerland

Museo del Tessuto, Prato, Italy
Museée des Tissus et Des Arts Décoratifs, Lyon, France
Musée du Louvre, Paris, France
Musée Guimet, Paris, France
Musée National du Mali, Bamako, Mali
Musées Royaux d'Art et d'Histoire, Brussels, Belgium
National Museum of China, Beijing, China
Nationaal Museum van Wereldculturen, Amsterdam, Netherlands
Petrie Museum of Egyptian Archaeology, London, UK
Seattle Art Museum, Washington, USA
The Cleveland Museum of Art, Ohio, USA
The Fowler Museum, Los Angeles, USA
The Kyoto National Museum, Kyoto, Japan
The Metropolitan Museum of Art, New York, USA
The Michael C. Carlos Museum, Emory University, Georgia, USA
The Museum of Fine Arts, Boston, Massachusetts, USA
The Museum of Natural and Cultural History, Oregon, USA
The Museum of World Culture, Gothenburg, Sweden
The State Hermitage Museum, St. Petersburg, Russia
The George Washington University Museum and The Textile Museum, Washington, DC, USA
Turpan Museum, Xinjiang, China
The National Museum of the Archaeology, Anthropology, and History of Peru, Lima, Peru
The University of Pennsylvania Museum of Archaeology and Anthropology, Pennsylvania, USA
Victoria and Albert Museum, London, UK

Glossary

Brocade: Term used for both rich patterned fabrics, usually in silk, and weaving technique where polychrome discontinuous wefts are 'brocaded' or inserted as supplementary wefts that float over a ground weave. These wefts were also fancy yarns such as gold and silver or chenille yarns. Brocaded fabrics were woven on drawlooms.

Damask: A term used to signify both the weave structure and fabrics woven using contrasting warp faced and weft faced weave structures juxtaposed next to each other. Usually in silk, damasks can be woven with the same color for both warp and weft or with contrasting colors of warp and weft. When woven in a single color, light bounces off the shiny and matt surfaces or different warp and weft effects differently and the pattern becomes evident. Sometimes, damasks can be woven with 2 contrasting colors – one for the warp and another for the weft. Classic damasks are woven with warp faced satin and weft faced sateen, creating a reversible fabric.

Discontinuous-warp-and-weft: Unique method used by Pre-Columbian weavers where the warp and weft threads do not go all the way across the fabric in one length. The warp ends are stretched out in short lengths going back and forth in the area that a color is desired, often achieved by using scaffolding weft yarns. Weft of the same color as the discontinuous warp is woven within these constrained warp areas, while interlocking with adjacent warps and/or wefts to connect the various pieces of the cloth. Like tapestry weave, this technique allows for pure color. However, a balanced tabby weave and lighter cloth can be achieved.

Dovetailing: A term used for certain types of tapestry weaves. When discontinuous wefts are used for tapestry weaves, they can cause slits to appear in the fabric. This is because each separate weft goes back and forth in a specific area to make the pattern. To avoid having slits, adjacent weft threads are turned around a common warp end. Several types of interlocking exist to avoid slits.

Double cloths or weave: A term used to signify both the weave structure and the fabrics that made using this structure. 2 sets of warp and weft interlace with each other to create 2 distinct layers that are woven simultaneously. Triple cloths have 3 sets of warp and weft.

Drawloom: looms equipped with cords that allowed a draw-person to manually pull up pre-defined groups of warp ends so that the weaver could weave complex patterns.

End: An individual warp thread.

Gauze: Also known as leno. Weave structure in which two or four warp ends are pulled out of their normal parallel alignment and crossed to the side. These displaced ends are kept in place with wefts and allowed to go back to their original positions. This creates an open weave structure.

Ikat: Warp and/or weft yarns are resist dyed before being woven. Usually woven in a tabby structure, twill ikats are rare.

Jin warp faced compound weave: This refers to a warp faced tabby weave structure where the patterning is achieved by using multiple warp ends. Most *jin* textiles with this structure have a short pattern repeat in the warp direction, directly linked to the number of pattern harnesses. Hence, a pattern with 84 harnesses will have 84 picks (see below) while lifting the colored warp ends. Warp faced textiles have a high number of ends per inch.

Lampas: A term used to signify both the weave structure and the rich silk fabrics that were made using this structure. Two warps (main and binding) are used to weave several wefts—main, pattern or discontinuous brocading wefts. The ground is often a warp faced weave created by one set of warp and weft. Weft-faced pattern is created by both continuous and discontinuous multicolored wefts that are bound by the binding warp. These fabrics were woven on drawlooms. Early lampas ground weave structures were mostly tabby weave. However, most lampas created in 17th century France employed satin weave structures.

Picks: An individual weft thread.

Samite: A term used to signify both the weave structure and the rich silk fabrics that were made using this structure. Two warps (main and binding) are used to weave 2 or more patterning wefts using a twill weave structure. It is a weft faced compound weave.

Satin: A term used to signify both the weave structure and the smooth shiny fabrics made using this structure. Satin is the third of the 3 basic weaves as the interlacement of warp and weft are relatively more complex than Tabby and Twill. The structure seeks to dissimulate binding points made by the weft under long warp floats. A good quality satin will have a high number of fine silk ends per inch to contribute to the smooth and glossy effect desired. A weft faced satin is known as sateen. The most common satin weaves require a minimum of 5 ends and picks or 8 ends and picks. Large satins are also employed on the reverse of lampas to bind wefts.

Supplementary warp: Supplementary warps are usually used for patterning and follow a different weave structure from the ground fabric. They add to the structure of a fabric. However, if a supplementary warp is pulled out of a fabric, the fabric will still hold together due to the weave structure of the main warp and weft. Supplementary warps require special tension and usually have warp beams or other devices separate from the rest of the ground warp as they are brought up on the face of a fabric when being patterned and float at the back of the fabric when not in use.

Supplementary weft: Supplementary wefts can be both continuous going from one selvedge to the other or discontinuous, being employed only in specific areas of the fabric. Both are employed to bring more color and to pattern to the face of a fabric. Like supplementary warps, they are an addition to a ground made of a main warp and weft and can be bound by the main warp or a binding warp.

Tabby weave: Also known as plain weave. Tabby is the first of the three basic weaves as it is the simplest form of warp and weft interlacement. Every end passes over one pick and under one pick, while every alternate end, does just the opposite—it lies under a pick and passes over a pick. A minimum of 2 ends and 2 picks are required to weave a tabby structure. Numerous derivatives exist including balanced tabby where both warp and weft are equally visible, basket weaves where the same principle of interlacement is applied to groups of warp and weft, and warp and weft faced versions.

Tapestry: A term used to signify both the weave structure and the fabrics using this structure. It is usually a weft faced weave with a single warp and discontinuous colored wefts that are woven back and forth in a specific area instead of going from selvedge to selvedge. Tapestry weave is usually woven using the tabby weave structure. However, one can also find balanced tabby weave where both warp and weft are equally visible, or a weft faced twill weave tapestry structure. As weaving with each color weft is done following the shape of a particular motif as opposed to a strict order of the shed row by row, this technique gives the weaver a lot of freedom in interpreting the pattern—motif and color.

Taqueté: A French term used to signify both the weave structure and fabrics made using this structure. Two warps (main and binding) are used to weave 2 wefts (main and pattern) using the tabby weave structure. It is a weft faced compound weave. As there is no English equivalent, the Centre International d'Étude des Textiles Anciens (CIETA) included this term in their 1964 vocabulary.

Twill: A term used to signify both the weave structure and the fabrics made using this structure. Twill is the second of the 3 basic weaves as the interlacement of warp and weft creates distinctly diagonal lines (wales) that can have either an S direction or Z direction. Twills can be created with a minimum of 3 ends and 3 picks or more (usually 4 ends and 4 picks). They have short floats of warp and/or weft. Twills can be warp faced, weft faced or balanced.

Velvet: A term used to signify both the weave structure and fabrics made using this structure. Two warps (main and pile) are used to weave this structure. While the main warp is used to construct a sturdy tightly woven ground, the pile warp is woven with temporary wefts of smooth metal rods. These rods allow the weaver to create loops (pile) of even height. Grooved rods, allow the weaver to cut the pile (cut pile), or leave the piles uncut (uncut pile), or not raise the pile warp on the face of the fabric at all (voided). These are the three main variations of pile. A multitude of variants exist in terms of methods of cutting or warp treatments.

Warp faced: In warp faced fabrics, the weft is hidden under a high number of warp threads (ends) that lie close to each other.
Weft faced: In weft faced fabrics, the warp is hidden under a high number of weft threads (picks) that lie close to each other.

4

Types of Conservation of Textiles in the Museum: Their Importance and Scope

K. B. Binita* **and B. Sunita†**

Department of Textiles and Apparel Designing, College of Community Science, Assam Agricultural University, Jorhat, Assam, India

Abstract

Textile conservation is a relatively new field and since the 1970s, museums have begun to research the conservation of textile specimens. Conservation aims to reduce the loss of our cultural heritage and protects precious objects from deterioration caused by the environment as well as physical and chemical effects. In addition, it also manifested in the loss of tensile strength, loss of surface texture, fading of dyes, and yellowing of the white fabric. Conservation is usually meant as a two-way action that includes preventive conservation and curative conservation. The art of conservation depends on the types of collected objects, which include tapestries, carpets, quilts, clothing, flags, furnishing materials and accessories. Many of these artifacts require specialized care. Presently, the cleaning of museum textiles is a growing area with new developments to diminish the deterioration of artifacts. The option of cleaning process mainly depends on the nature of the fiber, construction of the fabric, state of the fabric and type, and amount of soiling present. Besides cleaning, storing textiles is one of the most important aspects that should be performed with great care considering all the textiles elements that affect a textile in storage and the construction details, its state, and size determine the best storage method. Textile's conservation in museums varies enormously that valued for significant historic interest, aesthetic appeal, and cultural importance. Conservation of textiles in museums also offers the opportunity for documentation, research, publication as well as the service of society that they can have a quick look at the rarest and best artifacts in the world.

***Keywords*:** Museum, textiles, preventive conservation and curative conservation

4.1 Introduction

The heritage of our cultural artifacts was collected from different sources were stored in museums to provide necessary conservation for display as well as exhibits to disseminate the knowledge for common people [1]. Conservation is a process that aims to decrease the damage and loss of our cultural heritage from the environment and slows down physical

**Corresponding author*: Baishyakalitab@rediffmail.com
†*Corresponding author*: sunitaboruah182@gmail.com

Seiko Jose, Sabu Thomas, Pintu Pandit, Ritu Pandey and Vandana Gupta (eds.) Handbook of Museum Textiles: Volume I Conservation and Cultural Research, (87–100) © 2023 Scrivener Publishing LLC

and chemical deterioration which is a continuous and natural process [2]. In general, degradation of textiles articles results in loss of the surface texture, tearing strength, discoloration of dyes, and staining of white fabric. Indeed, science has suggested ways in which the natural lifespan of these objects can be extended [3].

The museum should have a conservator for better conservation of cultural heritage. The conservators with textiles specialization should start the documentation of cultural heritage by visiting the collections and recording the number, type, and condition of the objects. Literally, conservation has widely signified the preservation of a vast area of cultural heritage, from historical research to planning and technical interference. Museum implements and practices various preservation techniques to conserve entity, which exclusively objects specific to protect the damage caused by humidity, temperature, light sources, and from attacks of organisms like fungus, insects, rodent, etc. and demanding conservation work, for which special equipment and experience is essential [4].

Textile conservation is a quite innovative, intricate, challenging, and multifaceted discipline. The conservator tries to keep all the objects in their original state by treating them with great care so that materials will be conserving in good condition that ages well and if possible, using reversible methods. They meticulously document the textiles, their condition, and treatment through reports and photographs. In reality conservation of cultural property means curative measures to be taken to eliminate the defects already present in the textile articles and protect them from further deterioration by maintaining certain conditions for their better preservation [5, 6].

4.2 Importance of Conservation

Textiles are an important part of our culture and their conservation in the museum is of utmost importance for future references. The future generation will refer to the rich cultural significance and intrinsic work of textiles of today to find out the history behind its cultural articles. Nearly all textiles' items are not designed to last infinitely. They are treasured for their historic significance, aesthetic value, and cultural significance. Conserved objects are an important resource and a source of knowledge. The superlative conserved textiles artifact depicts the diverse impact of cultural influences, geographical factors, climatic conditions, and trade for centuries. Their artistry, color, and sheer beauty, enjoyed by both monarchs and common people across the country and the world with the skills and information passed down from one generation to the next. Collections of museums offer a rediscovery of the peoples, their distinctiveness, migrations, evolutions, and ideas that have formed and created the world. The documents store and record its aesthetic and scientific creations and offer a springboard for further development. They provide a personal attachment to textile and understanding culture through objects used with a view extending from the ancient past to the recent technological advancement. Every stretch of the world has its own motif, design, and fabric feature that entwines its history. The challenge is to conserve these heritage collections that reflected the richness of diversity, as well as pass on the past forward and enrich the future [5, 7, 12, 27].

4.3 Principles of Conservation

The intricate process of conservation and preservation of the museum collection is of great significance. Proactive and prudent planning, research and preventive strategies along with prompt intervention are quintessential to creating or maintaining the ideal conditions for preserving the museum collections from factors contributing toward their deterioration.

4.3.1 Determining the Need and Laying the Foundation for Conservation

Before starting conservation work, a location needs to fully identify with its purpose as an institution for the collection and its policies. The "place" of an artifact in a collection will determine its conservation needs. Efforts are directed toward preserving the content of archival collections with items that are important for their historic and cultural content. The nature of the collection is an important determinant of its conservation.

The policies and procedures adopted and enforced by a collection-holding organization are a defense to protect the collections and historic structure from inadvertent or accidental damage. Referred to as "intellectual controls," they are written and managed by the curatorial department or administrative staff.

4.3.2 The Conservation Assessment and Process

In the conservation process, the first step is to complete a conservation assessment, a tool designed to help institutions learn about their current ability to care for their collection and fulfilled the current standards for professional practice through both physical and intellectual controls. A Conservation Assessment offers prioritized recommendations for both short- and long-term collections care activities and forms the baseline for the development of a long-term conservation plan. The final assessment report should have an executive summary attached to it, which can be shared, with board members, community leaders and the press if desired.

4.3.3 Obtaining a Conservation Assessment

Conservation Assessments are performed by experienced assessors and can be arranged through a technical assistance program of the Institute of Museum and Library Services (IMLS) known as the Conservation Assessment Program (CAP) who is managed by Heritage Prevention. Conservation Assessments are also completed outside of the CAP program.

4.3.4 Selecting an Assessor or Conservator

The best way to select a conservator is through personal references from colleagues and also by using the AIC referral list on the AIC website.

4.3.5 The Assessment as a Planning Tool

The Conservation Assessment report should be used to develop the Preservation Master Plan where staff from different departments should be drawn together to form a Conservation Planning group to blend institutional priorities and integrate the assessment recommendations. The Conservation/Preservation Plan should be in written form and should elucidate assumptions, rationales, and goals.

4.3.6 Conservation Collection Condition Survey

A Conservation Collection Condition survey is a tool that examines individual objects for their specific conservation needs. The process involves examining each object in the collection or portion of the collection for condition, treatment needs, housing needs, exhibition recommendations, and storage methods. For example, a textile conservator can be engaged to evaluate the textiles collection or any objects and reengage to survey the object that needs preservation.

4.3.7 Object Treatment

After conservation assessment, preservation planning, and a collection survey, it is finally time for individual object treatments. The examination report should describe the object, its materials, manufacture, maker/artist, appearance, and condition. After evaluation of the report, the treatment proposal must be ready for professional practice. The treatment proposal needs to be signed and approved come with a clear estimate of the time and expenses for the work.

4.3.8 Risk Assessment and Management

This is a process of identifying vulnerabilities, analyzing risk, and implementing controls. Tools are used to evaluate the risk and help to manage its impact. These tools have been used extensible in finding increasing application in the preservation of cultural heritage [7].

4.4 Types of Textile Articles Conserved

The most important function of a museum is to collect and conserve the textiles articles of heritage since it is a custodian of the heritage of society or country. Diversity is the most important feature of folk costumes. However, there are striking differences in costume both in the cut and construction patterns, different construction techniques are present, such as sharply pleated dresses, simple tunics, characteristic sleeveless underdresses, aprons of trapezoid-shaped, quilted coats, padded head gear, and snugly fitted waistcoats with unfastening sleeves. The garments are decorated with equally rich metal and golden thread embroideries, polychrome wool and silk embroideries, golden thread bobbin laces; woollen tassels, sequins and elaborate metal fastenings, with variations in the types of materials used in making the costumes for each region. It was noticed that in certain regions, the folk costumes were not at all influenced by western fashion.

There are varieties of rich cultural articles conserved in a museum collected from different regions, states and the country. These wide ranges of well-off traditional artifacts include clothing, rich tapestries, flags, curtains, upholstered furniture, carpets, quilts, dolls, and accessories, such as parasols, gloves, hats, and fans. Further, it was also observed that there are various types of traditional textiles of different regions of India that were conserved in the museum including Phulkari, Kantha, Balucharisaree, Jamdani dupatta, Jamdani angrakha Kurta, Mashru, Banarsi dupatta, Chikankari kurta, Caps, Doshala, Sherwani, Brocade saree, Jackets, Odhani, Lehnga, Asawali saree, Candova, Paithani saree, Chakla, Patola saree, Carpet, Lehnga, Shawl, Moon shawl, Coga, Kalamkari and Pahari rumal with scenes, painting on cloth, cotton print, tribal veil cloth, Toran, Book cover, rumal, banner, painting on the cloth of 19th century, etc. [8–11].

4.5 Methods of Conservation

According to the International Council of Museums—Committee for Conservation (2008) that conservation describe all the measures and actions taken to protect textile material, cultural heritage for ensuring its ease of access to present and future generations. There are several elementary tenets of conservation that add the need to preserve the important and appropriate information of an object. In the museum, generally a bi-partite action of preservation techniques followed in conservation that includes preventive conservation and curative conservation [12].

4.5.1 Preventive Conservation

Preventive conservation is an applied form of conservancy for mitigation of deterioration and damage to artistic or cultural property through the formulation of different protective measures. Since these are indirect procedures and measure does not interfere with the materials and the structure as well as the original appearance of the items. We can consider one of the main objectives of preservation is for avoiding and minimize future deterioration or loss. These methods are carried out for the items which are in the fragile condition that they could be lost comparatively within a short time. The simple meaning of managing nature is that all the potential risks to the collection with the omnipresent environmental parameter like relative humidity or temperature; phenomena that are periodic and rare or simply access, handling and use by collection staff.

There are rare pieces of textiles with great cultural significance; need to conserve as reference material to designers for inspiration, weavers for design development, students, historians and art lovers for further research in their respective fields. The Azerbaijani carpet conserved in Azerbaijani Carpet Museum has a traditional culture with ancient motifs akin to those seen on ceramic and metal items of the Bronze Age and centuries-long historical development that have worldwide popularity [13].

The Coptic textile of the Coptic period is the most interesting artifact, as it acknowledged the features of the daily life of Egyptians from Greek-Roman period to the Islamic period. "Kabaty" is one of the most significant techniques for ornamentation of textile [14]. The condition of the collected textiles items in respect of storage, display and transit should be carefully supervised by the services of a qualified conservator, and which minimize deterioration by providing a suitable environment [15, 16]. Some of the examples of preventive conservation include proper measures and procedures for registration, environmental management (light, humidity, pollution and pest control), handling, storage, packing and transportation, safety, needful planning, public consciousness, legal agreement and staff education [17, 18]. Miller *et al.* (1989) conducted research intended at characterizing the physical and chemical degradation and changes in the properties of naturally aged silk aged between 20 and 400 years from various museums. Inferences after analyzing the data shed light on the fact that the weighting of silk with metallic elements of the highest percentage of tin degraded silk [19]. Luxford *et al.* (2009) Researcher conducted studies on factors that led to the deterioration of silk causing embrittlement of the textile leading to splits and tears and eventually a powdery fragile fabric [20].

4.5.1.1 Climate

The climatic condition put a profound effect on the preservation and storage of textiles material, mainly affected by humidity and temperature that are interrelated. Humidity has been found responsible for chemical deterioration, mechanical and biological reactions. According to the Midwest Art Conservation Center, the different fibers and components of textiles are more sensitive to conditions like heat, cold, humidity and fluctuations. The structural changes that occurred in fiber with a difference of humidity and temperature are swelling of textile fibers due to the absorption of water in a high relative humidity and shrinkage caused due to low relative humidity. The fibers of woven textiles are very close together and lead to abrasion with changes in temperature causes wear and tear of the textiles. Hence, it is vital to keep the environment under control in a museum. A moderate temperature of (about 18 °C) and relative humidity of about 55% is usually maintained in most museums Higher humidity encourages insect pests to thrive, whereas, very low humidity would discourage the growth of insects but not completely preventable. Climate of museum can be controlled through air conditioning, small humidifiers, use of silica gel, adjustments of the ventilation system, ceiling and exhaust fans [17, 18].

4.5.1.2 Light

All lights, irrespective of the source, natural and artificial, make a threat to certain textile artifacts in the museum. The damage of textiles, particularly colored textiles, depends on the length of light exposure, the intensity of the light and the amount of ultraviolet radiation. Among the natural fiber, flax is the most resistant to light damage (visible and UV), and silk is the most vulnerable.

In the case of colored textiles, some parts of the visible electromagnetic spectrum must absorb and reflect by the dyes. When light is absorbed, in turns absorbed energy by the textile, which causes the breakdown of fiber bonds and results damage of the structure. According to the law of electromagnetism, the higher energy is produced at wavelengths shorter than 500 nanometers was sufficient to damage the textiles compounds. Usually, ultraviolet radiations of less than 300 nm are rarely found in normal light since they are absorbed by the earth's atmosphere in case of sunlight and case of artificial light, by the glass envelopes of lamps. Light in the wavelength range of 300 to 500 nm is mainly responsible for photochemical degradation fading of colors, alterations of hues, loss of flexibility, becoming weak and brittle. Additionally, general yellowing and browning of textiles indicate the poor state of textiles (wool, silk, cotton, linen, jute, rayon, and nylon). The scientifically established 50 lux and 15 lux are recommended light levels for displaying artifacts to reduce damage to objects by light and minimum intensity of light needed for visibility and good color retention. When we want to control the illuminance, it is better to have artificial lighting up to 50 lux [18, 21].

Study conducted on the conservation of rare painted ancient Egyptian textile objects from the Egyptian Museum in Cairo to evaluate the reactivation consolidation technique for reinforcement of ancient Egyptian textiles. The inferences of the study suggested that the rate of fading and degradation of textiles and other organic materials on display is influenced by the intensity and wavelength distribution of the light source, to which they are exposed and by other environmental factors [22, 23].

4.5.1.3 Insects

Mainly, insects and microorganisms are the biological agents causing damage to textiles. To prevent the textile material from insect attack, the object should be kept in a place with great care.

Silverfish are usually white or gray having a greasy feel and shiny appearance with flat scales over the body. The temperature between 60°F and 80°F and humidity more than 55% are suitable condition for their survival in a dark place. The materials with starch and glue are mainly damaged by the silverfish as they attack the surface material of the objects and leave irregular holes. The larva does the actual damage, as it eats ravenously through many types of substances especially protein materials, like woollen textiles, feathers and fur.

Cloth moths in fact in the larvae stage attack the wide range of proteinaceous materials of different fibers, such as silk, wool, fur, and feathers, including clothing to floor covering, and it also spoils the articles by feeding the surface with irregular holes and leave frass and webbing that is difficult to remove.

Carpet Beetles damage textile artifacts extensively made of silk, wool and fur without preference and make the clean neat holes with a fine powder of the same color as the object.

Cockroaches that are commonly found in all tropical countries, also attack artifacts stored in the museum and they eat both protein and cellulosic fibers and excrements which leave stains in presence of moisture.

Termites destroy rapidly and silently the cellulosic fabrics of plant origin and cause almost destruction of the material by digging out the surface of the materials and leaving intact an external thin cover that hides the devastation underneath. Also, eat boxes and paper—regular and archival. Precautionary measures, such as properly treated timber, air conditioning, and integrated pest control, can protect collections and prevent damage from insect attack [18, 24].

4.5.1.4 Microorganisms

To prevent and eradicate the problem of textile caused by the growths of microorganisms, a blend of mechanical action should be attempted. Generally, textile materials made of natural fiber are more prone to biodeterioration and finishes, such as starch, protein derivatives, fats, and oils, promote microbial growth. Moreover, soil also plays an important role in the growth of microorganisms. Researchers felt the need for a study intended at understanding the potential of a polymer to act as a food source before deciding upon an appropriate polymer for a given application. Resins of both natural and synthetic origin may be destroyed by biological agents and stimulate the growth of microorganisms on textiles. Therefore a suitable polymer and resins applied safely and successfully in treating ancient Egyptian Linen textiles. The inferences from the study revealed a noticeable decline in tensile strength, color change, and substantial damage of surface among all consolidated samples after fungal deterioration [25].

The microbial growth on the surface of the fabric produces stain, develops an odor, and decreases the performance properties of the textile substrate. Microbes include a variety of microorganisms, such as fungi, bacteria, moulds, algae, and mildews.

Fungi release microscopic spores in enormous quantities into the air which remain in the atmosphere at all times. Since they do not synthesize their food; they damage the material, degrading and causing stains on the textiles as they grow on them. Most of them are aerobic but some species are anaerobic, requiring no air for growth. Algae develop greening and darker spots on textiles if exposed to continuous sources of water and sunlight for a longer period.

Moulds and mildews spore growths break down the cellulose and cause a stain that cannot be easily removed if stay for long period on textiles. They emerge as irregular gray, black, or green shapes on fabric and even discolor fabric and emit a musty odor. While handling moldy textiles, the use of dust masks, goggles, disposable gloves are recommended. Vacuuming and dry cleaning treatment control the growth of microorganisms [18, 25–27].

4.5.1.5 Dust, Soil, and Other Contaminants

Textiles in the museum are also affected by different types of pollutants become threatened and losses of originality are a very common problem these days. Mainly in the field of textiles, they are at risk from pollutants, such as sulphur dioxide in the air emitted from automobiles and industry and dust, dirt, etc., cannot be neglected. The simply cosmetic problem for clothing is dust. Dust is a fine airborne pollutant that may be a mixture of various materials, such as fibers, soil particles, fragments of human, animal skin and hair, soot, ash, mould spores, paint fragments, and pollen. Freshly settled dust on the textile surfaces can be removed, but during long periods it is impossible to remove as it gets embedded between fibers. Dust can also act as a breeding ground for pests by providing them with nourishment. Periodical cleaning of artifacts result diminish dust, soil, and other contaminants [28].

4.5.1.6 Disaster

Disasters pertaining to textiles, such as infestation with biological agents, floods, fire, theft, etc., make them susceptible to permanent damage. Prompt attention with regard to the preservation of textiles after a disaster will greatly reduce any permanent damage. Textiles become weak when wet and to reduce the risk of dye transfer separate whites from darks. Silt or debris on the textiles should rinse with clean, cool water and be carefully blotted with absorbent toweling to remove moisture and subsequently kept flat to dry, covered with clean, thin, cotton sheets. Textiles may be freeze (to prevent mold growth) for examined and soil may be removed with a soft brush. The fire, soot and smoke damage need to be addressed for a proper conservator. The fire prevention system must be planted in storage and display areas. A catalog of materials and artifacts will help in easily reachable artifacts during an emergency. Nation Institute for Conservation has come up with a decision-making process of an Emergency Response and Salvage to help the Heritage Preservation [29].

4.5.2 Curative/Interventive Conservation

Curative conservation is applied directly to an object or a group of objects to protect damage and stabilize their condition against further deterioration. These actions are carried out for articles having fragile conditions that result in loss within a short time. Moreover, they treat the defects present in the artifacts and preserve them in stable environmental conditions. With new technological developments cleaning historic textiles is a budding field. Cleaning is one of the methods to preserve the textile artifacts and in this regard conservator has to decide the appropriate cleaning method depending on the structure and component of artifacts. The various options accessible with a conservator are Surface cleaning, Wet Cleaning, Dry or Solvent Cleaning, Vacuuming and Stabilization [30].

4.5.2.1 Surface Cleaning

Surface cleaning removed dirt, dust, grit, soot and other foreign substances from the textile surface by use of different tools, such as brushes, dry cleaning polyurethane sponges, vacuuming, etc. Sponging techniques are of significance so far as minimizing the damage potential to the textile during cleaning [31]. Periodical Surface cleaning is performed prior to preservation processes depending on the textile products, condition of the artifacts and the preferred result [32].

4.5.2.2 Vacuuming

Textiles that are splitting, fragmenting and powdering that have loose threads or broken seams and areas of loss, should be vacuum clean rather than surface cleaned. It is effective for discouraging biological agents and other physical contaminants. Fragile or three-dimensional textiles are cleaned with a soft brush directly attached to the nozzle, while strong or big pieces of textiles are vacuumed with a net or flexible plastic screen [30].

4.5.2.3 Wet Cleaning

Water is used for cleaning textiles as it contributes an important role in dissolving both inorganic and organic types of polar soil and acidic deterioration of natural products that enhance the aesthetic improvement of the textiles. Moreover, water improves the flexibility and softness of the textile by acting as a "plasticizer" for the polymers of fibers and reduces strains, creases, wrinkles, etc. Deionized water is more suitable for wet cleaning.

The dyed textiles may change the color because of color bleeding hence; anionic sodium alkyl sulphate surfactant and nonionic ethoxylated nonylphenol surfactant do not cause significant color changes for non–PH-sensitive natural dyes. Fibers in poor condition may further descent result in excessive weight loss, wrinkling and dimensional changes due to swelling and shrinkage. Felting of woollen textiles may result due to the wet cleaning process, and it has a considerable effect on velvet, padding, feathers, and furs. Care must be taken to wash the woollen textiles with minimal agitation and under a temperature of 40°C to prevent felting [32].

4.5.2.4 *Solvent or Dry Cleaning*

The organic solvents used in dry cleaning facilitate in swelling up of dirt and assist in eliminating soil that is greasy, fatty or oily and treat textiles that has possibilities of color bleeds as well as a composite artifact, made of different types of materials. The use of organic solvent cannot dissolve the polar deterioration products of fibers and the discoloration as well as whiteness of the textiles cannot be obtained. The creases or wrinkles of the museum textile are not affected by solvent cleaning but sometimes synthetic adhesives get dissolved in organic solvents, such as white spirit, benzene, petroleum ether, diethyl ether, turpentine, toluene, acetone, ethyl acetate, etc. [18, 30].

4.5.2.5 *Stabilization*

The deterioration of textile artifacts can be minimized by a stabilization process that includes reinforcing, supporting and reuniting different parts of textile in order to achieve visual completeness or structural soundness. In the conventional method, sewing and mending techniques are used while in other methods adhesive is used. In addition to this consolidation and pressure mounting are adequate treatments under certain conditions. Mending of textiles can be done with appropriate sewing techniques using suitable threads and supporting fabrics. For example, the size and color of thread and weight, along with backing fabric have gained more significance. For stabilization of textiles, support fabrics should be selected based on the quality of the fiber, yarn and fabric as well as the properties, such as dimensional stability, weight, color and color fastness, hand, sheen, texture, finish, etc. [18, 30]. Ragauskiene make a study on preserving and secure method of adhesive lining for prehistoric textiles and inferred that brushing with dilute acrylic dispersion is the best method for the lining of ancient textiles that enables to conserve and retain its valid [33]. Leorna's [34] experimental research intended at investigating the abrasive nature of the backing surfaces for supporting historic textiles for conservation, display and storage subsequently shed light on the fact that no significant difference prevailed in the number of fibers removed for the face and back of the fabrics. The least abrasive nature of fabric was observed while experimenting with Muslin as the plain weave fabric is light in weight.

4.6 Storage, Display, and Handling of Museum Textiles

The storage materials used in museum textile for exhibition cases and frames must be carefully selected to make sure of environmental stability. Some of the storage materials, such as cardboard, wood, plastic, and metals, are considered unsatisfactory because they release volatile chemicals or acids. Heavy-duty aluminium foil, polyethene plastic, acid-free cardboard boxes and cotton sheets are used as a barrier of stable material that protects in storage [35].

Figure 4.1 Beaded fabric bands are provided with padded support that facilitates their safe storage and handling. Handles at either end of the support facilitate handling without having to touch the objects.

Source: Government of Canada, Canadian Conservation Institute. CCI79408-0020

In the late and postmedieval periods, strong-smelling cedar wooden boxes were used as they have a long tradition of repelling insects [36].

Storage cases and showcases for display arrangements ensure that they are made of appropriate materials, and prove to protect the textiles; such consideration with regard to the construction of display and storage cases would make sure a buffer against fluctuations in ambient humidity [35, 36].

Textiles are often susceptible to damages owing to mishandling, negligence, are often broken at the creases when they are kept rolled. Caution must be exercised to keep one's hand clean and must wear cotton gloves while handling textiles, as any oils, perspiration, and skin creams may be readily absorbed by fibers, often resulting in permanent stains and susceptible to physical damage. Textiles transported in protective packaging are ideal storage options (Figure 4.1).

4.7 Scope of Conservation

The scope for conservation has a far-reaching impact across different fields of application and recent advances in conservation through the digitalization of textiles have aided research by curators, students, teachers, scholars and specialists that will foster ease and effectiveness with regard to studying disparate images in new contexts. Moreover, the dissemination of images of unique local collections on global research platforms will encourage scholarly use of resources. Exploration through the exposure of other collection materials related to those on exhibit will increase and information about the museum and its significant collections will be enhanced. International museum professionals and researchers may find digital images useful in investigating links, similarities and contrasts with works at other institutions.

4.7.1 Education and Knowledge Dissemination

The conserved textiles articles in the museum offer an opportunity for the visitor's to observe the wonders and get knowledge on them. It supports people as an informal education centre for disseminating knowledge on art, history, science and technology through material and nonmaterial evidence. The knowledge dissemination to the society is another important function of conserved textiles of the museum to aware of the cultural heritage of different countries.

4.7.2 Lecture, Seminar, Workshops, and Research

From time to time museum organize some lectures, seminars and workshops on conserved textile articles by inviting renowned experts in the field of textile heritage and culture. These are the scholarly activities that help museum professionals to improve their potential for work as well as research and development.

Documentation and displaying objects are essential to conserve and publish the information for the use of the museum and also to serve research scholars. Research scholars in the field of history or heritage get the opportunity to refer to various artifacts of the museum for which the museum keep ready all the references.

4.7.3 Photography and Publication

There is a lot of literature on conserved textiles that are available in the museum for visitors like guide books, brochures folders and posters. In addition, catalogs on specific textiles collections, journals, art albums and monographs are published by museums, which are useful for scholars.

Photography, pattern-making, drawings and written descriptions are all good methods of documentation. Photography of museum textiles is done by professionals with expertise in the conservation of museum textiles; costume curators and conservators must communicate the proper procedures and the desired result. These mediums resorted for the conservations of textiles can help disseminate knowledge about the conservation through the documented procedures in the journal, books, folders, posters and photos to substantiate the significance of these procedures for conservation.

4.8 New Approaches in Conservation

Digitization

Digitization is an encompassing approach to preservation. In digitization, the information is converted in digital format which makes it easier to preserve access and share. The preservation and conservation strategies are enhanced by digitization which helped in the handling of fragile originals can be minimized. The workflows of digitization can be great substantial benefits in various fields and it can facilitate exchange, sharing and co-creation of information for cultural heritage conservation.

Although there were difficulties in assessing the references of textiles items due to time and distance constraints in present senior, the existence of digitization enables to easier access of fine art items, designs, structure and fusions of cultural aspect development for worldwide. It eases the spreading of the art information to thousands of users in any part of the world at a particular time. In present-day use of digitization in traditional textiles provide an opportunity for an easy and most accessible approach to enhance knowledge about the uniqueness of textile articles and set up them on a database accessible to a large group audience [37, 38].

4.9 Conclusion

The conservation of textiles in the museum develops a scope of work related to conserving the artifacts and provides service to the community through preserving textiles heritage and functions accordingly to fulfill its mission and vision. These actions are carried out when the items are in such a delicate condition or weakening at such a rate that they could be lost in a relatively short time. The chapter discussed the importance of conservation and focused on preservation principles. The types of article conserved in museum are enlisted and reflect the deterioration and damage of cultural heritage due to diverse climate, light, pollution, insects, microorganisms, disaster, and so on, as well as inappropriate storage, display, and handling. An understanding of the composition and structure of these objects leads to prevent and minimize the damage of artifact. Different techniques of surface cleaning were clearly explained to enhance the knowledge regarding vacuuming, wet and dry cleaning. The discovery and innovation of new techniques and procedures in the recent decades, such as the digitalization of textiles, have heralded in a new dawn for conservation of textiles capturing the fancy of many enthusiastic professionals to work toward the preservation of textiles, which, in recent years, try to uphold our rich heritage written in textiles.

References

1. France, F.G., Preservation of textile cultural heritage. *Text. Mag.*, 3, 24–26, 2004.
2. De la Torre, M., Values and heritage conservation. *Herit. Soc.*, 6, 2, 155–166, 2013.
3. Kim, K., *Conserving, preserving and restoring your heritage*, Dundurn Press & Ontario Geological Society, Canada, 2010.
4. Leene, J.E., *Textile conservation*, Butterworths, London, 1972.
5. Thirty, M.C. and Wyman, D.C., Textile conservation. *AATCC Rev.*, 3, 87–106, 2008.
6. Lennard, F. and Ewer, P., *Textile conservation*, Routledge, United Kingdom, 2010.
7. Priciples of preventive conservation chapter- 3, in: *Museum Textile Services, the Premier Textile Conservation Studio in New England, Specializes in the Preservation of Fabric-Based Materials for Cultural*, pp. 12–17.
8. Gillow, J. and Barnad, N., *Traditional indian textiles*, p. 6, Thames & Hudson Pvt. Ltd, London, 2002.
9. Nilsson, J., *In search of scientific methods for conservation of historic silk costumes*, Wydawnictwo Politechniki Krakowskies, Krakow, Poland, 202-211, 2010.
10. Amin, E. and Rashed, S., Preservation and restoration of a piece of textile at the Egyptian Textile Museum. *Egypt. J. Archaeol. Restor. Stud. EJARS*, 3, 1, 29–37, 2013.
11. Gillow, J. and Barnad, N., *Traditional indian textiles*, p. 6, Thames & Hudson Pvt. Ltd, London, 2002.
12. Zervos, S. and Alexopoulou, I., Paper conservation methods: A literature review. *Cellulose*, 22, 5, 2859–2897, 2015.
13. Taghiyeva, R., Traditions of the Azerbaijani carpet. *Textile Conservation Winter Colloquium. Presentations and Research*, 2018.
14. Thomson, G., *History of tapestry from earliest until the present day*, Hodder & Stoughton, New York, 1977.

15. Caple, C., *Preventive conservation in museums*, Routledge, United Kingdom, 2012.
16. Corr, S., *Caring for collections, a manual of preventative conservation*, pp. 21–26, The Heritage Council, Ireland, 1999.
17. Kathleen, D. and James, D., Managing the environment: An update on preventive conservation, vol. 15, p. 2, Getty Conservation Institute Newsletter, Los Angeles, CA, 2000.
18. Koka, V., Role of museum environment for textile conservation and restoration. *Int. J. Home Sci.*, 4, 2, 281–282, 2018.
19. Miller, J.E. and Reagan, B.M., Degradation in weighted and unweighted historic silks. *J. Am. Inst. Conserv.*, 28, 2, 97–115, 1989.
20. Luxford, N., Thickett, D., Wyeth, Preserving silk: reassessing deterioration factors for historic silk artifacts, in: *Book of Abstracts, Combined (New Zealand and Australia), Conference of the Textile Institute*, p. 39, 2009.
21. Evenson, J. and Patricia, C.C., *The effects of light and ageing on selected quilting products containing adhesives, resources*, pp. 1–5, International Quilt Study Center at Digital Commons @ University of Nebraska, Lincoln, 2004.
22. Hamlyn, P.F., Talking rot and mildew. *Textiles*, 19, 46–50, 1990.
23. Abdel-Kareem, O., Conservation of a rare painted ancient egyptian textile object from the Egyptian Museum in Cairo. *J. e-PRESERVATIONScience*, 5, 9–16, 2008.
24. Hamlyn, P.F., Talking rot and mildew. *Textiles*, 19, 46–50, 1990.
25. Abdel-Kareem, O., Microbiological testing of polymers and resins used in conservation of linen textiles. *J. Conserv. Restor. Art Archit.*, 202–211, 2000.
26. Ramachandran, T., Antimicrobial textiles - An overview. *IE (I) J.-TX*, 84, 42–46, 2004.
27. Thirty, M.C., Protecting the textile past. *AATCC Rev.*, 37–41, 2004.
28. Brooks, M.M. and Eastop, D. (Eds.), *Changing views of textile conservation*, p. 4, Getty Publications, Log Angeles, CA, 2011.
29. Francis, K., *Disaster prevention, preparedness and recovery: Special concerns for museum textile collections. federation files*, p. 34, The Federation of Histoniial Services, Troy, New York, 1990.
30. Singhal, D. and Bhagat, S., Cleaning of textile artifacts in museums. *Int. J. Appl. Home Sci.*, 2, 7&8, 247–252, 2015.
31. Maik, J. and Cybulska, M., Archaeological textiles- a need for methods of analysis and reconstruction. *J. Fibres Textiles*, 15, 5-6, (64–65), pp. 185-189, 2007.
32. Agrawal, O.P., *Care and preservation of museum objects*, National Research Laboratory for Conservation of Cultural Property, New Delhi, 1992.
33. Ragauskiene, D., Matulionis, R., Makuska, SEM studies of museum textile lined with acrylic adhesives. *J. Sci.*, 1–6, 2003.
34. Leorna, P.S., Abrasiveness of certain backing fabrics for supporting historic textiles. *J. Am. Inst. Conserv.*, 30, 2, 179–185, 1991.
35. Singh, P., Sharma, E., Fatima, N., A study on conservation of textiles in various museums of Uttar Pradesh. *Eduved Int. J. Interdiscip. Res.*, 1, 7, 1–11, 2014.
36. Clare, S., Digitized historic costume collections: inspiring the future while preserving the past. *J. Am. Soc. Inf. Sci. Technol.*, 60, 9, 1939–1941, 2009.
37. Keene, S., Preventive Conservation in Museums, Chapter 8, pp. 112–135, 1996.
38. Conway, P., The relevance of preservation in a digital world, in: *Preservation of Library and Archival Materials: A Manual*, 1999.

Webliography

Government of Canada, Canadian Conservation Institute. CCI79408-0020.

5

Fashion and Textile Museums Across the Globe

Arpana Kamboj* **and Surabhi Mahajan**

Department of Apparel and Textile Science, Punjab Agricultural University, Ludhiana, Punjab, India

Abstract

This chapter deals with different fashion and textile museums across the globe, which holds a prominent place in the heart of not only tourists but also students, researchers, fashion specialists, and all those who have a passion to learn about evolution of fashion trends. These museums play an important role in defining the textile and fashion developments of the past and presenting their memorable journey of change to witness the current fashion scenario. The location, establishment, history, recent collection, and preservation have been highlighted for the benefit of people seeking information on these museums. The prominent textiles museums of US, Europe, and Japan are covered in this chapter:

- Victoria and Albert Museum, London, England
- Fashion Museum, Bath, England
- Metropolitan Museum of Art, New York, USA
- Musée De La Mode Et Du Textile, France
- Palais Galliera, France
- Kyoto Costume Institute, Kyoto, Japan
- MFIT, New York, USA
- Museo del Traje, Madrid, Spain
- Fashion Institute of Design and Merchandising, Los Angeles, USA
- Kent State University Museum, Kent, UK

Keywords: Textile, museum, fashion, garments, collection

5.1 Introduction

Fashion museums are the reservoirs of collection of different textiles and artifacts from the glorious past so that people can witness them in the contemporary times to have an idea about the rich history in fashion. Due to this reason, such museums are also known as the "Fashion Storehouses." There are many fashion museums across the globe but major ones have been discussed in this chapter. The names of the museums has been presented in Table 5.1. Few of

**Corresponding author*: arpanakamboj@gmail.com

Seiko Jose, Sabu Thomas, Pintu Pandit, Ritu Pandey and Vandana Gupta (eds.) Handbook of Museum Textiles: Volume I Conservation and Cultural Research, (101–122) © 2023 Scrivener Publishing LLC

them are located in our country and abroad. These are visited by students, teachers, and tourists for gaining information and for entertainment purpose too. Preservation of old items is a difficult task, and museum authorities try their best to show the actual artifacts and textiles to the visitors. Hence, museums should be given additional weightage and made more popular. Visiting the museums is a lifetime experience and creates memories to cherish but for those

Table 5.1 Textile museums of the world.

S. no.	Museums	Location	Year of establishment	Number of preserved items
1	Victoria and Albert Museum	Cromwell Rd, London SW7 2RL, UK	1852	Over 2.27 million [3]
2	Fashion Museum	Somerset BA1 2QH, Assembly Rooms, Bennett Street Bath, England.	1963	Over 100,000 [9]
3	Metropolitan Museum of Art	1000 5th Ave, New York, NY 10028, United States	1870	Over 2 million items [15]
4	Musée De La Mode Et Du Textile Museum	Rue de Rivoli, 75001 Paris, France	1905	Over 81,000 [22]
5	Palais Galliera Museum	10 Av. Pierre 1er de Serbie, 75116 Paris, France	1977	Nearly 200,000 [26]
6	Kyoto Costume Institute Museum	103, Shichi-jo, Goshonouchi Minamimachi, Shimogyo-ku, Kyoto, Japan	1978	Around 40,000 [27]
7	Fashion Institute of Technology	227 W 27th St, New York, NY 10001, United States	1969	Around 50,000 [31]
8	Museo Del Traje Museum	Av. Juan de Herrera, 2, 28040 Madrid, Spain	1973	Over 160,000 [36]
9	Fashion Institute of Design and Merchandising Museum	919 S Grand Ave, Los Angeles, CA 90015, USA	1978	Around 215,000 [38]
10	Kent State University Museum	515 Hilltop Dr, Kent, OH 44240, United States	1982	More than 20,000 [44]

who cannot go physically and see the beauty of preserved collections, secondary sources, like books, leaflets, brochures, etc. giving an elaborate review of such museums can prove to be really helpful. The present chapter is a humble attempt in this direction.

5.2 Victoria and Albert Museum, London

The Victoria and Albert Historical museum situated in London is the world's biggest museum of applied and ornamental workmanship. It has been named after Queen Victoria and Prince Albert. It is a red-block palace full of sculptural delicacies, extravagant tiling and artistic creations. The gallery has more than 140 displays, and its collections range from 5,000 years of human creativity exhibited in many mediums for visitors to applaud and appreciate [1]. Its assortment is well arranged into different classes like style, theater, furniture and design.

5.2.1 History

This history of the Victoria and Albert Museum is very fascinating. From its early formation as a Museum of Manufactures in 1852, to the time when foundation stone was laid by Queen Victoria in 1899, to today's state-of-the-art galleries, the museum has constantly progressed in showcasing world-class unique pieces of art and design as shown in Figure 5.1. It was founded in May 1852, but it was only in 1857 that the museum was shifted to its existing location. The land was occupied by Brompton Park House and there were few iron galleries, which were later dismantled and used to construct the Museum of Childhood, which is still a component of the current Museum. The first building to be established in 1857 and still forms an important part of the museum is the Sheepshanks Gallery on the eastern side of the garden. The next major development was the Turner and Vernon galleries to preserve the beautiful collection. These are now used as the picture galleries and tapestry gallery respectively. In 1862, two courts, namely North and South, were started. They are now converted to galleries for displaying products during exhibitions. Toward the northern side of the museum, the Secretariat Wing is positioned which was also built in 1862 and consists of the offices and boardrooms and is not open to the visitors [2]. To the east of this wing, many additional galleries have been established for display of important items. The Museum endured the Second World War with

Figure 5.1 Victoria and Albert Museum [3].

few indemnities. A few galleries were redesigned in the 1990s and were further enhanced in 2002 when selected Victorian decorations were restructured. As a part of 2006 renovation of the museum, numerous galleries have been redesigned and renewed. The Exhibition Road Quarter was opened in 2017 with a novel entrance providing easy access for visitors.

5.2.2 Collection

The historical center embraces the assortment of materials and style, which incorporates in about 75,000 individual and set of articles that span a time of over 5,000 years, from Pre dynastic Egypt to the present day [3]. Practically, all materials are displayed in the assortments of the gallery, including woven, printed and woven materials, ribbon, artworks, and floor coverings. Among the principally rich showed regions are the early woven silks from the East, European and Chinese embroideries, English middle age weaving, Safavid rugs, Indian materials, etc. The primary materials that forms an essential part for the Gallery assortment is a trim gathering credited in 1854. The Historical center acquired a significant number of materials in the range of 1860 and 1883, which are at the core of the gallery's initial material assortment [4]. The assortments contain over 1,000,000 index records and unlimited pictures alongside exclusive materials displayed in wooden exhibits and assembled by their enrichment. The early pieces integrate uncommon sixteenth century kids' *kaftans* from the Stool Domain and seventeenth century European dresses of people. It has a specific forte with principally European elegant garments and embellishments for all kinds of people. The assortment additionally unites a wide variety of frill from across the world and also has footwear and caps. The assortment is remarkably elite and showcases old time ensembles to the elaborative dresses of Alexander McQueen [5].

5.3 Fashion Museum, Bath, UK

The Fashion museum in Bath is known for its unique galleries. The historical style exhibition hall known as The Gallery of Outfit is one of the world's amazing historical center of memorable and elegant dresses [6]. The Design Gallery was set up by Shower and North East Somerset Committee and is currently directed by the Legacy Administration Segment. The Style Exhibition hall includes an important assortment of frill and different attires. The gallery assortments are essentially isolated into menswear and ladies wear, dress for day and evening, old and contemporary fashion, etc.

5.3.1 History

The Fashion Museum in Bath is located at the splendid Assembly Rooms which are maintained by the National Trust since May 23, 1963. It was founded by a writer, designer, and scholar, namely Doris Langley Moore [7], who donated her well-known private collection of costumes to the city of Bath. From 1974, the Museum of Costume also included the Fashion Research Centre positioned at a building owned by a trust but in 2003, the Fashion Research Centre collections and activities were transferred to the Museum at the Assembly Rooms only. The Fashion Museum changed its name in 2007. The Museum is fully credited and labeled as a collection of international eminence by Government under various

Figure 5.2 Fashion museum [6].

schemes administered by the Arts Council, England. But now, it is being obligatory made to move for accommodating a new heritage site in Figure 5.2. The National Trust, which owns the building, plans to create a memorable event celebrating the city's Georgian history and is enforcing the museum to leave its underground site by March 2023 [8]. Bath and North East Somerset Council has stated that they may not be able to reposition the museum for another five years and the collection of over 100,000 items will be packed if a new location is not found soon.

5.3.2 Collection

This historical center has been named as an assortment of extraordinary importance. There are numerous unique and wonderful articles displayed in the Style Gallery ranging from gloves from the Shakespeare time to the present contemporary assortment of various things. Greater part of the valuable items has been given to the Exhibition hall by people and associations and they depict parts of the historical backdrop of design [9]. The assortment incarnates the progression that occurred from 1600s to the present day. The earliest pieces are woven shirts and gloves. The English design famous in 1970s and 1980s is included in the exhibitions as a bandage dress, skirt, coat, pants, shirt, short and long evening dresses, *kaftans,* layered dresses, etc. The Victorian dress and the old games outfits are an additional element in the assortment of the Exhibition hall. Other noticeable features include a sizeable gathering of woven petticoats from 1742 and work by twentieth century English style designers.

As far as contemporary and present day show of the exhibition hall is concerned, each year, the Style Gallery greets an expert designer from Design Industry to choose a "Dress of the Year" that emulates the look of the design and catch the visitor's creative mind [10]. The designers whose work has been addressed include John Bates, Jean Muir, Mary Quant, Giorgio Armani, Ralph Lauren, Alexander McQueen, Donatella Versace and many more. The "Dress of the Year, 2020" comprises of seven outfits in addition to one film by a popular style contributor and instructor, R. Webb. It incorporates the "feature making look" worn by Naomi Campbell on a tour to New York in Walk 2020, camel cashmere cape by Riccardo Tisci for Burberry with Tyvek hazardous materials suit, facemask, and pink latex gloves.

There is similarly a huge and varied photo assortment, including the Sunday Times Design File and nineteenth century Cartes de Visite of almost 400 types of apparel, embellishments and materials possessed by Woman Ottoline Morrell. The Design Historical center also possesses a few of evening dresses by Madeleine Vionnet, which represent her creative development methods and fascinating style.

5.4 Metropolitan Museum of Art, New York City

The Metropolitan Exhibition hall of Workmanship known as "The Met" is the biggest craftsmanship gallery and world's biggest display of products in the United States. The Exhibition hall is located at two destinations having a broad assortment of artifacts, design and elegant remains from middle age Europe. The Met has always endeavored to be much more than just a storehouse of rare and beautiful items. Everyday, art flourishes in the galleries of the museum through its exhibitions and events discerning new ideas and astonishing connections in time and between cultures [11].

The permanent collection of museum consists of works of art from Egyptian ancient times and paintings and sculptures by almost all the European masters, as well as widespread collection of modern art (Figure 5.3). Other major collections of the Museum include arms and armour, photographs and the Robert Lehman Collection. Different articles of Africa, Oceania, America, Greek, and Roman also form an important part of the collection. Islamic, medieval, modern and contemporary artifacts and musical instruments also catch a lot of attention.

5.4.1 History

The Metropolitan Gallery of Workmanship was approved on April 13, 1870, to build up and keep a historical center and Library of Craftsmanship authorizing and nurturing the exploration of expressive arts and the use of workmanship for advancement of general information on related subjects [12]. This representation was subsequently amended by the 1893 demonstration, section 476, which articulated that the assortments ought to be opened and made accessible to the visitors.

Figure 5.3 Metropolitan Museum of Art [12].

The organizers included many finance managers and investors and among them, Theodore Roosevelt Sr., father of Theodore Roosevelt, (26th leader of the United States) was the major one, to act as a driving expert and mastermind to open the gallery to the American public. The historical center was first opened in 1872, in an assembly situated at 681 Fifth Road. John Taylor Johnston was a famous designer of that period and he donated his craftsmanship assortment for display in the exhibition hall. He also joined as the first president of the cenete, and George Palmer Putnam was honored as the establishing director. Craftsman Eastman Johnson was made the Fellow patron of the exhibition hall along with Frederic Edwin Church. The previous official, Luigi Palma di Cesnola, was entitled as the first Chief and he served from 1879 to 1904. Under his direction, the Met's property, at first comprising of a Roman stone coffin and 174 European canvases rapidly propagated and was difficult to be assembled in the accessible space. In 1873, when the Metropolitan obtained Cesnola's assortment of Cypriot ancient pieces, the historical center moved from Fifth Road and was set up in Mrs. Nicholas Kruger's estate called the Douglas House at 128 West 14 Road. Nonetheless, the new premises ended up being temporary only as the increasing assortment required even more space [13]. So, the center moved to the current assembly in 1880. In 1954, the gallery released a movement of shows to launch a beautiful corridor to add workmanship discourses. The "Shows and Talks program" has grown to 200 occasions each season currently. From its commencement until 1968, the program was coordinated by William Kolodney and from 1969 to 2010 by Hilde Limondjian, both of whom were famous contributors for the gallery products.

During 1960s, the Metropolitan administration was extended to integrate the Executive of the Leading group of Trustees from the home of Robert Lehman who turned into the main Administrator of the Board at the Metropolitan during the 1960s. After his demise in 1969, the Robert Lehmann Establishment donated around 3,000 masterworks to the Metropolitan Gallery of Craftsmanship, which opened to general public in 1975. This centennial was commended with various shows, symposia, addresses, visits, and get-togethers for a considerable time from October 1969 to 1971.

In January 2018, the historical center President, Daniel Weiss, declared that the extremely old arrangement of entry would be replaced with a $25 charge for unfamiliar guests. In September 2020, the exhibition hall named Patricia Marroquin Norby as a Debut Partner Custodian of the Historical center of Local American Craftsmanship. In May 2021, The Met introduced a panel on the Fifth Road exterior identifying Lenapehoking, the country of the Native Lenape people groups.

5.4.2 Collection

The museum stocks an enormous assortment of craftsmanship crossing a wide exhibit of the historical backdrop of human culture. The historical center is popular as a home of prestigious artworks and excellent things like porcelain, instruments, relics, outfits and even lethal implements. It is extremely durable assortment contains more than 2,000,000 works, divided between 17 curatorial divisions [14]. The exhibition hall began with an assortment of 174 European canvases however currently; it stores in excess of 2,000 European artworks that date from the twelfth to the nineteenth hundreds of years individually. This assortment incorporates representative works from experts like Vermeer, Rembrandt, Caravaggio and Raphael. The gallery has around 20 super durable assortments. Two distinct assortments focus around antiquated workmanship, including the enlightened volumes and stained glass of the orders. Various societies are addressed, including model from Asia, artifacts from Egypt, Roman containers, African covers and articles from Islam [15]. Different assortments recollect embroideries and bands for the Antonio Ratti Material Center and photos, present day craftsmanship, drawings, prints and artistic creations, and figure and woodwork from America's commencement in different applications.

One of the popular occasions of this Exhibition hall is its "Met Celebration" formally called the Ensemble Establishment Function or the Outfit Organization Lead and furthermore known as the Met Ball [16]. It is a yearly fund raising occasion, which indicates the outset of the Ensemble Establishment's yearly style show. Every year, the occasion has a topic. The presentation launches the feel for the proper dress of the evening since visitors are trusted upon to pick their style to coordinate with the subject of the show. The subject for 2021 is "In America: A Dictionary of Style," which is exploring into the cutting edge jargon of American design.

5.5 Musée De La Mode Et Du Textile, France

The Musée De La Mode Et Du Textile (Museum of Fashion and Textiles) is currently a well-known department of the Musée des Arts Décoratifs, Paris (Figure 5.4). This museum is about interior design, furniture, objects, wallpaper, tapestries, ceramics, glassware, and toys from the Middle Ages to the present day and inhabits about 9,000 square meter area displaying around 6,000 items. At first, decorative art was on the threshold of fine art but now, it is more of an accepted art form and the collection consists of refurbishments of decorative art from different periods including an Asian, a Gothic and a Contemporary section [17]. Today, the number of departments of arts and crafts has increased due to many developments.

5.5.1 History

The collection of materials initially had an abode with the Association Centrale des Expressions Decoratifs and the Musée des Expressions Decoratifs formerly opened in 1905 with a massive assortment of materials, like silks, weavings, ribbon, embroideries, etc. [18]

Figure 5.4 Musée De La Mode Et Du Textile Museum [18].

In 1948, one more assortment of style and materials was established by the Association Française des Expressions du Ensemble (UFAC) on an initiative by a student of history, François Boucher. It was due to the initiatives taken by the guardian, Yvonne Deslandres, that the UFAC over the years became home to probably the biggest ensemble assortments on the earth. In 1981, the Association Centrale des Expressions Decoratifs and the Association Francaise des Expressions du Outfit assented to unite to make the Musée des Expressions de la Mode, however, it took an extra five years before this center was accessible to general society inside the Structure de Marsan in the Louver [19]. The assortments and presentation displays were then drifted to the Rohan wing of the Louver in 1997 and were spread in more than two floors. This gave more space for show of the remarkable assortment of the museum.

Les Expressions Decoratifs is, likewise, connected with a café in the past called Le Saut du Loup and presently renamed LouLou, which exists inside the Louver Royal residence and is open through historical center and the merry go round garden [20].

5.5.2 Collection

The Musée des Arts Décoratifs showcases one of the world's largest collections divided into five sequential sections (the Middle Ages and Renaissance, the 17th and 18th centuries section, the 19th century section, Art Nouveau and Art Deco section and the Modern and Contemporary section) and five thematic divisions (graphic arts, jewellery, toys, wallpaper, glassware). The assembly of this museum was earlier known as the Musée de la Mode de la Ville de Paris, the Victoria and Albert Museum in London and the Metropolitan Museum in New York [21]. This unique collection occupies an integral place in the field of the decorative arts and represent a wide range of techniques like woodwork (ornamental carving, furniture making, wood trim), metalwork (gold, silver, iron, bronze, pewter), ceramics, glass work, leather work, painting, and arts, such as straw carving, bead embroidery, and painted metal [22].

The museum presented collections showcasing the historical costumes from the French Regency period to the present and textiles from the 7th century onwards [23]. Important French fashion designers and textile artists whose works are held in the collection include Christian Dior, Pierre Balmain, Paul Poiret, Madeleine Vionnet, Sonia Delaunay and many more.

5.6 Palais Galliera, France

The Palais Galliera is officially known as the Musée de la Mode de la Ville de Paris (City of Paris Fashion Museum) and formerly known as Musée Galliera. Paris has been always famous for its fashion history [24]. The museum is accommodated in a palace established in the 1800s in the Renaissance style on the Duchesse de Galliera's orders (Figure 5.5). The construction of the museum was completed in 1894. Palais Galliera is one of the important 14 museums in the city of Paris, which since January 1, 2013, have been amalgamated into the public institution Paris Musées.

The architect Léon Ginain created his design on a genoa palace held by the Duchess Galliera. The structure is covered in cut stone in the Italian renaissance style and is supported by a steel underframe built by the Eiffel Company. Giandomenico Facchina designed the

Figure 5.5 Palais Galliera Museum [25].

mosaic flooring and domes. Painting is represented through the sculptures on the Avenue du President Wilson concealment.

5.6.1 History

Maria Brignole Deal De Ferrari, the Duchesse established that she needed to formulate a historical center to hold her work of expressions. As per her desire, a land was gifted to the French state. The Duchess passed an order in December 1888, preceding the historical center was finished, yet in May 1889, her beneficiaries gave around 1.3 million francs to the City of Paris to complete its development. The engineer who was accountable for this development was Léon Ginain. He finished the exhibition hall in February 1894, which was formally professed by the city [25]. Without a trace of the Galliera craftsmanship assortment, for which it was planned, the City of Paris utilized the gallery for temporary presentations. On Walk 1, 1895, President Félix Faure unveiled the debut show, which included pictures of women and trim. In 1902, it was consigned as a historical center of modern expressions and later worked as a division for impermanent contemporary workmanship presentations.

5.6.2 Collection

The Palais Galliera is an exhibition hall of design and style history. It safeguards extravagant assortments which mirror the encryptions of apparel in France from the eighteenth century to the current day. It empowers to show the development of style over three centuries and reflect the changes of the outline from the flying dress of the 1730s to the vanguard expressions of ongoing occasions. The assortment is arranged into different offices are as per the following:

5.6.2.1 Eighteenth Century Dress Office

This is one of the world's driving assortments of eighteenth century dresses. It involves people's articles of clothing dating from the late seventeenth century to the year 1800, along with kids' wear and theater outfits.

5.6.2.2 *Nineteenth Century Ensembles Division*

This office relates to the historical backdrop of design among France's privileged societies from the Napoleonic time to around 1906, when Paul Poiret's "Directoire" style dresses started with the twentieth century style [26]. The assortment is comprised of garments for ladies, while men's and kids' wear also form an essential part. The assortment incorporates various pieces of clothing used by exquisite individuals from the upper working classes.

5.6.2.3 *Fashion of the Principal Half of 20th Century*

From the start of the twentieth century to the Christian Dior's New Probe in 1947, the division has accumulated numerous things. The assortment of popular plan houses, like Callot Soeurs, Chanel, Chéruit, Doeuillet, Doucet, Nicole Groult, Jacques Heim, Lucien Lelong, Jeanne Lanvin, and many others, have been displayed.

5.6.2.4 *Haute Couture*

This is one of the most generous, representative and in-depth assortments and an ideal impression of the imaginative variety of Paris' couture houses like Dior, Balenciaga, Chanel, Grès, Fath, and Balmain during the 1950s who were the forerunners of Yves Holy person Laurent, Courrèges, Paco Rabanne and Pierre Cardin during the 1960s. With prepared to-wear high fashion during the 1970s, the division's elements keep on showing a realistic representation of the creativity of ongoing periods.

5.6.2.5 *Contemporary Office*

The Contemporary office assortment has been assembled through gifts from couture houses, designers and other eminent personalities including press attachés, style chiefs, models, writers, editors and picture takers.

5.6.2.6 *Extras Office*

Caps, shoes, handbags, scarves, gems, fans, strolling sticks, parasols, umbrellas, gloves, buttons, clasps, etc. are completely found in the Frill office, whose assortment covers a period extending from the late seventeenth century up to the current day.

5.7 Kyoto Costume Institute, Japan

Kyoto Costume Institute is located in the city of Kyoto and was initiated in the year 1978. The Kyoto Costume Institute (KCI) methodically collects and preserves exceptional western apparel as well as documents and artifacts. The KCI digital collections present image and text information for roughly 300 objects that is arranged in a chronological order [27]. The Institute also conducts researches, exhibiting or publishing its verdicts for conveying useful information to the promising fashion learners.

Figure 5.6 Kyoto Costume Institute Museum [28].

Through study on western clothes, KCI explores various aspects related to clothing. Relevant documents from previous times reveal a good deal when observed in the context of human civilization. Clothing highlights different features of each era, as well as ways in which its civilization operated and changes that were evident in styles and dresses. Stuffs from KCI's collection are given for addition in exhibits fashioned by other institutions, in addition to being exhibited in the institute's own shows. They work closely with other art museums and their collections have been displayed in the famous Victoria and Albert museum in London and the Metropolitan museum of Art in New York and Musée des Arts de la Mode Et Du Textile in Paris (Figure 5.6).

5.7.1 History

KCI was set up in the year 1978 and the actual model was created related to the Outfit Establishment of the Metropolitan Historical center of Workmanship, New York. In 1980, development was made in the Public Gallery of Present day Workmanship, Kyoto and from that point forward, there was first issue of Dress Review and The Secret Story related to FIT, New York [28]. From April 1989 to September 1999, the assortment of KCI surpassed 10,000 pieces and later on in October 2006, 50th issue of Dress Review was published. The assortment continued to develop since then. KCI has limited scope displays in the foundation consistently related to the Public Exhibition hall of Present day Workmanship in Kyoto and different establishments across the world.

KCI allocates a release dedicated to themes in design study twice in a year, Through Taschen, KCI distributed "Design: A Past filled with the eighteenth century to the twentieth century, Assortment of the Kyoto Ensemble Foundation," which recollects 400 photos of things for the assortment.

5.7.2 Collection

KCI gathers western outfits mostly underpants, frill and other attires. Since its footing, KCI has made its assortment and examination discoveries accessible to the general public through presentations, talks, distributions, and computerized chronicles. It is especially famous for its assortment of products that exhibit how Japanese culture and style, just as Japanese creators, have significantly affected western garments.

KCI's assortment right now descends from the seventeenth century to the current day with possessions of about 13,000 different type of apparel and 20,000 reports. The establishment has grown due to the gifts from top style houses like Chanel, Christian Dior, Louis Vuitton, and was also given an endowment of around 1,000 arrangements of apparel from Comme des Garçons. The assortment is displayed in a controlled environment. The establishment additionally spans to predict the changes in shape and style.

KCI shows a few pieces at the KCI Display; with carefully picking up pieces from the assortment [28]. One more credit to KCI is the association of displays on topics of design, which are held at galleries all throughout the world. It handles the arranging and opening of such shows, delivering outlines of dress according to alternate points of view and has impelled new interest in design displays in Japan. The computerized chronicle of KCI is one of its classes. It incorporates digitized interpretations of property offering help for crafted by fashioners and specialists. Text and visuals are accessible for many things, including outfits, accessories, etc.

5.8 Museum of Fashion Institute of Technology, New York, USA

This museum is an important part of specialized fashion museums, all of which have been acquainted by the American Alliance of Museums. It is located at Fashion Institute of Technology and is known for its pioneering and high-quality special exhibitions. Also, MFIT has won an American Alliance of Museums (AAM) Media and Technology MUSE award [29]. These awards recognize outstanding achievement in Galleries, Libraries, Archives and Museums (GLAM) media and celebrate scholarship, community, innovation, creativity, education, accessibility and inclusiveness (Figure 5.7). The Museum at FIT is supported by a Couture Council which helps to make it possible for the museum to support world-class exhibitions that educate and inspire, build and conserve its extraordinary permanent collection and organize free public programs to serve the FIT's students as well as the interested public.

5.8.1 History

The historical center was established in 1969 and shows were accessible to the public in 1975 after the gallery was relocated to its present area in 1974 [30]. Starting around 2003, Dr. Valerie

Figure 5.7 Museum of Fashion Institute of Technology [31].

Steele has filled in as the Overseer of the gallery at FIT and as a head custodian in 1997. The State College of New York (SUNY) framework has made the historical center in 1969 to help starting facilitating displays in 1970s by utilizing assortment on credit from the Brooklyn Gallery of Craftsmanship. The late Richard Martin and Harold Koda, coordinated original style displays like Design and Oddity, Three Ladies, Athletes and the Geeks. A protection lab was also settled to focus on the fast developing assortment of dress and other materials.

In 1993, the Leading body of Trustees of FIT, taking note of the meaning of the Plan Research facility's assortments and shows, changed the foundation's name to The Exhibition hall at FIT. In 1997, Dr. Valerie Steele joined the staff of the Gallery and she curated shows like London Design, which got the principal Richard Martin Grant for Greatness in Outfit Displays from The Ensemble Society of America [31]. Dr. Steele turned into the Manager of The Gallery at FIT in 2003 and the historical center was granted authorization by the American Partnership of Galleries in 2012. In 2003, The Historical center at FIT held its first yearly Design Conference regarding the show Femme Fatale: Style and Visual Culture in Paris determined on boosting style information.

5.8.2 Collection

The historical center is made out of a lab and visual studio in an area exceeding about 14,000 square feet and extra rooms, three study halls and diverse office regions. The historical center's task is to propel information on design through shows, distributions and public projects [32]. The exhibition hall has an assortment in excess of 50,000 pieces of clothing and embellishments spanning between the 18th century to contemporary times with strength in current ladies' design. The outfit assortment incorporates style by planners like Chanel, Azzedine Alaa, Balenciaga, Comme des Garons, Dior, Galanos, Halston, Charles James, and so on. The menswear assortment highlights around 2,000 pieces of clothing ranging from formal to dynamic wear, including suits, covers, vests and garbs. Additionally included categories are swimwear, undergarments, outerwear and knitwear.

Among the 15,000 accessories, there are more than 4,000 pairs of shoes alone, including illustrations by Manolo Blahnik, Ferragamo, Perugia and Roger Vivier. The millinery collection contains more than 3,000 hats by famous milliners. There are also many examples from designers, such as Christian Dior, Balenciaga, and Jacques Fath. The handbag collection includes fine specimens by luxury houses, such as Herms and Gucci, as well as samples by Roberta di Camerino, Judith Leiber, and Bonnie Cashin.

There are also 30,000 textiles, dating from the 5th century to the present, including the work of designers, like William Morris, Salvador Dali, Raoul Dufy, and Junichi Arai. The textile collection includes apparel and home furnishing fabrics, embroideries, shawls and also a number of ribbon sample books [32]. The J.B. Martin Velvet Room records hand woven and production velvets spanning 125-year period. This museum often credits different objects from its permanent collections to other institutions for use in exhibitions.

A small archive of fashion photography also contains works by Louise Dahl-Wolfe and John Rawlings. The museum is divided into three sections. Special exhibits are held at the lower level gallery, while on the main floor, there is fashion and textile history gallery, which exhibits a selection of around 200 historically and aesthetically significant artifacts from the museum's permanent collection.

5.9 Museo Del Traje, Spain

The Museum of Garment Ethnologic Heritage Research Center (*Museo del Traje—Centro de Investigación del Patrimonio Etnológico*) was established in 2004, although the history of its collections dates back to 1925. The Museum is located in a recent building designed by Jaime López de Asiaín and Ángel Díaz Domínguez and is surrounded by beautiful gardens. The museum's collections is a combination of products from the Middle ages and the recent works of contemporary fashion designers from Spain. The mission of the Museo del Traje museum is to maintain, protect and promote fashion and clothing collections while also providing the visitors with cross-disciplinary information through activities based on a solid and accessible discourse [33].

The exhibition hall features the noteworthy development of the apparel, exploring its specialized, social, logical and inventive implications through a variety and a consistent change in dressing rehearses. This supports in uniting the material examples and useful components from the early times and report to a current time driving the Museo to record the advancement and feature accomplishments of contemporary style plan. The significant purposes of the gallery are nurturing research in the Spanish ethnological legacy, turning into a public and global benchmark community and changing over the exhibition hall into a Middle for Ethnological Legacy Exploration.

5.9.1 History

The Museo del Traje exhibition hall has a long and intrusive history. Its starting point lies in the Notable Ensemble Show in 1925. The Count of Romanones anticipated in making the brief display into a durable settlement during his feature address at this occasion. He set up a Leading group of Trustees of the Gallery two years after the fact to assume responsibility for the assets of the presentation, getting them from the state, and structure them into the new Territorial Ensemble Historical center [34]. In 1934, the Count thought of the idea of setting up another establishment to get the loved assortments a separate place (Figure 5.8). This new gallery, the historical center gained the assortments of Ethnography and People Craftsmanship at the Institute of Training and covered a broad series of items given by the Provincial Trustees. After

Figure 5.8 Museo Del Traje Museum [34].

a few changes, the assortments were moved in 1983 into the structure which was around then involved by the Spanish historical center of Contemporary Craftsmanship Museo Español de Arte Contemporáneo. In 1993, the Exhibition hall of the Spanish Town and the Public Gallery of Ethnology were united to shape the public historical center of human sciences [35]. At last, in 2002, after a wide conversation about the historical center's future, it was chosen to promote them in the public presence according to an advanced point of view.

5.9.2 Collection

The Exhibition hall has accomplished an ideal situation at Madrid's social scene as a result of imperativeness and the nature of its assortments. These assortments are wide and different, comprising authentic, contemporary and conventional dresses, decorations and a huge quantity of things acknowledged with the Ethnological Legacy.

It is extremely durable show permits the visitors to go through the historical backdrop of design from eighteenth century to the twenty-first century, utilizing recorded ancient photos and banners to give a live setting [36]. This experience is advanced by the assorted transitory shows and exercises intended to carry the historical backdrop of apparel to a spread-out public identified with altogether different subjects, like style, photography, toys, and so on. Some interesting instances of presentations offered are Modachrome, Mujeres de Blanco and Trajes de novia. The Gallery of Ensemble designs all sort of studios, courses and exercises, just as shocks for exceptional dates, like the Worldwide Historical Center Day.

5.10 Fashion Institute of Design & Merchandising, California

Fashion Institute of Design and Merchandising has four campuses: San Francisco, Orange County, San Diego and the original Downtown Los Angeles location. The FIDM Museum and Galleries on FIDM's Los Angeles campus represent a unique resource for contemporary and historic costumes and textiles (Figure 5.9). The square-foot gallery spaced in a huge space spotlights special exhibitions year-round. The museum is situated on the ground floor of the Los Angeles campus of the Fashion Institute of Design and Merchandising, and it includes a collection of varied costumes, accessories and textiles dating to eighteen century to modern times including couture, ready-to-wear and film/theater costumes.

Figure 5.9 Fashion Institute of Design and Merchandising Museum [37].

The FIDM Gallery houses the early Hollywood Ensemble Assortment from the City of Los Angeles, Branch of Parks and Entertainment. It has the yearly Film Ensemble Plan presentation at the exhibitions [37]. The Annette Green Fragrance Gallery is in on-going occupancy on the second floor at the FIDM Exhibition hall and Displays.

5.10.1 History

Tonian Hohberg, the current FIDM President, made the Style Organization of Plan in 1969. Hohberg projected an astonishing learning climate for understudies where they could dominate workmanship. The Style Plan Division was acknowledged in 1973. In 1978, the not-for-profit FIDM Exhibition hall and Library, Inc. was officially settled. The underlying property was improved by a liberal legacy of French high fashion outfits from famous contributor Betsy Bloomingdale [38]. In 1985, Rudi Gernreich granted the originator's record to the FIDM Historical center. The plan of an advanced Los Angeles site developed in 1990 and it included open hoarding for the historical center's shows. The FIDM Display opened in 1993, giving a super durable show space. From California active apparel to Hollywood ensembles, the Gallery started showing various displays.

A curatorial office was set up during the 1990s, the exhibitions were extended and the historical center shop was started. In the interceding years, the gallery moved up to a computerized data set, built cutting edge reduced capacity regions, and gained the Annette Green Scent File.

5.10.2 Collection

The museum's collection quality, uniqueness, and explanation have evolved and expanded toward educational exhibitions, documentary, and catalog publications, blog presence, and social media activities [39]. More than 15,000 fashion objects dating primarily from the 18th century to the present, and 200,000 special collection artifacts have been displayed at the FIDM Museum. The museum focuses on the high quality of design and most of the collections are dedicated to women dresses. The collection also includes folk dress, ready-to-wear, menswear, and nonwestern garments.

Accessories are another important component that are displayed individually or used for mannequin dressing, allowing visitors to have a better knowledge of fashion adornment. Interior textiles, swatch books, and embroidery samples complement the fashion holdings and ensure a rich understanding of related disciplines. The museum also provides online exhibition feature, as well as access to an online permanent collection, like the Study Collection, Rudi Gernreich Archive, Textile and Jewellery Collection, Hollywood Costume Collection, Fragrance Collection, and various special collections like documentary materials, including books, patterns, magazines, etc.

5.11 Kent State University Museum, USA

Kent State University Museum is a house to important collections of fashion and decorative arts. It contains astonishing collection of historic dresses, fashion, textiles and arts [40].

Figure 5.10 Kent State University Museum [41].

It showcases works by many of the world's great designers across the globe. It has eight galleries featuring work of artists and designers displaying first-hand experience of historic culture and contemporary fashion as well as costumes representing many of the world's cultures. A broad collection of American glass, fine furniture, textiles, paintings, and other decorative arts combine to give context to the study of design.

The museum serves both the university and the community through exhibitions and its public programs (Figure 5.10). The museum store is located in close proximity where jewellery, books, apparel, stationery, and other fashion-related merchandise are available for purchase. Additionally, the store features selected student and faculty made merchandise from the School of Fashion Design and Merchandising.

5.11.1 History

The Kent State College Exhibition hall holds an exceptionally captivating history. The President, Brage Golding, supported the vision of building up both an exhibition hall and an educational program [41]. He convinced both the collaborators of Jerry Silverman fuse, Shannon Rodgers and Jerry Silverman for an incredible gift and the gallery activated with significant grant of the assortment of design, memorable outfits, artworks and ornamental expressions that included 4000 ensembles, 1000 bits of workmanship and 5000 volumes of reference library [42]. The Ohio's foundations of Shannon Rodgers and other concerned individuals from Kent, Cleveland and Akron group opined the possibility of development of this generally rich exhibition hall in 1982. It was viewed as one of the best demonstrable outfit assortment focus in the United States in 1960s, with more than 40,000 items. The Gallery opened in 1985 in Rockwell Lobby, the first College Library, which was refurbished to fulfill the most noteworthy historical center guidelines [43]. Implicit 1927, the structure reflects Beaux Expressions plan components which were carefully reinstated under the management of Ted Curtis, previous Kent State College architect from Whitley and Whitley. The Higbee Display exhibited components of the first structure including the marble, woodwork and crystal fixtures. From the eighteenth century to the present, this historical center houses perhaps the broadest collection of style plan.

5.11.2 Collection

The current collection is to the tune of over 20,000 objects and augmented by major gifts from Borowitz, Paige Palmer, Martha McCaskey, Wilkinson and many other profuse donors. The museum acquired the archives of several designers, including photos, sketchbooks and press kits. These archive span the last half of the twentieth century and represent many sector of the fashion industry [44]. Among the designers whose archive are represented are Charles Kleibacker, Shannon Rodgers, Jerry Silverman and Pauline Trigère and George Stavropoulos.

5.12 Conclusion

The fashion and art museums constitute an important space and a showground dedicated to displaying varied objects from the past to the present to discover their beautiful journey. The 10 important museums discussed in this chapter reveal the glorious expedition of historic costumes, contemporary clothing trends, paintings, murals, accessories including hats, scarves, footwear, jewellery, etc. Their stories of origin and growth covered in the chapter are an inspiration in themselves to understand how small ventures perceived with the idea of common well-being of one and all grow at the end. The generous donations of designers and celebrities have added to the vast collections of all the museums. These museums are not only vital in attracting thousands of visitors but also act as information reservoir where fashion-related knowledge can be produced, negotiated and transmitted. Worldwide fashion exhibitions, which are a blazing trend in these museums, provide an opportunity to many fashion designers to promote their collection. The visitors can take advantage of the magnificent assembly of these museums and at the same time learn to appreciate the hard work put in by the museum authorities in collecting, archiving, conserving, and displaying the artifacts.

References

1. Victoria and Albert Museum, Comics in the National Art Library, 2021, www.vam.ac.uk, (accessed 31 august 2021).
2. Alexandra, W., Catalogues of architectural drawings in the V&A, in: *A. W. N. Pugin. and the Pugin. Family*, p. 77, 1985.
3. Victoria and Albert Museum, 2021, https://www.vam.ac.uk/collections/the-va-story/, (accessed 29 august 2021).
4. Victoria and Albert Museum, 2020, https://www.vam.ac.uk/shop/fashion.html, (accessed 29 august 2021).
5. *Museum, Victoria and Albert*, The Oxburgh Hangings | Elizabeth Talbot | Mary Queen of Scots | V&A Explore The Collections, Victoria and Albert Museum: Explore the Collections, 2021.
6. Fashion museum, 2021, https://en.wikipedia.org/wiki/Fashion_Museum,_Bath, (accessed 7 august 2021).
7. Fashion museum, 2021, https://www.fashionmuseum.co.uk/collection, (accessed 12 august 2021).

8. Thomas, D., Paris's fashion museum gets a new director, pp. 289–294, The New York Times, 2018.
9. The fashion museum, 2021, https://www.fashionmuseum.co.uk/about. (accessed 26 july 2021)
10. Lester, R., Dress of the year, pp. 141–142, Antique Collectors' Club, 2013, ISBN 9781851497256.
11. Barnet, P. and Wu, N.Y., *The cloisters: medieval art and architecture*, p. 176, Metropolitan Museum of Art, New York, 2005.
12. Peck, A., Period rooms in the Metropolitan Museum of Art, p. 275, Metropolitan Museum of Art, 1996.
13. Pyhrr, S.W., *Arms and armor: Notable acquisitions 1991-2002-The Metropolitan Museum of Art*, pp. 6–10, Yale University Press, New Haven, 2003.
14. Brenson, M., *90 works by Paul Klee donated to Met Museum*, p. 189, German, Switzerland, The New York Times, 1984.
15. Metropolitan Museum of Art, 2021, https://www.metmuseum.org/blogs/metkids/2016/history-of-the-met/.
16. Husband, T., *The art of illumination: The limbourg brothers and the Belles Heures of Jean de France, Duc de Berry*, Metropolitan Museum of Art, New York, 2008, ISBN 978-1-58839-294-7.
17. Musée de la Mode de la Ville de Paris, 2021, https://en.wikipedia.org/wiki/PalaisGalliera, (accessed 10 october 2021).
18. Musée de la Mode de la Ville de Paris, 2021, https://www.eutouring.com/musee_galliera.html, (accessed 10 october 2021).
19. Musee de la Mode et du Textile in Paris, 2011, https://en.wikipedia.org/wiki/Mus%C3%A9e_de_la_mode_et_du_textile, (accessed 09 october 2021).
20. Hrvol, F. and Carol, A., *Owen Jones: Design, ornament, architecture, and theory in an age in transition*, pp. 94–98, New York: Rizzoli, Random House Incorporated, 2006.
21. Musee de la Mode et du Textile in Paris, Mode/textile, 2021, https://madparis.fr/francais/musees/musee-des-arts-decoratifs/collections/departements/#article3700, (accessed 01 october 2021).
22. Musee de la Mode et du Textile in Paris, 2017, https://madparis.fr/en/about-us/collections/, (accessed 09 october 2021).
23. Musee de la Mode et du Textile in Paris, 2021, https://www.eutouring.com/museedelamodeetdutextile.html, (accessed 15 october 2021).
24. de Andia, B., *Le 16e: chaillot-passy-auteuil: métamorphose des trois villages*, pp. 39–43, Délégation à l'ActionArtistique de la Ville de Paris, Paris, 2004.
25. Hubert, D., *Auteuil et passy, de l'annexion à la grande guerre*, p. 142, L'Harmattan, Paris, 2008.
26. Miren, A., *Nouvelle directrice du Palais Galliera*, p. 45, lejournaldesarts.fr. Le Journal Des Arts, 2017.
27. Kyoto Costume Institute, 2021, https://artsandculture.google.com/partner/kyoto-costume-institute (accessed 21 november 2021).
28. Kyoto Costume Institute, 2021, https://artsandculture.google.com/exhibit/the-kyoto-costume-institute-kci-kyoto-costume-institute/IQKyC5hlsoXMLQ?hl=en, (accessed 25 november 2021).
29. Jackson, K., Fashion Institute of Technology, in: *The Encyclopedia of New York City*, pp. 392–93, New York, Yale University Press, 1945.
30. Mzezewa, T., Fashion Institute of Technology's Library gets a makeover, pp. 458–460, The New York Times, 2018.
31. The museum at FIT (MFIT), 2021, https://www.fitnyc.edu/museum/about/history.php. (accessed 27 july 2021)
32. Steele, V., Suzy, M., Fred, D., Robert, N., Fashion Institute of Technology Museum, fashion designers: The collection of the museum at FIT, p. 14, Köln, Taschen, 2012.

33. Museo-del-traje, 2011, http://everymuseummadrid.blogspot.com/2011/09/museum-of-costume-museo-del-traje-10.html, (accessed 11 september 2021).
34. Museo-del-traje, 2021, https://www.gomadrid.com/museums/museo-traje.html. (accessed 11 september 2021)
35. Museo-del-traje, 2021, http://www.emporis.com/building/museodeltraje-madrid-spain, (accessed 21 september 2021)
36. Museo-del-traje, 2014, http://www.saatchi-gallery.co.uk/museums/museum-profile/Museo+Del+Traje,+C.i.p.e./413.html, (accessed 11 september 2021).
37. Fashion Institute of Design & Merchandising, 2021, https://en.wikipedia.org/wiki/Fashion_Institute_of_Design_%26_Merchandising_Museum, accessed 13 july 2021.
38. Fashion Institute of Design & Merchandising, FIDM/Fashion Institute of Design & Merchandising, San Francisco Campus, Petersons's, 2020, (accessed 15 july 2021).
39. Cordaro, C. and Clark, B., *Dressed for success*, vol. 4, pp. 36–38, Philadelphia Phillies, Phillies, 2014.
40. Gigenbach, C. and Walton, T., p. 251, Kent State University Athletics, Charleston, South Carolina, Chicago, Portsmouth, 2008, New Hampshire and San Francisco, Arcadia.
41. Kent State University, A history of Kent State University, 2019, (accessed 5 november 2021).
42. Grismer, K., *The history of Kent Kent*, p. 187, Kent Historical Society, Ohio, 2001.
43. John, E. and McGilvrey, P., *Kent State University special collections and archives*, Kent State University, 1890-1945, 2005.
44. Kent State University special collections and archives FAQ, Ohio, Kent State University, 2021, https://www.library.kent.edu/about/departments/copyright, (accessed 27october 2021).

6

Documentation of Museum Textiles

Simmi Bhagat* and Radhana Raheja

Department of Fabric and Apparel Sciences, Lady Irwin College, University of Delhi, New Delhi, India

Abstract
Museum textiles are a source of valuable historical and sociocultural information. There are many activities performed by the museum professionals to provide details and status of each object in their custody. The written documents along with their images provide essential information of the artifacts. Documentation is one of the essential activities of collection care processes to be performed by the conservator. A good documentation can be credited for successful curatorial, conservation, educational, and managerial activities of the museum. There are various guidelines available on documentation processes for museums by international organizations. Museums across the world develop their own collection care policies and follow different formats of documentation process depending on the availability of resources, size, and nature of the collections. Thus, as a conservator, it is important to understand the purpose, advantages of documentation, its implementation, various methods, and formats.

***Keywords*:** Documentation, textile conservation, documentation systems, collection assessment, written records, photographic records, documentation types

6.1 Introduction

From the very moment of arrival of an object in the museum's 'collection, a variety of collection care activities are performed by the museum personnel. The evidence of historic objects can be greatly attributed to the written records that give a detailed description about the artifact. Textiles being organic in nature will at some point deteriorate and disintegrate. What keeps their legacy alive is their detailed descriptions, which provide evidence for their existence [1]. From the written and photographic records of national and international exhibitions to the accounts of various travelers and courtiers, documentation has given a vast database of information of the sociocultural importance and techniques of historic textiles, which otherwise would have been lost over the years [2]. Noting the importance of such evidence, documentation of objects becomes of prime importance.

**Corresponding author*: simmi.bhagat@lic.du.ac.in

Seiko Jose, Sabu Thomas, Pintu Pandit, Ritu Pandey and Vandana Gupta (eds.) Handbook of Museum Textiles: Volume I Conservation and Cultural Research, (123–142) © 2023 Scrivener Publishing LLC

Documentation is a process that involves the systematic interaction of people, events and documents to create the records of the organization. It was during the First International Congress of Architects and Technicians of Historic Monuments held in 1931, the Athens charter was adopted. It viewed documentation as an inseparable part of conservation. At that time, documentation was understood as a record that captured and described an existing state of historic monuments and the changes made to those monuments. Initially, it was devoted to the historic monuments and subsequently to antiquities and art objects within the museum. It provided the basic concept of documentation that remains as an unchanged core in all formats and types of conservation documentation applied to all types of objects. In 1969, an International Committee for Documentation (CIDOC) was created by the International Council for Museums (ICOM) to study documentation and reference materials. This working group continues to provide advice on good practices and developments in documentation to the museum community [3].

In the Code of Ethics of Canadian Association for Conservation of Cultural Property, documentation is described as one of the essential activities to be performed by professional conservators. It states "The conservation professional shall document his/her work by recording all essential details of the conservation of a cultural property. The extent and type of documentation will vary with the nature of the cultural property and conservation work required. Documentation is part of the history of the cultural property and shall be produced and maintained in as permanent a manner as is practical. Documentation shall be available for appropriate access when this access does not contravene confidentiality" [4]. The documentation of artifacts, as we understand now, aims to provide a detailed description to the insurers and answer queries regarding the object at hand. Good documentation is a vital aspect of modern conservation. Not only does it help to distinguish original work from the fake ones, it also helps to provide absolute identity and legal ownership of objects in matters of dispute and ambiguity. From the perspective of curation, documentation becomes a fundamental source of information. It helps to identify potential themes for exhibitions, streamline the planning of exhibition galleries and aid in efficiently planning exhibitions by providing the necessary information in an organized manner. Along with being a vital tool for curation, documentation also proves to be a useful instrument in the areas of conservation, education and management of collections. These records provide a detailed understanding of the object to be conserved, along with being a source of information to trace and link missing pieces of information and enable effective management of collections.

6.2 Functions of Documentation

A lack of documentation or poorly documented collections has resulted in the loss of important details, which were an authentic source of information. Thus, to ensure the provision of accurate information, a good documentation becomes imperative, which provides the following functions, covering the aspects of curation, conservation, education, and management:-

1. Curatorial aspect: A systematic documentation report of an object helps to identify the potential themes and topics for an exhibition. When it is time

to select objects for display, these reports provide information regarding the theme, size, material, time period of the objects, which makes the planning of exhibition much more systematic, efficient, and time-managed. The classified nature of the inventory greatly shortens the exhibition planning time.

2. Conservation: When the need for conservation of an object arises, it becomes important to understand the object in its totality. The nature of material, the technique of manufacturing, the time period, details of previous conservation work, and scientific analysis of the material, all become the base for a conservator to formulate conservation treatments. These details are provided by the condition report. Addition of the treatments performed by the conservator to the previously documented details ensures comprehensive information of the object, which then becomes a baseline for any other future treatments required.
3. Education: In the field of research, authentic sources of information play a pivotal role. A good documentation provides a researcher with the information required. Written information supported with photographic documentation gives the researcher the scope to study the object and develop a deeper understanding, which in turn would produce valuable and quality research articles.
4. Management: When a museum acquires an object, it is important to determine the legal ownership, and terms of agreement among the transactors. Whether the object has been bought, gifted, loaned, or donated, once the terms and conditions have been recorded, not only does it make it convenient to identify the origins of an object, but also reduces the chances of legal troubles in the future. In the case of theft or loss, documented evidence helps to recover and trace the object even from a foreign country under the provision of the UNESCO convention. Lastly, a well-managed collection makes the organization, storage, movement, auditing, and insurance of objects more easily manageable, without any ambiguity. At any given point, movement of an object to or from storage, period of display, handling by various personnels, time and duration of loan, among other management related matters can be traced.

6.3 Features of Documentation System

According to ICOM Code of Ethics, 2006—"Museum collections should be documented according to accepted professional standards. Such documentation should include a full identification and description of each object, its associations, provenance, condition, treatment and present location. Such data should be kept in a secure environment and be supported by retrieval systems providing access to the information by the museum personnel and legitimate users" [5].

The documentation process, thus formulated, should serve as an effective tool for preservation. It should aim to preserve all known historical, social, technical, and aesthetic information about the object or a collection that can help satisfy the needs of all the stakeholders, which include the curator, the researcher, or the public. To achieve these aims, the documentation system should possess features, which comprise a comprehensive record

- Museums will keep on acquiring textile objects from time to time, so the documentation system formulated should be capable of adding on any number of entries. It should not impose a limit on the size of total records and should also allow for cross-referencing. Ensuring that each object has a unique code, specific to itself eliminates ambiguity and makes for an efficient documentation system.
- The developed system should be uncomplicated and easy to use and maintain. This will save time in documenting timely records. It will also ensure ease of access by others who wish to refer to the records.
- The system followed by the museum should be set down in a manual for future curators to use. Once a well-researched comprehensive system has been developed for a museum, it should be maintained for continuity.
- The documentation system should permit to protect confidential information. While the information pertaining to the history, technique, method of manufacturing and other research-related details should be accessible by all, administrative information should be kept confidential, only accessible to the authorities in charge [6].

6.4 Collection Management Policy

Management of collections is one of the vital functions of museums. A well-managed collection reduces ambiguity and provides required information about objects in details. The Museums Association states "Museums enable people to explore collections for inspiration, learning and enjoyment. They are institutions that collect, safeguard and make accessible artifacts and specimens, which they hold in trust for society." An audit report of Controller and Auditor general (New Zealand) 2006, reveals that past collections in the museums are not planned nor strategically acquired rather they have evolved from social and cultural tastes and curatorial preferences. Thus, a planned collection strategy will help to manage every aspect of the collection with ease. To achieve that it is thus necessary that collections in museums are based on clear purpose and articulated guidelines [7].

The mission statement of the museums often addresses the collection activity. It focuses on expanding knowledge, fostering appreciation and developing creativity. In order to be able to perform this task with ease and efficiency, it becomes important to organize the collections in a manner that can be understood and accessed by all. The Collections Management Policy helps to achieve this objective. It acts as a blueprint document that guides in managing museum art collections for their future development, presentation, and care. The policy needs to be consistent with the mission of the museum. It should be able to showcase how the collection will be useful in helping the museum achieve its goals, as well as how it will further develop the collection. A collection policy should also be able to not only explain the classification of its collection but also specify the legal and professional procedures and standards that are followed to acquire and remove the objects from the collection [8, 9].

Keeping in mind the concept behind formulation of a policy, the following are the elements commonly found in a collection management policy:

1. Mission, vision and History: This information will provide an understanding of the museum's role and the basis on which it will take its decisions.
2. Statement of Authority: This section will enumerate the details pertaining to the government body, committees, staff, and the role of each member. It will also provide details of the people who are responsible for legal matters.
3. Legal and ethical issues: The code of ethics of a museum will be listed in this section. It will also state national and international laws, which ensure protection of cultural property.
4. Scope of collections: This section entails the history of the collections and details of how they are being used. It will also give guidelines of how the collection can grow keeping in mind the mission of the museum.
5. Categories of collections: Common categories include permanent collections, collections for education purposes, collections for temporary exhibitions, etc. It can also include information regarding categories, such as acquired collections, accessioned and deaccessioned, collections regarding conservation treatments, and so on.
6. Acquisitioning and accessioning: Museums acquire objects through various means and it becomes important to accession them accurately, in a manner that explains their method of acquisition and the category to which that object will belong to.
7. Deaccessioning/Disposal: In the event of deaccessioning or disposal of an object, it becomes imperative to be transparent in the transactions to ensure that the museum and its employees are aware of the decision that has taken place. This eliminates confusion regarding the presence or absence of an object from the collections. The process of deaccessioning is also a way for the museum to refine its collection, keeping in accord with the mission and vision of the museum. The process must specify the reason for disposal and mention the method of disposal as well.
8. Loans: Whether a museum may be giving a collection for loan or taking objects on loan, this section will outline the conditions and formalities of loaning. It will also include the history of loaning with other museums and the policies that were followed for the same.
9. Objects in Custody: The objects, which are found in collections, but are either abandoned or unclaimed, will fall in this category. This section of objects in custody will mention how the museum plans to deal with such objects.
10. Conservation/Care: This section will provide information on how the collection will be taken care of in terms of preventive and curative conservation. Storage management specifications with respect to temperature, relative humidity, pest management and disaster planning will be mentioned. If there are any curative treatments required for objects in the collection, those too will be mentioned in this section, along with their plan of execution.
11. Insurance and Risk Management: The measures that are being taken by the museum to safeguard its collection against disasters, and the security measures to be taken in case of any disaster or natural calamity will be outlined in this section. This section will also mention the insurance procedures and policies to be followed for collections in the case of a disaster.

12. Documentation and Records: In-depth details regarding the condition of an object and its history, its technical specifications and value will be included in the records.
13. Access: Certain collections are accessible to the public, others for museum personnel only. The rules, regulations and reasons to access collections by different people will be mentioned here.
14. Appraisals: While the collection is meant for the public, appraisal of objects by the public can be detrimental. The policy for appraisal, as well as the authorities who can appraise collection, is mentioned for clarity.
15. Intellectual property: To ensure the museum is able to efficiently protect its collection, it should be aware of the laws placed by the government for their protection. Whether it may be regarding trademarks, commercial usage of information provided by the museum, photography restrictions, or any other such rules, which will ensure safeguarding of intellectual property, the museum should mention these details in its policy.
16. Review/Revision: With changing times, it is a possibility that the rules and regulations put in place by the museum too need to be revised. This section will address the procedures and intervals at which reviewing of the policy will take place and how the revisions are to be incorporated in the existing policy.
17. Glossary: A glossary of terms is formulated to familiarize the employees, volunteers and other workers with the terminologies used in the collection management policy by the museum [10].

6.5 Assessment Standards

6.5.1 Collection Assessment

An accurate and intensive documentation is required for all textile artifacts. It is stated in ICOM Code of Ethics 2006, "Every aspect that provides intrinsic information and historical evidence about the object should be recorded" [5]. Prior to documentation of the objects, museums prepare a collection assessment report. It is a general survey of collections that consists of the details in context of the geographical location of the museum and its surroundings. The overall assessment of the collection may include the following elements:

- examination of the museum building that houses the collection;
- evaluation of environment within and around the museum that includes exhibition and storage areas;
- assessment of storage and display conditions in terms of environment, furnishing, and packaging materials;
- review of handling, loan policies, and emergency preparedness [11].

This collection assessment format will help in identifying areas that need improvement and further assist in developing a program that will achieve those improvements. It focuses on assessing the condition of the entire collection. It provides details on agents of deterioration and their potential adverse effects on the collection. According to Keene (1996), the condition survey of collection is used for collection management purposes. This document can be a summarized version of an object condition report. However, some museums may prefer to prepare a comprehensive and a detailed collection assessment report. It provides details on risks from environmental conditions, packaging and display materials, and further protective actions needed for the collections care. Based on these reports and available resources, a list of problems with suggestions for improvements is recommended for the collection [12].

6.5.2 Assessment of Objects

Subsequent to collection assessment is the assessment of objects. It includes all the details of the journey of the objects even before it becomes part of the museum. All the evidence in the form of photography, films, video, tape recording, oral history recording, and even the rituals and folklores attached to the objects and its condition are documented and preserved. Canadian Code of Ethics mentions "The conservation professional shall prepare an examination report (alternatively called 'condition report') that shall identify the cultural property, include all relevant information on its structure, materials, history and condition, and provide the date of the examination" [5]. A detailed examination is the first step in assessment of the object. It is an essential step for any new acquisition, objects given on loan or on return from a loan, and also conservation treatments prior to display or long-term storage. The object documentation is a brief or a detailed diagnostic report that includes scientific and analytical tests along with all laboratory examination of all the components of the object. For performing any conservation procedure, often, conservators prepare a detailed documentation of the object's condition and examination report. The format of the assessment depends on the purpose and importance of the object. This document provides recommendations for undertaking any treatment for improvement of the object. It helps in formulating the methodology for proposed treatments. All the conservation work on the object till completion is documented as it is an integral element of the permanent records of every treated object. The materials and methods that are used for examining objects depend on the availability of resources, object condition, and also the custodians' preferences [13].

A conservator should know that the condition report of any textile artifact should include specific information. The details of the report can vary according to the purpose and information requirement; however, the following sections are essential in textile artifact documentation:

I. Object description—it includes all details in terms of iconography, artist, dating, dimensions of the artifact and collection to which it belongs or the owner of the artifact.
II. Technological survey—it provides an overview of the preservation state of the artifact. It includes study of material composition and techniques. Analytical findings of fiber structure and construction using microscopic images, construction, and fabrication through photography, dye identification, and material composition through SEM and FTIR are also recorded in this survey.

III. Condition and damages—this section describes the condition of the object through careful examination of its state. The damages occurring on the object are interpreted and documented by the conservator. Correct terminology is used to describe and catalogue the damages. If required and permitted, scientific analysis may be carried out on the object and results are recorded. The concluding remarks of the conservator are mentioned to describe the overall state of the object.
IV. Conservation/Restoration—any type of conservation and restoration performed can affect the future condition of the artifact. Hence, every detail is completely documented. The improvements required in the object are described in terms of preventive or curative conservation. Often, the required changes in climatic conditions, storage conditions, packaging or display may help to improve the state of the artifact. Additionally, a conservator may specify specialized treatments in the form of cleaning, humidification, consolidation or display and storage supports for specific artifacts. Any previous conservation or restoration treatment performed on the artifact is also acknowledged in this section. This will help the conservator to assess and monitor the condition of the artifacts [14].

6.6 Types of Documentation

As mentioned in the previous section ,documentation consists of all the records pertaining to the history, examination, scientific analysis, condition of the object, conservation treatment plans formulated and executed, and guidelines for storage and display. The purpose of recording this kind of information is to provide an accurate information of the condition of the object and to have a record of the evolution of the deterioration of the object. This kind of information helps in formulating conservation treatments, which can be helpful in prolonging the life of the object. Once the cause of concern or the sensitive aspects of a textile are known, better care can be taken of the object. Such documentation also helps retain information regarding the techniques and methods of construction, which may be lost over a period of time to the degradation process of the textile. When objects are borrowed or loaned, such documentations help in establishing the condition of the object before and after receiving the textile and also provides a guideline as to how the object is to be transported, stored and the methods that can be used for display by other museums, without risking the condition of the object.

To elaborately document these features, written records along with photographic evidence aid in understanding the characteristics of an object, along with providing clarity and overcoming language barriers that may be faced with only written records. Thus, documentation of objects can thus be bifurcated into two forms; written description and photographic records.

6.6.1 Written Description

A well-formulated written description should include the following details:

- Accession, location, collection: this includes details pertaining to the accession number assigned to the object, its location in the storage and the collection that it belongs to.
- Historical significance: details regarding the provenance, importance of a costume or textile with respect to its usage, historical evaluation and any sociocultural importance associated with it that provides better understanding of the artifact.
- Technique, method of construction, and surface ornamentation: this includes an analysis of the weave, fabric construction method in terms of weaving, knitting, felting, crocheting, lace making etc., and the kind of embellishments that have been used and their method of attachment which may include embroidery with threads, metal, applique work of any other form of surface ornamentation. It also includes an analysis of the seams and stitches used for the construction of the textile.
- Scientific analysis: analysis of the fiber, of the fabric, fugitive nature of the dyes used, investigation of the paint layer and binding media in printed textiles and determination of metal components, which are a part of the composite textile.
- Condition of the object: this involves a detailed description of the damages including tears, holes, missing threads, abraded surfaces, and missing ornamentation components. It also includes any stains, microbial growth, insect attack, or any form of visible degradation.
- Suggested conservation treatment plan: based on the condition of the object and scientific analysis, a tentative plan of treatment is formulated and written in this section.
- Executed conservation treatment: sometimes, the proposed treatment may differ from the executed treatment plan based on how the object responds to the treatment. The reasons for change in the plan and the final treatment is explained.
- Guidelines for storage and display: the necessary support required and the best-suited method for the object is enumerated for minimal handling of the object.

6.6.2 Photographic Records

Written and photographic documentations support one another to produce a comprehensive report regarding the collections present in a museum. Digital images provide a clarity in understanding and interpreting the information present in the written records. Photographs for the abovementioned details of written description provide support in understanding the information and enhance the quality of documentation in terms of-

- Historical evidence and cross referencing: Images of the textile found in photographic collections, published books, journals, catalogues would help in understanding the significance of the object. It also aids to cross-reference certain descriptions mentioned in written records which may not be backed by photographic evidence. Visual documentation can provide information to scholars

- Scientific analysis: Apart from the detailed description, microscopic images and images related to other scientific investigations like X-ray radiography, 3D imaging, UV and IR imaging and multispectral imaging, would help to understand the deterioration process of the textile as well as confirm certain conclusions drawn from these. Research on the languishing craft of warak printed textiles of Rajasthan done at Lady Irwin College as a part of the SHRI-DST project initiative involves understanding the materials used in the manufacturing of these textiles in order to formulate appropriate conservation treatments. The SEM-EDX analysis (Figure 6.1) of the silver leaf used in traditional printing provides the information and authenticity of the materials used in printing. The researcher will be able to formulate conservation procedures appropriate to the nature of materials detected, such as the methods and materials that would be compatible with silver.
- Techniques, method of construction, and surface ornamentation: Photographic details of the fabric, such as method of construction and the types of surface ornamentation, can bring clarity in terms of materials used. There can be many terminologies associated with one particular technique or ornamentation, and providing photographic details can help understand the techniques better. Figure 6.2 below displays a 'warak' printed motif on cloth.

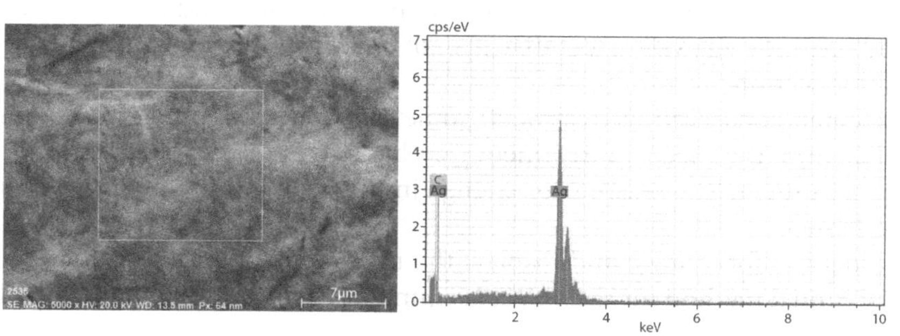

Figure 6.1 SEM-EDX analysis of silver used in warak printed textiles.

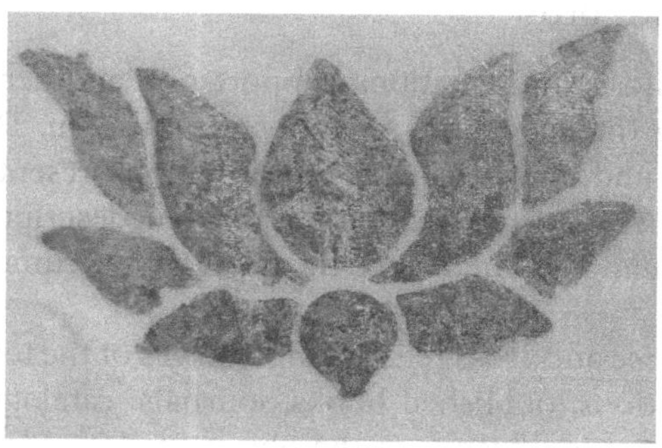

Figure 6.2 Warak printed motif.

Further the microscopic image (Figure 6.3) shows the different layers of the printed textile such as the base fabric, the binding medium and the silver leaf. This gives details regarding the process of warak printing whereby the fabric is first printed with a binding medium and then applied with metal leaf.

- Condition of the object: A complete front and back image of the object is the foremost requirement while photographically documenting the condition of an object. Sometimes it becomes difficult to explain the color, size, nature and location of damages, which can be better understood with images. In Figure 6.4, the complete front and back images of this object (Namda of Kashmir)

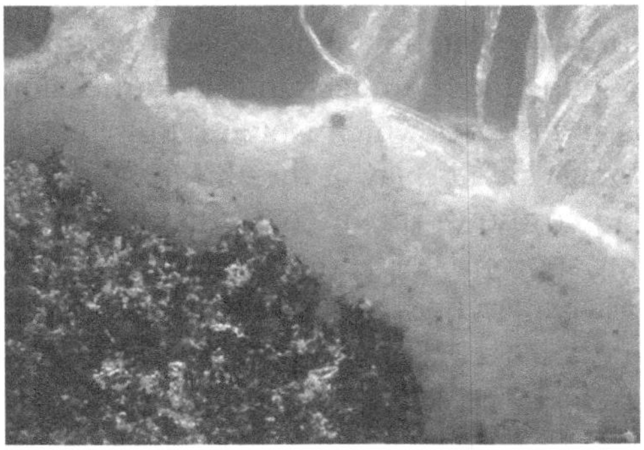

Figure 6.3 Microscopic image of the print.

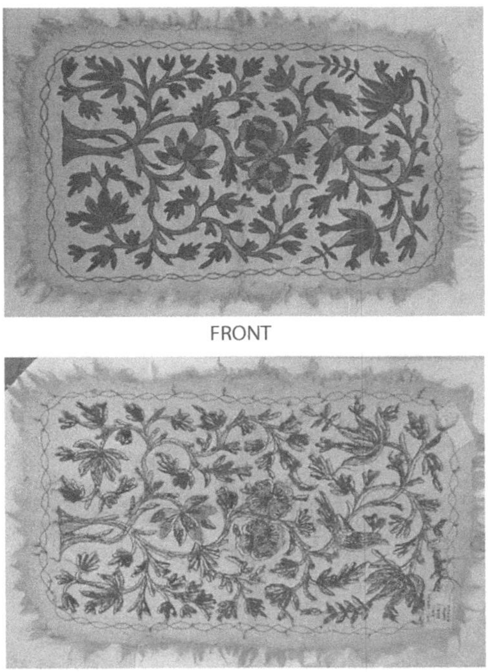

FRONT

BACK

Figure 6.4 Front and back of Namda.

helps in understanding the precise location of damages (Figure 6.5) observed and better explains the specifications of the object.

- Executed conservation treatment: A before and after images of the treatment given to the object can visually depict the changes that have occurred and it can help in identifying the materials used for conservation.

  The following example (Figure 6.6) shows the treatment given to the object to repair the holes present by providing a support material and using appropriate conservation stitches to stabilize the object.

- Guideline images: Pictorial representation of how the object should be stored and displayed, giving details regarding appropriate padding and support to be given at the stable locations eliminates the excessive handling of objects.

  A diagrammatic representation of the object (Figure 6.7), like the one seen below provides a guideline to understand the dimensions of the object and the areas of folds and creases which would require extra paddings and protection during storage and display of the object.

  Upon display of costumes (Figure 6.8), the diagrammatic representation helps in developing appropriate padding and custom support for the costume to ensure the fabric experiences least strain, which would ensure the longevity of the object.

(a) MANUFACTURER LABEL

(b) CORRODED NAIL AND RUST STAIN

Figure 6.5 Damages marked as (a) and (b) and further detailed images of the damages.

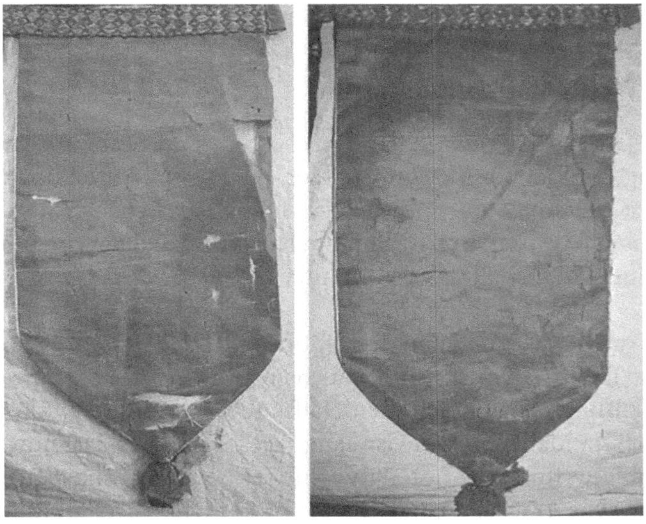

Figure 6.6 Before conservation and after conservation.

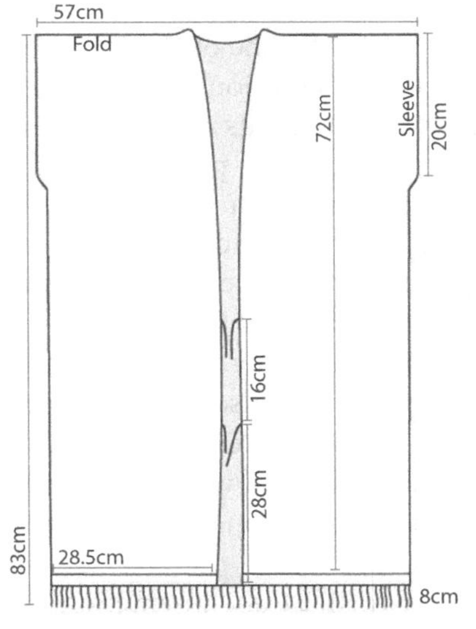

Figure 6.7 Diagrammatic representation of costume.

Figure 6.8 Display of costume.

6.7 Formats of Documentation

6.7.1 Styles of Written Documentation

Depending on the use of the documentation report, it may be for curatorial reference, for exhibition planning, for loan, or for conservation purposes, written documentation can be prepared either as a narrative/descriptive essay style, or as a checklist, or as a combination of both. These styles can be prepared in a manual traditional paper document format or can be recorded in a digitized format using computer technology.

- In the narrative style, detailed description of each segment is provided. This style is flexible as a conservator can record in his/her style but it is also a time-consuming method. All the details are handwritten by the conservator in an understandable yet concise language. Since it is a handwritten document, legible writing is a necessity. This style permits subsequent additions easily; however, traceability of information is time-consuming. Free-style forms allow for much more detailed evaluation of the work being done to the object or its environment and allows for the observations, explanations, and conclusions of the conservator to be recorded. MDA describes this style as "the ideal recording medium" for the object documentation.
- The checklist style of documentation form is a very structured form that includes a premade list of statements and words pertaining to the condition of the object, which are checked as per the status of the individual object. This style has become more popular in use for several reasons. It is simple to read and quick to document. The conservator needs only to check off the information that applies to the object. The information and terminology used in the form are standardized. This provides easy documentation and retrieval of information. As cited by Moore, this style, however, is not very flexible to any changes. The form also has to be extremely detailed in order to cover all possible reasons of deterioration and damages. Textile collections being diverse and composite will require a much exhaustive checklist to cover the different conditions in which documentation should be recorded.
- In the combination style form of documentation, both checklist and narrative styles are used for certain sections, for example, a narrative form may be applied to when describing the historical significance of the object, while a checklist may be used to list the damages seen in the object. This can be considered as an ideal format for documentation as it can draw upon the best of both styles [14].

6.7.2 Manual and Digitized Documentation

When analysis and examination of artifacts is recorded by manually writing the observations and inferences, it is known as manual documentation. Manually writing the details poses the advantage of being able to note down details while examining the object. It helps to explore answers to leading questions that one may have in the process. Manually preparing visual records gives the flexibility to include details, which may not be visible in photographs. It also gives the option to focus on specific details, such as visual records depicting only seams and stitches, or only damages, or only surface ornamentation. It helps to provide a clearer understanding and a detailed documentation of the object at hand.

While manual documentation has its advantages, it also becomes cumbersome to manage them when a museum has documentations of hundreds and thousands of objects. Also, while accessing a particular object, one would have to go through many registers and folders to locate the documentation of that particular object, which is a time-consuming process. As per conservation standards, retrieval of data should not exceed 3 minutes. This is where digitized documentation is useful. Manual documentation also poses the risk of perishing, which is handled by digital documentation whereby all the information is fed into a database, which stores the information for as long as required, and information is accessible in a shorter period of time. It is an easier and faster way to write and edit documents through computers. It allows for photo documentation, much bigger storage and update of the sections. Overall collection management of museums is possible through the digitized systems. The documentation system can be created according to the end users. Researchers have worked on MS-ACCESS to create a digitized documentation system for a museum where selected artifacts have been documented in layers for Curators, Visitors and Researchers [15]. In India, JATAN software, created by Centre for Development of Advanced Computing, Pune, has been used by many museums across the country to digitize imprints and is accessible to the public [16].

Another advantage of digitization of documentation is that the information will never be deteriorated and lost till the material on which information is stored is kept updated.

6.8 Case Study

An artifact condition report from Heritage Textile Collection of Lady Irwin College, New Delhi [17] is presented here to provide an understanding on documentation format and the extent of details-

CONDITION REPORT OF A COSTUME.

Collection Name: Lady Irwin College Heritage Textiles Collection	**Accession Number:** 2008.1.65
Dimensions: 43 × 49 cm	**Title/Name:** Jacket
Provenance: Unknown	**Period:** Unknown
Storage Location: Lady Irwin College Textile Heritage Storage Unit	

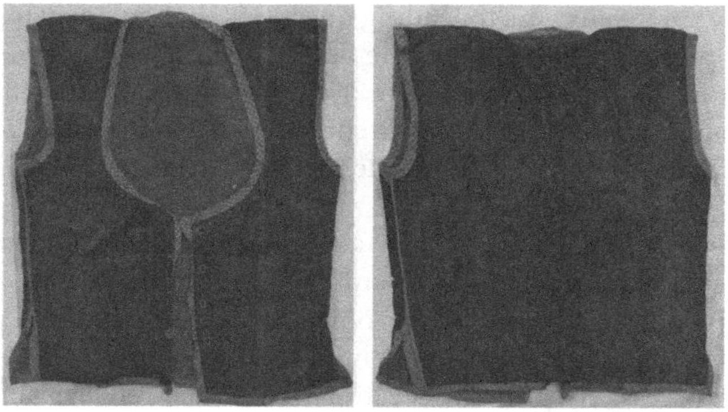

Front Back

Graphic Representation

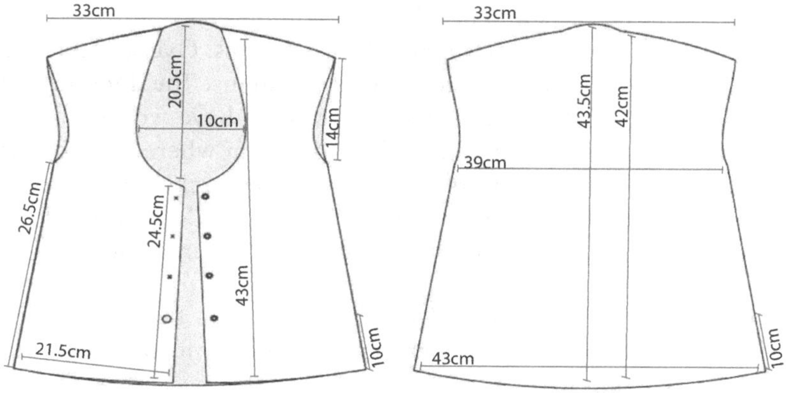

Front Back

I. **Brief description:** Sleeveless jacket in deep wine color with round neckline. This type of jacket, known as a *koti,* was a popular upper body, outer garment in Rajasthan and surrounding areas, and was worn by both women and men.

II. The **accession number** is written on muslin and handstitched on all four sides on the inside of the garment

III. **Material:**
- The base fabric of the jacket is a wine colored velvet fabric
- The jacket consists of a red cotton lining
- Presence of cotton piping on the neckline, armholes, placket and hemline of the jacket
- Presence of metal buttons on placket

IV. **Construction:**

The garment has been machine stitched, which confirms the garment was constructed after the 1800s. It is completely lined with a red cotton fabric and consists of cotton piping to finish the edges of the garment. The front open even hem left over right placket consists of four buttons placed equidistant from one another.

V. Deterioration

- Physical:
  - Loss of pile in several areas of the object, especially along the fold lines
  - Multiple tiny holes on the inside of the object in the cotton lining, most likely insect exit holes
  - Three large areas of tears and fabric loss (more than 2 cm) in the cotton lining; two in the front right panel, one in the back panel
  - Severe deterioration of the green edging, missing in several areas
- Stains:
  - Several dark brown colored stains on the cotton lining, mostly on the lower area of the back panel
  - One rust stain on the green edging on the back, right armhole
- Fading:
  - Color of velvet pile appears to have faded
- Can it take needle and thread: Yes
- Condition of ornamentation/fasteners:
  - Three of the original buttons are missing
  - Remaining metal button is heavily corrode

Deterioration: Front lining

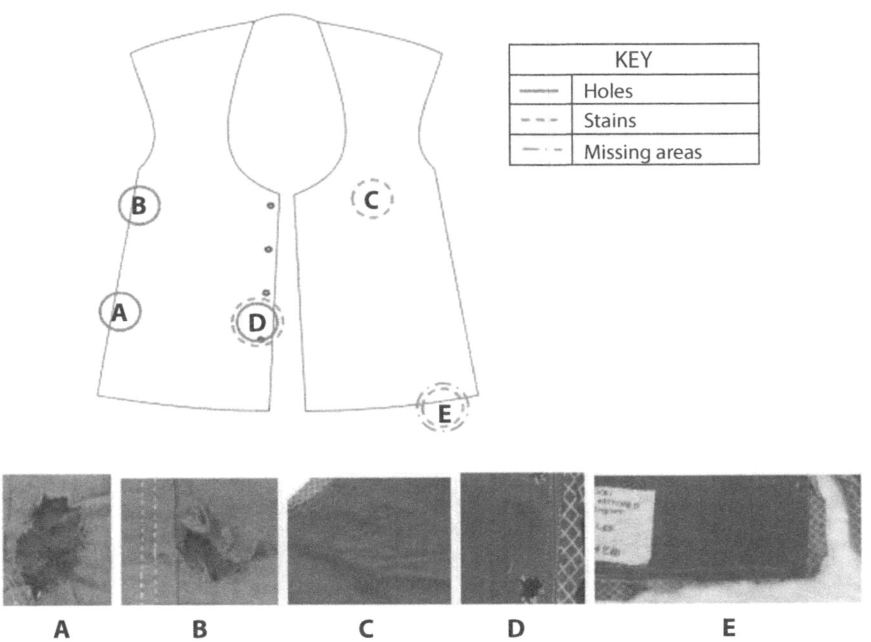

Deterioration: Back

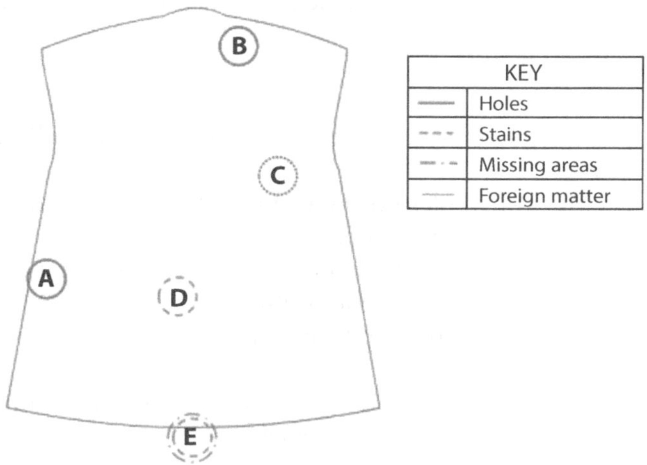

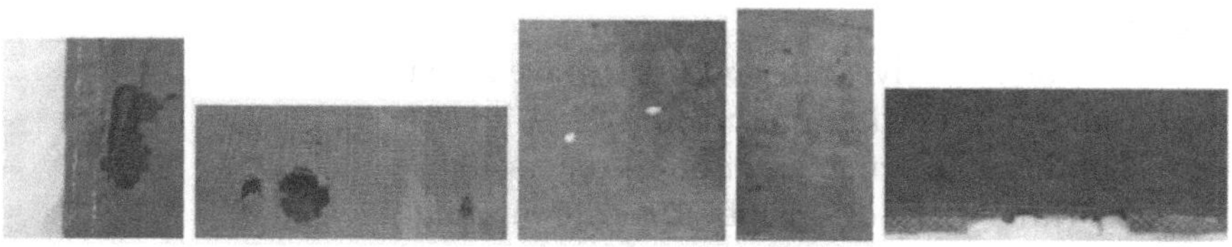

A B C D E

Fiber and fabric analysis

Microscopic evaluation

Thread Count

Fibre Identification

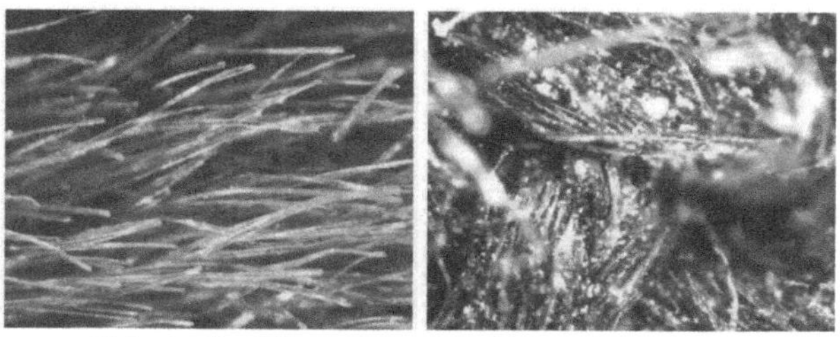

Velvet pile: Silk Velvet base: Cotton

Weave Analysis—Plain Weave at the base

X		X		X		X	
	X		X		X		X
X		X		X		X	
	X		X		X		X
X		X		X		X	
	X		X		X		X
X		X		X		X	
	X		X		X		X

VI. **Seam and stitch analysis**
Plain Seam used on the side seam, covered with the lining. All edges finished with edge facing of cotton fabric. Top stitching along with facing at a distance of 0.5 cm on the velvet part of the garment.

VII. **Treatment proposed**
- Object should be isolated and sent for fumigation to deal with possible insect infestation.
- Treatment should be carried out for removal of stains.
- Areas where there is fabric loss, both in the green edging and lining of the object should be stabilized.

VIII. **Packing and storage**
- **a.** Since the object is structurally strong, it can be hung on a padded hanger
- **b.** Alternately, it can be stored flat, while providing padding, to soften the folds and to maintain the shape

IX. **Documented By:** ABC Student

X. **Date of Documentation: December 8, 2018** [18]

6.9 Conclusion

Documentation is a vital step toward the conservation process in the museums. Documentation helps to preserve the information contained in the object. It is an exhaustive activity performed by the conservator. There are national and international professional bodies that have established code of ethics and standards for practice to document conservation work. Conservators keep records of all the information of the past and all the treatments performed by them as this may have an impact on the future treatments and the stability of the object. The data related to objects can be preserved in traditional paper documents or digitized versions, depending on the availability of resources with the organization. Across the world, preference for digital format is due to the ease, speed, and efficiency of the system.

References

1. Orlofsky, P., Textile conservation, in: *Conservation Concerns: A Guide for Collectors and Curators*, K. Bachmann, (Ed.), pp. 79–83, Smithsonian Institution Press, Washington, 1992.
2. Good, I., Archaeological textiles: a review of current research. *Annu. Rev. Anthropol.*, *30*, 209–226, 2001, http://www.jstor.org/stable/3069215.
3. Anonymous, (2011, November 11). The Athens Charter for the Restoration of Historic Monuments-1931. ICOMOS. France. https://www.icomos.org/en/167-the-athens-charter-for-the-restoration-of-historic-monuments. Accessed on February 15, 2021
4. The Canadian Association for Conservation of Cultural Property, The Canadian Association of Professional Conservators. Code of Ethics and Guidance for Practice of the Canadian Association for Conservation of Cultural Property and of the Canadian Association of Professional Conservators. CAC-ACCR and CAPC-ACRP. Canada. 2000. https://www.cac-accr.ca/download/code-of-ethics/

5. International Council of Museums. ICOM Code of Ethics for Museums. ICOM. France. 2017. https://icom.museum/wp-content/uploads/2018/07/ICOM-code-En-web.pdf
6. https://www.metmuseum.org/-/media/files/about-the-met/policies-and-documents/collections-management-policy/Collections-Management-Policy-11_10_20.pdf.
7. Museum Documentation Association (1999). *Documentation, why bother?* [online]. MDA. United Kingdom. Accessed July 30, 2000. http://www.mda.org.uk/bother.htm
8. American Alliance of Museums. *Developing a collections Management Policy* [online]. AAM. Arlington, VA. 2012. https://www.aam-us.org/wp-content/uploads/2018/01/developing-a-cmp-final.pdf. Accessed February 20, 2021.
9. https://oag.parliament.nz/2006/heritage/part3.htm.
10. https://www.aam-us.org/wp-content/uploads/2017/12/Developing-a-CMP-2018.pdf.
11. Szczepanowska, M.H., *Conservation of cultural heritage (key principles and approaches)*, Rutledge, Oxon, 2013, ISBN 9780415674751.
12. Keene, S., *Managing conservation in museums*, Butterworth-Heinemann, Oxford, 1996.
13. Moore, M., Conservation documentation and the implications of digitisation. *J. Conserv. Mus. Stud.*, 7, 6–10, 2001, http://doi.org/10.5334/jcms.7012.
14. https://cool.culturalheritage.org/coolaic/sg/bpg/pcc/05_written-documenation.pdf.
15. American Institute for Conservation (AIC), *AIC code of ethics and guidelines for practice* [online], 1994, Washington DC, https://www.culturalheritage.org/about-conservation/code-of-ethics. Accessed Feb 20, 2021.
16. Gaur, S., Bhagat, S., Dalela, N., Digitized documentation system for museum textile collections: A prototype. *Int. J. Appl. Home Sci.*, 4, 1 & 2, 51–56, 2017.
17. https://www.cdac.in/index.aspx?id=mc_hc_virtual_museums Accessed March 15, 2021.
18. *Collection documentation reports, costume (2008.1.65)*, Heritage Textile Collection, Lady Irwin College, New Delhi, India, 2019.

7

Ideal Storage Conditions for Museum Textiles

Simmi Bhagat[1] and Kanika Sachdeva[2*]

[1]Department of Fabric and Apparel Sciences, Lady Irwin College, University of Delhi, Delhi, India
[2]Conservation Scientist, London, UK

Abstract
Museums around the world are custodians of natural and cultural heritage. It is a well-known fact that most museum collections spend the majority of their lifespan in storage. All these objects in storage require organized facilities that provide stabilized conditions and easy accessibility to the objects. Storage of textile artifacts is a well-laid scientific process where all environmental factors, like humidity, temperature, light, etc., need to be suitable for every artifact. Apart from the environmental conditions, storage architecture and design are other important factors that can affect the safety of the objects. There are enough examples of loss of precious artifacts because of negligent and badly designed storage. Different textile artifacts, depending on their condition, design and composition, need specific storage materials and methods. There cannot be a standard material and a technique attending to the requirements of diverse artifact range. This chapter provides details of three basic things that are imperative to efficient textile collections storage—environment control, storage architecture and design, and storage materials and methods. While the chapter will provide basic essentials for collection storage, best practices need to be object and place specific.

Keywords: Museum storage, storage environmental conditions, museum storage architecture, museum storage design, museum textiles storage

7.1 Introduction

Museums are repositories of objects of cultural, religious, scientific and historical importance. They preserve, research and present collections to the public in the most safe and accessible manner for the purpose of education and enjoyment. It has been reported that most textile artifacts spend most of their lifetime in storage and are ideally taken out for display only once in 5 years [1]. According to ICCROM (RE-ORG), 95% of the museum objects are kept in storage [2]. Many studies have shown that poor storage facilities will put valuable objects at risk. The findings of an international survey conducted by ICCROM and UNESCO in 2011 identified that 25% museums find it difficult to circulate in storage,

**Corresponding author*: Dr.Kanikabhatiasachdeva@Gmail.Com

Seiko Jose, Sabu Thomas, Pintu Pandit, Ritu Pandey and Vandana Gupta (eds.) Handbook of Museum Textiles: Volume I Conservation and Cultural Research, (143–160) © 2023 Scrivener Publishing LLC

about 33% museums are unclear about who is responsible for storage and 50% museums have insufficient storage units. They developed RE-ORG, a method to reorganize storage in museums. This method helps in identifying the problems in all aspects related to storage, such as building and space, furniture and small equipment, management and collection. It has been successfully used across 23 countries.

The importance of collection storage has become an important focus for the museum community across the world. During the 34th ICOM General Assembly held in September 2019, a resolution was adopted on "Measures to safeguard and enhance collections in storage throughout the world" [3]. There is a need for appropriately designed, organized, and maintained storage facilities to stabilize and maintain the condition of objects.

Textiles are organic in nature and are prone to physical and chemical degradation due to many factors. The fundamentals of environment control in museum storage needs to be maintained as advised by experts. Ideal storage areas should have temperature and humidity levels that fall within the recommended range to prevent any physical and chemical alterations. Storage location should be thoughtfully planned according to the museum size and funding. Individual textile items and space availability will influence choices. Housekeeping and trained professionals are essential elements of storage maintenance.

Storage furniture and their materials need special attention. In the Indian context, the storage of textile artifacts becomes much more complex. This is primarily because of the fact that Indian textile artifacts could be made of multiple materials, with each having different characteristics. A typical 16th century skirt or bodice could be made of silk, with a muslin/cotton lining, gold/silver wire embroidery and at some places encrusted with stones, etc. When selecting storage units, it is important to choose materials that will not adversely affect any part of the textiles. A variety of materials, such as acid free cardboard, wood, or metal, are used for storage containers and devices. These should not be placed in direct contact with the textiles. Recommended materials for storage include baked enamel metal shelving units and acid-free lignin-free boxes. For storing textile artifacts, literature describes horizontal orientations that include flat, rolled, folded or mounted storage while vertical orientations include hanging storage.

7.2 Published Standards in Museum Storage

In 1998, Museum and Galleries Commission, U.K., published "Standards in the Museum Care of Costume and Textile Collection" [4]. The document extensively enlists various factors that can ensure a fool-proof storage arrangement for textile artifacts. However, museum collections have become more varied over the period of time. Also, technology has further enhanced the possibilities of making storage facilities more efficient than ever. To be able to systematically discuss various aspects of museum storage, the discussion can be divided into various sections like building architecture, environmental conditions inside storage area, storage design, safety systems, disaster handling & management of dust/dirt & pollution.

7.3 Storage Design and Architecture

Location & design of the building is one of the most important & least discussed factors in museum storage. The importance of a good storage facility has been rightly summed up by this observation, "Inadequate storage buildings in the short run are expensive buildings in the long end" [5]. It is important to note here that most museums are housed in heritage buildings not originally meant to store & display museum collections. However, there are enough examples of modern museum buildings that are planned and designed keeping in mind all requirements of education, storage & display.

There is not much documented literature available on norms for space allocation & architectural design of a museum building. What should be the space allocation and design of a model museum building? This question might not have a single correct answer! However, it is important that this is done keeping in mind the risk potential for the collections. Since most museums' housing textile artifacts might not be solely textiles museums, the storage space has to be allocated and designed taking into consideration utilities for different categories. For example, a National Museum would house collections ranging from sculptures, coin collections, arms & armours, paintings & textiles. While sculptures could be considered a very stable category that does not require much attention in terms of temperature & humidity control; textiles & metallic collections have very different & specific environmental requirements. Keeping this in mind, a museum of national scale would have multiple storage facilities, each catering to the requirements of a specific collection. However, there could be smaller local museums and private collections that cannot afford segregated storage for each artifact category. These museums would then resort to standard norms that could take care of majority but not all the requirements of each artifact. Although some museums resort to localized storage for each section, most museums have dedicated spaces for storage and exhibitions respectively. A document from Philadelphia Museum of Art [6] lays down specific guidelines for a museum storage:

- A storage area dedicated only to storage
- Storage units that would allow curators to view the collection easily, yet would close fully to protect the objects
- Controlled access (security) to the storage area
- Safe and easy object retrieval (without actually handling the object)
- The ability to view objects without removing them from their housing whenever possible
- Environmentally stable conditions (45% relative humidity, 70°F ± 2°)
- A gaseous fire suppression system, backed up by a wet pipe suppression system
- A system for objects to be grouped by:
  - Classification (costume/textile)
  - Country of origin (western/non-western)
  - Type (printed fabric, hat, sampler, etc.)

It is imperative that storage space & design is planned providing adequate protection against environmental changes, pollutants, pests, dirt etc. Another important consideration for storage buildings is a disaster management system and proximity to exhibition and conservation/restoration spaces. Additionally, it is important to enforce regular upkeep and monitoring procedures to ensure safety of the stored collections. Following sections would further elaborate upon the specific housing requirements of museum textiles and different solutions available to keep deterioration arrested for long durations of time.

7.3.1 Museum Storage Building and Space Allocation

Although there is some literature available about the ideal location of museum storage, not much data is available to map the actual storage location with proposed standards. While museums readily share information about the conservation and museology procedures followed, there are not many pointers about the storage locations in these museums. This could be primarily due to security reasons as museums are storehouses for priceless and timeless pieces of art & culture that need to be secured against unholy intentions.

Most available resources propose ideal museum storage chambers to be sufficiently chaperoned from all exterior influences of temperature, humidity, light, pollution etc. In addition to climatic buffering, these storage chambers need to be in close proximity to exhibition and conservation sites within the museums while restricting access to the few authorized personnel. As discussed earlier, not all museum buildings are ideal constructions. Many of these could be remodelled heritage buildings or buildings built with local limitations of geometry, access and more.

It is advisable for new museum buildings to be constructed in a way that they provide a stable environment to the artifacts. It should be noted here that frequent fluctuations in environmental conditions like temperature, humidity, etc. can be the biggest triggers for the process of degeneration. Artifacts housed in less-than-ideal conditions can be expected to live longer than compared to artifacts that are housed within ideal range but frequently fluctuating environmental conditions. The artifacts tend to reach a state of equilibrium with their immediate environment and frequent changes to these immediate environments could trigger the dormant process of degeneration. Therefore, it is imminent for museum buildings to effectively protect its chambers from external environment changes. It is primarily for this reason that most museums allocate basement spaces as storage areas. The basement spaces have the tactical advantage of being least impacted by outdoor temperature variations. Additionally, they can be safer in terms of limited access options. It is advisable to design storage chambers with high efficiency low-energy designs to make them more environment friendly.

However, in most scenarios, an ageing refurbished building is used mostly because of its historic and cultural value. A thorough assessment needs to be carried out of the building structure, electric wirings, plumbing network etc. before the building is deemed fit for housing delicate historic collections. All parameters of the building works should be adequately assessed and stabilized before huge funds are invested in setting up a museum.

7.3.2 Building Monitoring and Maintenance

Storage chambers/buildings need to be constantly monitored for any fluctuations in environmental conditions. As per the standards published by Museums & Galleries commission, UK (1998), [4] a new building should be intensively monitored both indoors & outdoors before housing the collection for the first time. The data procured can provide useful indicators about the suitability of the building. Additionally, these data can help in dividing the building into different zones that could be suitable for different purposes according to their natural environmental stability. These zones can be identified naturally or created artificially by air conditioning systems. In either circumstance, the devices used for environmental stability should be supporting the natural environment & not replacing the natural environment of the building.

Once the artifacts have been stored in the building, routine environmental monitoring should be done covering all four seasons. In olden times, it was customary to install thermometers or thermo-hygrograph for continuous monitoring of the storage chambers. With technological advancements, many museum storages are being monitored by computer-operated devices that collect data through intelligent sensors. It is important here to note that not just the environmental factors but human access inside storage chambers also needs to be monitored. As specified by some standards, every person entering the storage chamber can release approximately the same heat as a 60-Watt light bulb and humidity equivalent to a wine-glass full of water per hour.

Museum organizations take various measures to insulate buildings from exterior fluctuations of temperature & humidity. This should include measures like draught-proofing, thermal insulation etc. The buildings need to be sufficiently water-tight while addressing all possible sources of dampness like leaking pipes or water tanks, faulty guttering and missing roof tiles. It is advisable to take technical advice from building specialists so that harmful impacts of these measures like reduced air supply can be taken care of.

7.4 Environmental Conditions

The published handbooks on museum maintenance prescribe temperature levels at 21°C and humidity levels at 50% to 55% RH. However, it also talks of later recommendations of US engineers' Handbook that provides flexibility to temperature, between 16°C and 21°C. Most storage chambers in established museums are centrally air-conditioned with inbuilt climate control. In India, the storage chambers of the National Museum, New Delhi, are also centrally temperature controlled. However, many smaller museums do not have such facilities. These museums then depend on primitive measures like use of silica gel and placement of Thermo hydrographs in storage chambers to keep check on humidity.

7.4.1 Temperature and Relative Humidity

Standard and stable relative humidity (RH) is imperative for maintaining longevity of aged artifacts. High humidity levels can trigger fungal growth and low humidity levels can brittle the fragile artifacts [7]. All published standards advice RH levels between 50% and 55% and temperatures between 16°C and 21°C. However, there might be specific variations between products of different categories. Museums should take care of these specific requirements and it is advisable to assign storage chambers to specific products as per their temperature and RH requirements. It might not be possible for smaller museums to create multiple storage chambers with specific temperature/RH levels, the key to arrest decay lies in stabilizing the environment thus allowing artifacts to maintain long-term equilibrium with their surroundings.

Air conditioning the storage chambers is one of the most efficient means to maintain stable temperature/RH conditions. While this could be a de-facto option for established museums, smaller private & state-owned museums who do not have a robust revenue structure, might not be able to afford 24X7 air-conditioning because of its huge installation and running expenses. Additionally, centralized air-conditioning can be a potential risk as malfunctioning in one corner might pose risk to the whole collection. Thus, localized air-conditioning is a preferable option for museum storage. In absence of air-conditioning systems, such inner environment control can be achieved by humidifiers, dehumidifiers or simple heaters controlled by humidistats. In addition to the monetary requirements of an air-conditioning system, its maintenance costs and repair staff availability also needs to be considered before taking a decision.

7.4.2 Light

Light can be counted as the single biggest enemy of textile artifacts. It is a form of electromagnetic radiation that can be extremely damaging for textile artifacts, combined with other factors. Both visible & invisible components of light, i.e., Ultraviolet (UV) & Infrared (IR) have damaging effects on aged textile artifacts although their impact is different on different textile substrates depending on their age, fiber, types of dyes/finishes used, etc. It has been established by laboratory research that those certain fibers, weave types, dye molecules, etc., accelerate the process of degradation. This means that fabrics made of the same fiber but different weave type might age differently under the same duration and intensity of light [8, 9]. It is advisable to keep textile artifacts in dark storage, with illumination used only when required for the purpose of handling and maintenance. In this scenario as well, care should be taken to eliminate the UV component of light as it has no role in the illumination process and is invariably the most damaging component of light. UV light can be efficiently avoided by using ultraviolet filters with the light source. Natural light should be completely avoided in museum storage. Additionally, care should be taken to use the sources of light that emit as less heat as possible.

The damaging effect of IR light on textile artifacts could reflect in the form of color fading, fabric embrittlement etc. and these changes are completely irreversible. Therefore, it is paramount to restrict the duration and intensity of all types of light exposure to the artifacts. The light exposure to museum collections should be seen from a holistic perspective. This needs to take into account the unavoidable light exposures for conservation and exhibition purposes. Therefore, keeping collections in long periods of dark storage is an efficient means of reducing the cumulative damage done by exhibitions and stabilizing textiles towards a longer life-expectancy.

7.5 Storage Techniques

The guidelines for storage designs in museums are quite varied. More than providing direct instructions, it provides a framework of the dos and don'ts within which an ideal museum storage must be designed. Some of the important pointers laid out by the guidelines stated in section 1.2 are as follows:

- All cupboards and containers should be clearly labelled
- Never place objects on top of each other
- All objects should be padded or supported using inert materials so that their original shape is retained. This is specially recommended for accessories like shoes, hats, fans, gloves etc.
- Trims and fastenings must not get crushed
- Boxes should not be over-stacked or over-filled
- Storage furniture should be designed in a way that it is safe and easy to work with for individuals. This can be achieved using inert materials & avoiding sharp edges
- All storage cupboards should be dust-proof
- Folding should be kept to minimum
- Folds if any, should be layered with acid-free tissues or knitted fabric stockings
- Folded textiles should be periodically unfolded and refolded to avoid damage at folds.

7.5.1 Accession and Labeling

When a museum acquires an object, it is given an exclusive and permanent code, typically a number, that is known as the accession number [10]. This number is unique and establishes the object's individuality and links it to its written collection records.

Museums have a location register that provides the exact location of any object in the collection. If the object is in the collection storage areas, it is critical that its exact location be specified. In order to describe its exact location, the storage areas must be systematically arranged and both the storage rooms and the individual storage systems should be coded and clearly labelled. Labels are an essential part of any object in the display or storage within a museum. As the objects in storage are packed and not visible, their recognition without touching is only possible through their labels. Thus, besides being essential for documentation purposes, labels also facilitate in locating and recognizing objects without handling them.

There are three basic types of storage—flat, rolled, and hanging textile storage [11]. These can also be classified into Horizontal orientation that includes flat and rolled textile storage and Vertical orientation that includes hanging storage. Museum handbooks advise flat storage for small textiles, textiles with pile structure, textiles with fragile surface, such as paint, textiles that are brittle and stiff and textiles with heavy surface embroidery and beadwork. Large textiles like carpets and coverlets are advised for rolled storage, and dimensional textiles like heavy costumes to be placed on padded hangers

7.5.2 Flat Storage

This is the ideal storage for most textile artifacts for various reasons. Flat storage does not add stress/strain to any part of the textile as there is no rolling or folding involved. Also, the gravitational pull is distributed all over the surface of the textile, thus saving it

Figure 7.1 Flat storage drawers for textile artifacts.

from localized strains and weakening. Most hand-painted textiles, flags and regimental colors are laid out in flat storage (Figure 7.1). The textile is sandwiched between layers of acid free paper and placed in storage drawers made of non-corrosive material. The technique stands perfect for small size flat textiles and even costumes and made-ups to a large extent. However, it is not possible to store textiles of bigger size in this manner as it becomes practically impossible to create functional drawers beyond a certain size. Most organized museum storage sections would have numerous chests of drawers with thin sliding sections that contain one artifact each and can be completely slid out for easy removal and placement of the artifact.

7.5.3 Rolled Storage

This is a storage design where the artifacts have been nicely placed between layers of acid-free paper and rolled over non-corrosive rollers (Figure 7.2). This ensures that the artifact does not crease at any point of time, thus avoiding breakage at creases. Also, the fabric weight is distributed all through, with no undue pull of gravity at any one point. The technique stands just perfect for full length, open-width, unstitched textiles, like carpets, sarees, etc. It is advisable to use rollers with as large a diameter as possible, and the rollers should be longer than the width of the textile. It is important to ensure that these rollers have been made from acid-free materials and they are padded with inert materials. Once the textile has been carefully rolled over the roller beam, it needs to be then covered with sheets of acid-free tissue and dust-proof cover. However, this technique gets very complicated with costumes and made-ups as rolling over might place creases on parts stitched in various directions.

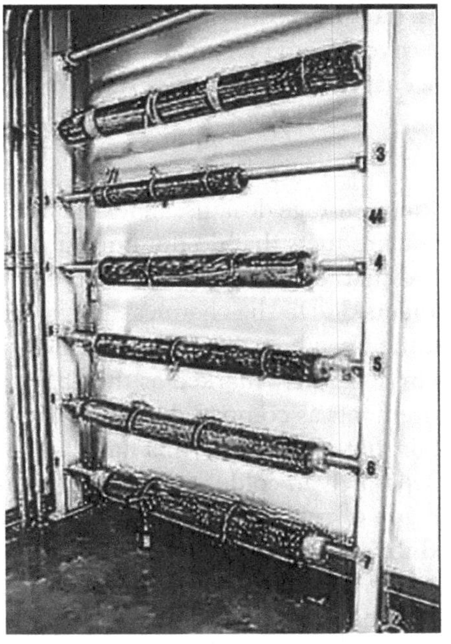

Figure 7.2 Rolled storage stands for textile artifacts.

7.5.4 Hanging Storage

Hanging storage is one of the most efficient storage designs in terms of space management. Additionally, it might be the only option for some design rich costumes that have significantly 3D designs and cannot be folded, rolled or laid out flat without compromising with the design/accessories. However, it cannot be widely used for most textile artifacts due to various functional reasons. Hanging storages concentrate the gravitational pull on certain central locations of the artifact, which can cause considerable wear and tear in those areas of the textile artifact. Although padding and support systems can be used to spread that pull over a larger area, it might still not be a viable option for already weak aged products. Keeping all this in view, hanging storage can be considered suitable for only very strong costumes using padded hangers that suit the individual features of the costume. This storage design should never be used for heavily decorated costumes or garments made from loose knits or bias-cut fabrics as these would gradually stretch out of shape.

Care should be taken to use wooden hangers and not wire hangers for hanging storage. Also, these hangers should be adequately padded with untreated polyester wadding covered with clean, washed calico. Polyester is the safest option for wadding/filling as it is stable, inert and does not absorb moisture like cotton or wool. It is advisable to create these padding covers as removable so that they can be washed occasionally. Care should be taken to design long neck hangers for costumes with high collars. It is also important here to understand that there cannot be a standard design or dimension for hanging storage hangers. Rather, each hanger needs to be designed and customized in-house taking into consideration the specific design features and susceptibility of the garment where extra support needs to be placed in areas that are already weak.

Like other storage designs, hanging costumes need to be protected from light, dust and crushing. Therefore, suitable covers from inert materials like well-washed cotton/inert bags or acid free tissues should be customized for each of these hanging costumes.

7.5.5 Special Storage

Apart from the above-mentioned storage designs, all textile museum storages need to make special arrangements for some products that do not find suitable space in any of the above mentioned due to their special characteristics. Artifacts containing fur, rubber, plastics etc. have special storage requirements due to the chemical activity associated with their degeneration process. Sometimes, they need customized storage designs; however, most of the time, these artifacts need to be separately marked as they have different maintenance routines or environmental requirements as compared to regular textile artifacts.

Textile exhibits that contain fur or feather trims have higher potential of pest infestations and need to be stored separately in colder temperatures in the range of 0-4 °C. Again, objects partially or wholly made of rubber might release Sulphur fumes in the process of degeneration and risk the exhibits stored nearby. It is advised to wrap these in barrier films and placed in oxygen-free storage if possible. Another common problem area is the plastics and lacquered leather. These can be found on costumes as accessories, buttons, patches, and in shoes handbags etc. Biggest problem with these artifacts is that each plastic has its own chemistry and thus a different deterioration curve. These objects are known to release plasticizers, organic acids and become sticky over a period of time. As the process is different from case-to-case, predictions are always not the best way out. It is therefore important to keep these pieces wrapped in acid-free packing and stored in well-ventilated rooms. Additionally, it is an established practice to store 3D artifacts on customized support platforms that can be used for both storage & display (Figure 7.3). This reduces the need for direct handling of artifacts and minimizes decay. This could include artifacts like headgear (hats, turbans etc.) or other textile articles like umbrellas, bags etc. As a standard practice, customized

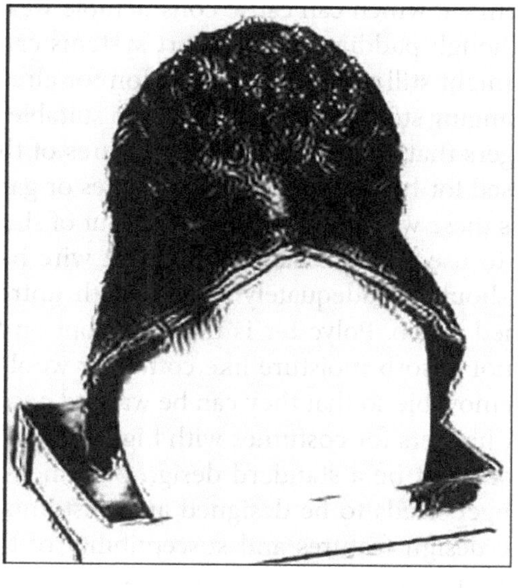

Figure 7.3 Customized support for a shaped headgear that works for both storage & exhibition.

shape supports are constructed for each artifact that provide support as well as help in dissipating the impact of gravity. The 3D storage mounts double as display mechanisms as well and thus ensure that the artifact does not need to be handled directly at any stage of display or storage [12–19].

Another interesting category of special textiles is the "Archaeological textiles." These artifacts are extremely fragile and require special care in handling. Archaeological textiles could be so sensitive that many a times these could disintegrate due to minimum handling required for research or exhibition procedures. Keeping this in mind, the textile fragments are often mounted on conservation safe frames that can be used for both exhibition and storage without any need for direct handling. This has necessitated three-dimensional mounting storage, which can be easily stored in boxes and can be made accessible without damaging the sensitive textile fragment [20–22].

7.6 Safety Systems

Museums are storehouses of timeless pieces of art that can be often priceless because of their heritage value, material or techniques. It is the prime duty of museum storage to keep these articles safe from theft and vandalism. It is desirable for museums to create and maintain a photographic record of all the objects with their specified location inside the storage areas. Museums have various systems in place to ensure security of the objects in their collection. Some of the most prevalent ones being physical protection by strengthening the building, perimeter alarms, invigilation, integrated electronic security systems and key security [23]. We will discuss the basic tenets of all these in the same order.

7.6.1 Location, Structural, and Physical Protection

The buildings meant to store museum artifacts should be designed in a manner that helps in defending or delaying an attempt of theft or vandalism. The building design and security layers around it should be able to delay unholy attempts enough so as to allow the alarm systems to react and gather support. Most importantly, the storage buildings need to have strong walls and roofs that cannot be penetrated easily. There should be a minimum number of doors and windows, just the minimum required for ventilation. Emergency doors should be equipped with quick-release functions. Also, access to the interior of storage chambers as well as the building roof should be restricted and regulated.

7.6.2 Perimeter Alarms

A perimeter alarm or intruder detector system needs to be installed across all openings of the storage building, such as doors, windows, roof lights, ventilation shafts, etc. Also, the system should be intelligent enough to reduce the false alarm rate. Systems relying solely on movement are liable to higher false alarms, especially in chambers designed for hanging storage. Ideal option is the alarm technologies that use a combination of movement and body heat parameters to detect untoward movements. Additionally, there needs to be an active, round-the-clock protocol to address the signals raised by these alarm systems.

7.6.3 Invigilation

Manual monitoring of the storage systems is necessary to ensure total safety. It is important to regulate access to museum storage chambers by restricting access to only authorized personals and creating a log of people entering or exiting the building. Also, any visitors, researchers requesting access for academic or administrative reasons, need to be accompanied by authorized professionals at all points of time. Invigilation professionals need to be extra vigilant during specific times like evening events or when an exhibition is being installed or uninstalled.

7.6.4 Key Security

Museums need to design and administer strict protocols regarding the use and possession keys to the storage areas. It is advisable to have only as many sets of keys as strictly required, and these should be actively logged and accounted for. Under any circumstances, museum staff should not be allowed to take internal keys out of the building. Modern electronic key systems can maintain auto logs of every single use and help in monitoring access to the storage chambers. Nevertheless, any theft or security breach should be immediately reported to local authorities.

7.7 Disaster Handling

Despite all measures taken, it is not unusual for museums to face unexpected situations like floods, fires, pest infestation to name the few. Disaster handling in museums essentially needs to be discussed under two categories, i.e., (a) Disaster potential management, (b) Disaster response. Museums need to have a well-laid policy and protocol for handling these situations and thoroughly train its professionals to handle these situations. Every museum needs to have a well-written detailed disaster plan document to lay out an action plan for disaster prevention and management. This plan document should essentially start with a discussion on *Risk Assessment* where a thorough and practical assessment of all imminent risks should be made. Although all museums need to be ready for all kinds of emergency situations, some museum storage will be more prone to certain disasters than others. For e.g. A museum or museum storage situated close to a major river needs to proactively plan for potential flooding than any other disaster whereas a museum located in a desert area faces almost negligible chances of such an episode. The disaster management document needs to take the assessment forward by laying out risk mitigation procedures. It needs to detail a clear policy in times of emergency in terms of items that need to be rescued on priority. This plan should contain updated emergency contact numbers as well as the contact details of public emergency services. It should also detail the responsibilities of specific personnel at the time of emergency and a road map to crisis management. It should contain a clear, confidential up-to-date plan of the building marking all areas, big or small. The plan should also contain a complete record of the collection with prioritized items marked in the document and the collection covers. Although this information is already available in museum registers, access to this information in the disaster plan document might save valuable time and facilitate quick action. It should further elaborate upon the safety policy for working in hazardous crisis conditions and an emergency security policy in case the premises have been damaged. All storage areas should be equipped with disaster boxes to be used during an emergency. One of the most important parts of this document could be the prescribed first-aid measures drawn up in consultation with conservators. The staff needs to be regularly trained on emergency disaster response drills.

7.7.1 Protecting from Fire

Risk of damage due to fire can be both direct or indirect like melting/burning of mannequins etc. Disaster management in terms of protection from fire can be discussed under two aspects: (a) prevention through design and practice, (b) emergency response at the time of disaster. A well laid out policy in both the scenarios and regular training of museum professionals to help them put the plan into practice complete the circle of protecting the artifacts from fire.

Prevention is an ongoing aspect of disaster management that is intertwined with every aspect of museum building, care and maintenance. Museum buildings should be architecturally designed and physically constructed in a manner that minimizes the risk of a fire hazard. The architecture of the building should essentially include an adequate number of fires exits and emergency outlets for ventilation. The building needs to be equipped with fire warning and safety and firefighting equipment. Adequate numbers of water sprinklers and carbon dioxide-based fire extinguishers (more preferable than water-based fire extinguishers for textiles) need to be placed and regularly maintained. Museum area storing collections should be rigorously insulated to provide at-least one hour protection from fire. This guidance is based on the assumption that it might possibly take firefighters that much of time getting into action. Also, a building should be planned in a way that high risk areas, such as laboratories, kitchen, boilers, or chemical stores are placed away from the storage chambers. All these factors need to be kept in mind while detailing the placement of fire sensors across the building. As far as possible, the materials used in construction of buildings and their interiors should be fire-resistant. All electric, gas and oil installations or equipment should be done as per prescribed standards and a copy of the wiring plan should be included in the disaster management document. Museum buildings should comply with all safety legislations and should be duly certified by local fire departments after regular inspections.

Artifacts that are known to be made of cellulose nitrate like bags, buttons etc. should be identified and monitored. These should be kept in special storage to prevent build-up of harmful gases. It is advisable to maintain a copy of all museum records & documentation in a separate building. Smoking should be forbidden in and around storage areas. Careful thought should be given while planning exhibitions & concerts. Additionally, any construction or repair work on premises that involves working with heat sources, such as blow torches or arc welding, should not be permitted in the vicinity of storage areas.

Above all, it is imperative that staff & volunteers are given regular training in preventing & responding to fire hazards.

7.7.2 Protecting from Floods

Interestingly, "Museum & Galleries Commission's Book of Standards" advises museums to make arrangements assuming, "If a flood can occur, one day it will." This statement emphasizes the readiness with which museums should keep themselves ready to face adversities. It also wants to explain that although calamities occur once in years or maybe decades, if caught unprepared, they might wipe out the efforts made in developing and managing a museum collection over the long preceding years. Although the occurrence of such incidents is rare, its damage potential is huge and therefore worth mitigating the risk by a suitable disaster management plan.

The policy towards flood management is not limited to flood caused by overflowing water bodies nearby, but includes any damage potential by aqueous functions like plumbing lines, fire lighting equipment, etc.

It is advisable to store textile artifacts in buildings that have possibly no pipe works, at least in areas where collections are to be stored. It is always expected that the storage cabinets are at least 150 mm (6 inches) above the floor. In buildings, where it is not possible to exclude the pipe-works altogether, one solution could be to run the pipe work at ground level instead of ceiling level or side walls. Nevertheless, adequate drainage should be provided in all chambers. Also, automatic cut-off valves and leak detectors and flood detection alarms should be adequately installed. Dangers of leaks from humidifiers should also be considered. Advice should be taken from local bodies about the likelihood of floods & buildings susceptible to floods should have adequate equipment like bund walls, stop-boards & sandbags etc.

The disaster response plan should contain detailed information about pipe-works & stop-cocks around the building. All drains should have non-return taps & all taps in sinks should be spring loaded, auto-turn off type. The plan should also consider the danger of water damage from fire hazards or flooding danger posed by central heating systems. Care should be taken to protect all racks and shelves with water-proof sheeting at all times, however, not using materials that can accelerate other forms of deterioration by creating microclimates. As valid for all other possibilities, the disaster boxes should contain adequate equipment for dealing with floods, including absorbent pillows and super slurpers [24]. Finally, regular and thorough staff and volunteer training stands imperative to success of any of these flood management plans.

7.7.3 Protecting from Pests

Biggest biological threat to museum collection apart from people is from rats, mice, birds, insects, fungi, algae and bacteria. Although some of these can be effectively barred from coming in contact with the housed collections by physical arrangements, some of them cannot be monitored by simply denying access. The threat is such that the introduction of one infested piece in the storage chamber can effectively trigger havoc for the whole collection. Therefore, it is imperative that all incoming objects, together with their packaging material should be inspected in a separate quarantine room for the presence of any biologically active agents. Some museums follow a policy of quarantine and sanitation by prescribed treatments immaterial of the inspection reports [25].

7.7.4 Day-to-Day Maintenance

Regular inspection and maintenance procedures further ensure safety of textile artifacts from these pests. Placing pest traps in storage areas is a time-tested strategy that should be followed. Museums should develop and follow an Integrated Pest Management (IPM) strategy that details methods of quick and effective eradication in case of an outbreak.

The biocidal remedial treatments to eliminate pests should be kept to minimal as these can be potentially harmful to the objects, environment, staff and visitors. Some of the old techniques include widespread fumigation of storage chambers with pesticides or use of naphthalene balls. However, these techniques carry long-term health risk for museum workers and also the fumes might cause color fading in textiles over a period of time. A range of new pest control methods have since been developed that are safer for both the workers and the stored objects. One of the most effective treatments prescribed is timed exposure of textiles to carbon dioxide or nitrogen chambers. This process effectively kills/deactivates the microorganisms and sanitizes artifacts for a very long time. Some manuals also propose heat treatments or freezing treatments for sanitization. However, these might not be conducive for textile artifacts given potential damage in terms of burn marks or color bleeding respectively. Many museums in India have been managing their collections effectively while using herbal remedies like *Neem* leaves, spices and condiments. Since textile artifacts have varied susceptibility, the remedial disinfection process should be duly discussed with the conservation scientists before practical application.

7.8 Managing Dust and Dirt

Dust and dirt need to be regularly managed in order to keep the collection in healthy condition. Dust can originate both from internal and external sources and if left unattended, it can fasten deterioration by attracting and holding moisture. These basic nuances are capable of fostering mold growth and attracting pests and other microbiological infections. Therefore, all storage areas should follow a regular cleaning regime, which is monitored by trained and experienced museum professionals. Good housekeeping is the key to preventing dust-initiated damage to artifacts. However, housekeeping protocols for a museum storage can differ from that of a regular public premise. Wet cleaning of museum storage floors can disturb the stability of humidity levels and further encourage deterioration of the products. It is advisable to use vacuum cleaners for cleaning the storage interiors while completely avoiding wet cleaning if possible. Also, care should be taken that vacuum cleaners have been properly maintained and not using the same attachments as used in some other pest infested areas.

Storage area windows, if any, should be close fitting and kept shut and floors should be covered or sealed. All costumes and textiles should be covered with dust sheets. Also, there should be loop-piled door mats and under door brushes at the doors to the store. Additionally, care should be taken to keep the curtains and dust covers clean by regular laundering. There should always be additional dust covers available when one set goes for laundry. It is important to use only those cleaning methods and equipment that are approved by the conservators and collectors. Special precaution should be taken to temporarily relocate the collections in case of building works in vicinity. It needs to be duly mentioned here that sometimes the dirt acquired becomes part of the history of the object due to certain circumstances. This aspect should be duly considered and discussed with the collectors before an object is cleaned.

7.9 Pollutants

Pollution is a widespread term used in different perspectives. Pollution can be broadly described as the introduction of harmful elements in the environment. Museum artifacts in storage suffer from pollution risks both from indoors and outdoors sources.

The impact of outdoor pollutants is directly proportional to the rising pollution levels of the city or local area. Harmful elements, such as sulfur dioxide, ozone and nitrogen dioxide, smoke, dust deposits from diesel fumes, etc., can cause fading, staining, and deterioration of both organic and inorganic objects. Needless to say, organic artifacts like textiles are impacted much higher as compared to the inorganic ones. The impact of outdoor pollutants can be significantly reduced by draught proofing doors and sealing windows. Additionally, mechanical ventilation equipment with scrubbers and charcoal filters can assist in reducing the impact of outside pollution.

A potent and continuous indoor source of pollution could be the materials in the immediate environment of the artifacts. Storage construction materials like boards, hard wood, fire-retardant coatings, paints and adhesives, and some artifacts themselves, like the ones made of wool fur might be constantly emitting small amounts of vapors and gases that might accelerate the deterioration of adjacent textiles. Although good ventilation minimizes this risk, enough time should be allocated to testing of all materials before they are used in the storage chambers. "Oddy tests" are routinely used by museums to test materials that come in close proximity to the stored objects. These tests use metal coupons that have very reactive surfaces. These metal coupons are placed in a closed vessel with a sample of the fabric, which is to be tested. Any resulting corrosion to the metal coupon indicates the type of pollutant being given off by the material. For example, if a silver coupon tarnishes, then a sulfur compound is being given off. The test takes up to 6 weeks to complete.

As noted earlier, the stored artifacts and their accessories could themselves become potent sources of pollution in a storage chamber. Metallic fastenings are one such notorious category. It is for this reason that only *Velcro* covered with unbleached fabric is considered safe for museum textiles. Additionally, objects containing rubber (such as elastics, carpet backings, raincoats, shoes, etc.) release sulphur into its surroundings that can cause adjoining fibers to disintegrate and metallic inclusions to erode. Plastics present as buttons, etc. release plasticizers over the period of time, which can cause similar damage to adjoining fibers. Metals present as buttons, sequins, metal threads can tarnish and corrode and their corrosion products can accelerate degeneration in the surrounding fabrics. Some combs, buttons, buckles might be made of cellulose nitrate that gives off corrosive acidic products at ageing. Also, glass beads used as embellishments on dresses can also be chemically unstable, as well as easy to break and may affect the underlying fabric by bleaching it or damaging the yarns. Interestingly, another hidden category of indoor pollutants is the remnants of chemicals used in the production, finishing, and treatment of fabrics from previous conservation treatments.

Another important source of indoor pollutants is the building works and redecoration works. These can introduce contaminants, such as dust, paints, solvent fumes, and large quantities of moisture, which are potentially harmful to the objects. Building and finishing materials emit particles like saw dust along with ammonia and water vapors for some time after application. Therefore, a newly built or redecorated building should be ventilated and completely dried using industrial dehumidifiers. Care should be taken to not store objects in newly decorated space until tests show that emissions have been reduced to acceptable levels.

7.10 Conclusion

Museum storage has become a central issue for museum policies as significant parts of collections are kept in storage. Textiles being fragile and hygroscopic in nature are extremely susceptible to damage from many factors. They need to be given special attention within the storage of museums. Improper environmental conditions surrounding the artifacts in storage spaces affect the rate of textile deterioration. Natural disasters, fires, and earthquakes being other conditions, which have been a major concern for the museum professionals. It can be safely concluded that stabilized storage conditions in terms of humidity and temperature play a decisive role in long-term preservation of textile artifacts. Storage spaces need to be inspected regularly to ensure minimal stress on the housed collections and keep them safe from weather, pollutants, pests, dust and dirt, and other external threats. Safe storage furniture materials that are inert and do not rust or oxidize and are also dust proof should be used for storage of textile artifacts to protect them from many damages that are associated with material types. Where technology has a huge role in maintaining optimum storage conditions, museum architecture plays a crucial role in maximizing the benefits of technology for the protection of cultural heritage. There cannot be a standard solution that can effectively take care of maintenance needs of varied textile artifacts. Each textile artifact in museum collections need to be carefully analyzed in terms of condition, components, etc. to be able to decide the best possible storage regime for it. While ensuring a longer lifespan for textile artifacts could be both labor- and resource-intensive, their cultural importance could well justify the time, effort and resources spent in their careful storage and maintenance.

References

1. Sachdeva, K., Suri, M., Bhagat, S., An overview of Textile Conservation Practices in Museums of Australia, in: *Conservation of Cultural Property in India*, 2009.
2. https://www.iccrom.org/sites/default/files/RE-ORG%20Leaflet%20A4%20-%20for%20printing.pdf, 2020
3. Kreplak, Y. and Mairesse, F., Museum collection storage. *Mus. Int., ICOM*, 73, 1-2, 1–4, Routledge, 2021.
4. *Standards in the Museum Care of Costumes & Textile Collections*, Museums & Galleries Commission, UK, 1998.
5. https://backend.orbit.dtu.dk/ws/files/131863954/2016_04_ACE_2016_New_Concept_for_Museum_Storage_Buildings_Christensen_Knudsen_Kollias.pdf, 2021
6. https://philamuseum.org/conservation/5.html?page=3, 2021
7. Sachdeva, K., Suri, M., Bhagat, S., Assaying changes in museum textiles on ageing-A short note, in: *Conservation on Cultural Property in India*, vol. 39, pp. 119–122, 2011.
8. Sachdeva, K., Suri, M., Bhagat, S., Effect of fabric construction on the longevity of aged cotton fabric. *Int. J. Eng. Res. Appl.*, 10, 9, 49–61, 2020.
9. Sachdeva, K., Suri, M., Bhagat, S., Role of fibre in degeneration of aged fabrics in museums. *IJERASET*, 531–539, 2021.
10. https://unesdoc.unesco.org/ark:/48223/pf0000042316, 2021.
11. https://www.spacesaver.com/portfolio-posts/designing-off-site-museum-collections-facility/, 2021.

12. https://collectionstrust.org.uk/wp-content/uploads/2017/02/Museums-Galleries-Commission-An-Illustrated-Guide-to-the-Care-of-Costume-and-Textile-Collections-2000.pdf, 2021.
13. https://www.canada.ca/en/conservation-institute/services/conservation-preservation-publications/canadian-conservation-institute-notes/storage-costume-accessories.html, 2021.
14. http://insidetheconservatorsstudio.blogspot.com/2017/10/conserving-storing-and-mounting-hats.html, 2021.
15. https://www.mfa.org/collections/conservation/feature_costumeaccessories_hatsandheadware_carved, 2021.
16. http://museum.wa.gov.au/research/collections/materials-conservation/care-wa-museum-collections/storage-19th-century-bonnets-2, 2021.
17. https://artsandculture.google.com/exhibit/conservation-in-action-preserving-hats-and-head-wear-at-the-mfa-boston-museum-of-fine-arts-boston/wAKCMTQv7r5NKA?hl=en, 2021.
18. https://americanhistory.si.edu/blog/2011/06/out-of-storage.html, 2021.
19. https://www.airforcemuseum.co.nz/blog/military-headgear-conservation/, 2021.
20. Peacock, E. and Griffin, E., Rehousing a collection of archaeological textiles. *Conservator*, 22, 68–80, 1998.
21. http://resources.culturalheritage.org/wp-content/uploads/sites/8/2015/02/osg008-012.pdf, 2021.
22. Childs, S.T. and Benden, D.M.A., Checklist for sustainable management of archaeological collections. *Adv. Archaeol. Pract.*, 5, 01, 12–25, 201.
23. http://ignca.nic.in/Museum_security2013/museum_security_proceedings01.pdf, 2021.
24. Amaral, J.R., Thinking inside the box: evaluation of polypropylene boxes for the storage of museum objects. *Conserv. Patrim.*, 34, 1–23, 2020.
25. Faheem, F., Management of pests risks in museums: A review. *Int. J. Adv. Res. Biol. Sci.*, 122–136, 2019.

8

Tools and Methods for Handling and Storage of Museum Textiles

Pratikhya Badanayak[1], Seiko Jose[2], Ragini Dubey[3] and Ritu Pandey[3*]

[1]*Department of Textile and Apparel Designing, University of Agricultural Sciences, Dharwad, Karnataka, India*
[2]*Textile Manufacturing and Textile Chemistry Division, ICAR- Central Sheep and Wool Research Institute, Avikanagar, India*
[3]*Chandra Shekhar Azad University of Agriculture & Technology, Kanpur, India*

Abstract
Textile museums exhibit exquisite historical evidence of our glorious past in the form of fabric, specimens, accessories, and robes. Textiles in the museum survived for so long owing to their proper care, handling, transportation, and storage conditions. Thus, we can understand the value of proper handling of museum textiles so that it is available to future generations for the viewing in the same condition as it is now. Storing ancient textiles is a challenge today, as they differ greatly in terms of content, structure, and adornment. It is quintessential to manage, store, and display costumes and textiles in a way that protects them from harm and does not expose them to unnecessary risks. In this context, certain tools are used for handling and maintaining such textiles, and this chapter unfolds the tools used for handling museum textiles. In addition, the transfer of costumes and textiles whether room-to-room or continent-to-continent should be carefully prepared and carried out with the necessary equipment and people. This chapter contains all the tools and protecting accessories to handle the textiles by museum personnel. Storage conditions, instructions for visitors, suggested personal clothing, care, and standard protocol during the transportation are written in detail. Few case studies regarding ancient textile conservation techniques, fabric, and chemicals used in museums is also presented.

***Keywords*:** Textile museums, museum tools, textile artifacts, handling tools, ancient textiles

8.1 Introduction

There are several functions performed by museums. They collect artifacts, handworks, and handicrafts of various materials, often including precious materials, from different cultural backgrounds. The museum's objects have been categorized, transcribed, and obtained from various places. Some are displayed in continual galleries or transitory exhibitions, while others are stored carefully. Other aspects that make the museum a source of knowledge

**Corresponding author*: ritupandey70@gmail.com

Seiko Jose, Sabu Thomas, Pintu Pandit, Ritu Pandey and Vandana Gupta (eds.) Handbook of Museum Textiles: Volume I Conservation and Cultural Research, (161–180) © 2023 Scrivener Publishing LLC

are preparing publications with well-researched information and making objects accessible to scholars. Nonetheless, a museum is an exhibition site and opens to the public for learning, amusement, or gratification. Textiles vary enormously in museum collections. Their historical interests, artistic attractiveness, and historical relevance are valued. Due to their broad appeal, textiles are often permanently displayed in most museums, particularly in historical museums. Textile museums are recognized for their historic attractions, artistic appeal, and cultural relevance. In museums, ancient or historic textiles, and crafted textiles, such as engravings, rugs, quilts, banners, apparels, curtains, upholstered chairs, dolls, and accessories, such as supporters, parasols, handcuffs, and bonnets, are stored. Museums bear a great deal of responsibility for the preservation, study, and display of cultural heritage. Particularly, historic textile is gaining more appreciation throughout the display of the material in museums. Culture and heritage conservation and its preservation is a multidisciplinary field that, on the one hand, requires close collaboration between archaeologists, art historians, curators of museums, and restorers. This is a prominent part, and in this area of study, a number of scientific papers have been provided [1].

The preservation of museum artifacts offers the choice among passive or active preservation methods. Passive conservation methods can cause damage or disintegration to the object. The use of integrative conservation methods to structurally preserve the artifact over time may lead to change in the mechanical and physical properties of the artifact [2]. In order to improve the quality of care and conservation of textiles and to guarantee continuous access to historical and cultural information, and the aesthetic pleasure it provides, necessary steps can be taken to manage, display, and store textiles safely.

8.2 Care, Maintenance, and Handling of Museum Textiles

Preservation, also known as conservation, encompasses all of the "dos and don'ts" that contribute to ensuring the life of the objects, whether they are on the window, archived or managed, studied or handled. In all conservation terms "care" and "monitoring" are used synonymously. Conservation in laboratory settings, on the other hand, refers to "radical treatment." This process is usually designed to make the object resemble the original, often by adding new material and to stabilize the state of the object [3]. Fundamentally, preservation and conservation of works of art are studies of the object's chemical, physical, and photochemical interoperability with other matter and aspects with which it comes into contact, and challenges to regulating the interplay between them in order to maintain the inherent quality of the object.

8.2.1 General Storage Factors

Long-term storage of historical items can cause materials to change structurally due to exposure to various conditions, such as UV irradiation, high moisture, and changing temperatures. Ultraviolet light promotes the oxidation of natural fibers, causing intermolecular bonds to rupture and microbial enzyme penetration. High humidity accelerates the attack and the degradation process consequently [4]. Although made of natural fibers, the microbial attacks and decay lead to discoloration, deterioration, and loss of structural strength.

Textiles, on the other hand, may also be the carriers of microorganisms, such as pathogenic or odor generating bacteria, and fungi [5]. Museum managers have been involved for a long time in the degree of fabrication in terms of antimicrobial protection of textiles to prevent biodeterioration [6, 7]. Light damage to specimens in historic houses and museums has risen in recent decades. With the availability of air conditioning to maintain an artificial temperature balance, there is little need to seal the sunlight into the room.

8.2.2 General Guideline in Handling

Fabrics should always be handled with care, and the use of protective plastic gloves is advised. The dressmaker's dummy is thought to be extremely useful during the accession process because it aids in providing a better fit for the product labeling. This could improve the way pruderies are handled. Unless the item has been registered, it will most likely be kept unless preservation is required and facilities are readily available.

8.3 Ideal Conditions, Temperature, Humidity

Depending on the current archive conditions, the storage practices were recognized. Ideal storage creates a nonchemical and physical environment in which objects are altered. The specifications are kept at temperatures ranging from 65°F to 70°F, with minimum lights ranging from 25 to 1000 Lux depending on the object type, maximum ultraviolet levels of 75 UW/lumen, and relative humidity ranging from 45% to 55% [8]. Even though microbes can attack almost any material, including synthetic polymers, historical textiles primarily made of organic materials are particularly vulnerable [6, 7]. Museums' physical conditions are currently being regulated, particularly in storage and exhibition space, including the establishment and maintenance of acceptable temperatures and relative humidity (16–18°C and 40-65 RH percent, respectively) [9, 10]. It is also a concern that natural fibers have varying moisture absorption and desorption rates, as well as uncontrollable local microclimate conditions, which may be the reason for contaminating the fibers.

8.4 Storage Units

Fabric fading can be accelerated by natural sunlight, as well as artificial lighting within period rooms, though Ponsonby [11] stated that opaque window shades can be used to completely darken the rooms. The museum's maintenance staff had to walk across the rugs and reach behind the curtains to remove the shades, which caused some abrasion damage to these textiles over time. Air conditioning and insect infestation protection should be controlled in the storage area. Spraying is best done with a vacuum and "fumigas 10" [12]. To monitor harmful light, direct sunlight should be avoided, ultraviolet filters can be installed over fluorescent lights, or low-ultraviolet pipes such as the Philips TL37 can be used. Storage areas should be inspected on a regular basis for insect infestation, lighting, and structural issues.

8.5 Storage Materials

The degree and the rate of degradation are dependent on the chemical and physical properties of the substrate, including the chemical structure, molecular weight, and crystallinity and the prevailing microbial pollutants. Many garments can be hanged and safeguarded with appropriate materials on padded hangers. For this application, the acid-free tissue or dacron fiber is ideal. If box storage facilities are necessary, very little hanging space should be necessary. It is made of nonsubstantial cardboard and measures 14 × 53 × 102 cm long [12].

8.6 Tools Used in Maintenance of Museum Textiles

The essence of the fabrics chosen and used by artists is one of the key characteristics of modern and contemporary textile art. The use of common fabrics, such as silk, wool, cotton, and linen, has spread to new and radical materials, such as plastic bags, rubber, and film, through modern synthetic textile fibers. Understanding, anticipating, and avoiding the degradation of these materials requires information that also spans conventional conservation fields.

8.6.1 Equipping the Workspace

The world's oldest dress, Tarkhan dress dating back to 3482 and 3102 BCE was excavated in the year 1912 to 1913 from one of the Egyptian mummies and transported to England's Victoria and Albert Museum for conservation. However, the packet containing the dress and other linen garments remained unattended till its rediscovery by the museum officials in the year 1977. The radiocarbon dating of the dress, done in the year 2015 confirmed its antiquity [13, 14] and exclusivity [15] in terms of pleated sleeves and designs. The use of conservation fabric was done on the golden linen V-neck Tarkhan dress as lining material before mounting it on the museum's mannequin (Figure 8.1). Proper handling, transportation, storage, and restoration work done by the art collectors and museum personnel made the Tarkhan dress a shining jewel in the Petrie Museum of Egyptian Archaeology at University College London. Handling precautions and equipment of the museum artifacts depend upon the collection scale and use [16]. Use of protective clothing is recommended for the art collectors and handlers as the collection and care can be unhygienic and messy work. Clean overalls, or aprons and clean white cotton gloves, vinyl and latex gloves, and dust masks are basic and useful requirements. Some of the basic items required for handling and restoring historical artifacts are listed in Table 8.1. Jewelries, such as rings, brooches, and chunky necklaces, should be removed before handling costumes or textiles as they present a risk of causing damage.

8.6.2 Housekeeping

The routine housekeeping practices are dusting and cleaning of the artifacts and the environment. Written cleaning procedures for protective clothing, gloves, dust sheets, and

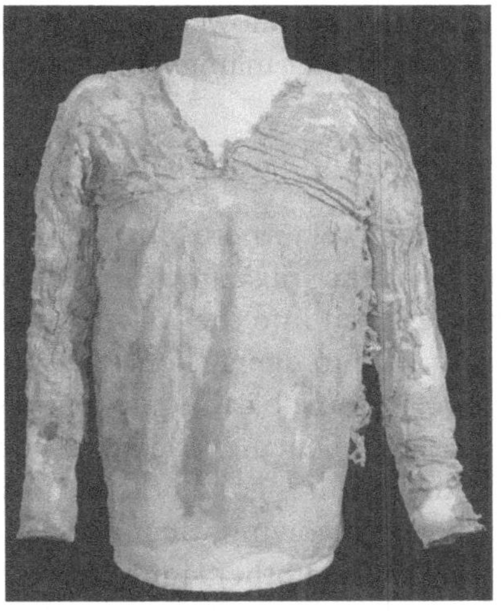

Figure 8.1 World's oldest dress (3000 BCE), the Tarkhan dress at Petrie Museum of Egyptian Archaeology at University College London [17].

Table 8.1 Basic equipments and tools for handling museum textiles.

Equipment and tools	Function
Mount making tools	Object display and safety
Labeling tools	Identification tagging
Sealing equipment, Trimmers	To protect from dirt, making customize enclosures
Suction table/conservation platform	Used for repairing work
Cleaning equipments	Glass cleaning cloth, sponges, suede, towels, steamers, deodorizers
Washing machine and hand-washing facility	Washing of dust sheets, overalls, gloves, etc.
Cylinder type vacuum cleaner with adjustable suction	Cleaning stores and displays, etc.
Dust sheets	Covering, lifting, and rotating of museum textiles
Shallow trays of various sizes for lifting, gloves, forceps, etc.	Segregating, preventing loss
Plastic sheets of conservation quality and papers	Lifting, turning, moving during the inspection
Thermo hygrograph, whirling hygrometer, lux meter, ultra-violet monitor, insect trap, deacidification spray, etc.	Humidity, lighting, temperature conditions, and pest activity monitoring in the museum

appropriate quarantine measures must be followed. Regular good housekeeping tasks practices, evaluation, and checks ensure proper maintenance of museum artifacts [18].

8.6.2.1 Cleaning the Collection and Environment

Cleaning is an important activity everywhere. Museum cleaning staffs usually takes responsibility for daily cleaning of the museum site but some cleaning activities especially in object store are best integrated into the collection care program. Ancient textile being organic in nature is susceptible to damage due to light, heat, moisture, pest, mold, and environmental pollutants [19]. Some basic equipment and tool for space cleaning is the adjustable suction, cylinder type of vacuum cleaner a good range of nozzles, and soft brushes. This equipment handling requires good practice to a deep clean storage area and display galleries at least twice a year involving vacuum cleaning space behind and under the furniture, plinth, and cases. Pest is another significant threat to the textile collection. There is a number of creatures that can cause damage to the fiber of textile material. Pest activity vacuum cleaner bags should be removed after each use and disposed of sealed in a polythene bag. Delicate and necessary content should be carefully removed before vacuuming the interior surface. Shaving brushes and bed brushes are useful for brushing scattered pieces of remains into the suction stream to avoid damaging the surface of display boards.

A useful guide from a conservator regarding costume displays in room settings may be sought. To seek further advice on historic interiors, National Trust Manual of Housekeeping may be contacted [20, 21]. Cleaning is necessary to protective clothing and dust sheet sailed overalls aprons and gloves should be washed after every use, preferably with lukewarm water using bleach-free nonbiological and biodegradable detergents. Dust-sheet used in close contact with objects should be boil washed without adding any detergent. Taking proper safety measures, incoming objects made of wool, fur, or feather and any object showing sign of pest infestation must be separate from other objects to carry out careful monitoring and treatment if infested. Quarantine also involves sealing objects in cleaning polythene bags or sheets [22]. It is best practice to build quarantine measures into your object entry procedure.

8.6.2.2 Basic and Best Practices for Checking and Monitoring in Museum

Making routine checks of the collection, environment, and managing the organization go hand in hand. Integrated managing and monitoring of the environment, checking for pest control, and maintenance of the hygiene in the museum are part of the inspection practice.

8.6.2.2.1 Premises and Facilities

Regular checks of galleries and stores are required to maintain the cleanliness of walls, windows, blinds, lighting, and the environment and thereby help in conservation measures. Regular checks of ancient textiles in a museum prevent the damage and ensure taking preventive measures. Faulty water pipes fitting and sources of pest infestation are the source of immediate danger that must be identified and avoided.

8.6.2.2.2 Environment

Routine monitoring of temperature, relative humidity, visible, and ultraviolet light are necessary basic practices to be followed at least weekly for ancient textile museums. Routine checks of visible light levels and their variability are necessary several times a day. Annual checks of the performance of ultra-violet filters should be measured. Inappropriate relative humidity, temperature, and pollution promote fungal growth in museums [2].

8.6.2.2.3 Pests

Continuous checking and monitoring of integrated pest management (IPM) program need to be implemented in textile museums. Monthly checks for evidence of cricket activities especially in spring/summer, including inspection of debris from cleaning under units, corners, and on windowsills must be routine practice.

8.6.2.2.4 Furniture and Basic Practice Containers

Annual maintenance is required to check the stability of racks, the functionality of locks, cupboard sliders, door closure, hinges, etc. Condition of storage boxes, tissue paper sheets, and spaces quality and requirements essentially need to be checked annually.

8.6.2.2.5 Overall Management of Museum Condition

Annual check of inventory records and regular update of large collections is essential. Full condition assessment of all objects going to and from display on a routine basis is necessary, depending on the degree of risk. As a basic practice, collection conditions and monitoring need to be commissioned every 5 years.

8.6.3 Materials and Supplies

Materials required to clean and maintain the collection needs to stock in any museum at all time. Table 8.2 presents a guide to the materials required for handling of museum textiles. Another prerequisite is to have conservation quality materials used for storage and display of the collection.

8.6.3.1 Handling

Museum collection needs some handling for regular cleaning regimes, storage, and display purposes. Handling is done by trained museum professionals following good working procedures (Figure 8.2). Gloves are always worn when in contact with textile objects. The need to touch objects is quite small if basic materials required for handling textile artifacts are used appropriately. Common materials, like sheets of tissue, fabric, forceps, and transparent sheets, are usually used to improve the quality of handling and eliminate the need of touching the objects while handling [23].

Table 8.2 List of useful materials for proper storage/rolling/packing of textile artifacts.

Material	Function
Unbuffered acid-free tissue paper sheets and rolls.	Packing and in interleaving
Unbleached, undyed cotton fabric ('calico'), in various widths and weights. Must be sterilized/boil washed before use.	General protection and support to be used in direct contact with textiles
Polyethylene foam ("Ethafoam," "Plastazote").	Support mounts
Woven tape for sewing in	Labels tying for identification marks
Spun-bonded polyester padding	General-purpose padding
Spun-bonded polyolefin sheet ("Tyvek").	General protection and support, tie-on labels
Thin polyester and nylon film	Protection and support, lifting, making lightweight tubes for storage.
Cotton hosiery stockinette tube	Covering of mannequins arms and costume hangers
Conservation and museum quality acid-free boards.	Mounting and framing of textile wall hangings for support
Acid-free corrugated cardboard ("Perma"/"Dur board")	Making boxes, trays, support forms, and boards
Polyethylene boards	Boxes for storage and transporting artifacts
Velcro' fasteners	Display of textiles
Laminated barrier film ("Moistop," "Seet," "Marvelseal")	Covering of textile artifacts
Wooden coat hangers, wishbone, and crescent types. Extra-long hooks	Costume display in store
Acid-free tissue paper ream	Required for packing and transport

8.6.4 Packing and Unpacking

The requirements for packing for long-term storage will be different from packing for moving and protecting an object. Packing material must be absorbent and able to withstand harsh environmental conditions of high relative humidity (RH) and temperature fluctuations during transportation and storage [25]. Packing and unpacking may be done by two different individuals at different places; therefore, instructional labels should be put on the packing for ease of unpacking. Textile objects should be unwrapped carefully as per the instructions to avoid any damage to the objects.

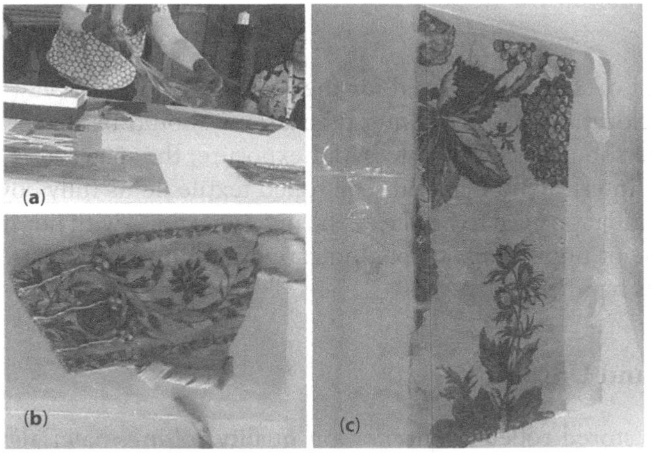

Figure 8.2 Handling and storage of 15 to 17 CE ancient textiles at Temple Newsammuseum, Leeds (a) Handling of textile using latex hand gloves (b, c) Storage and display of textiles covered in transparent nylon film to prevent atmospheric dust [24].

- Line the box with sheets of tissue paper (Figure 8.3).
- Allow space for cushioning. Never overfill a container.
- Arrange fragile and light items on top and heavier at the bottom of the packing boxes.
- Arrange the item using cushions, tissue paper, and snowballs.
- Labeling of the list of items in the box is always helpful during unpacking.
- Wrap each object separately using a tissue paper sheet.
- Use the tissue to lift the object into the box. Use tissue liner over the objects.

Figure 8.3 (a) Handling and packing of 15 to 17 CE ancient textile tapestry bag at Temple Newsam Museum, Leeds. The bag was wrapped in conservation quality tissue paper and stored in the same size cushioned box. (b) An 18th CE bridal silken brocade lehenga having all over moon motifs from India (private collection). The lehenga was basked and stored in wrapping paper after every use and largely untouched with water and any dry wash solvents. The wrapped lehenga was stored in a box with moth repellents dispersed to keep away the insects [24].

8.6.5 Moving

Museum items are occasionally moved for the reasons, such as display, photography, and research for its relocation. The following precautions should be taken while moving the objects: the textile should face down to avoid exposure; the material is rolled and placed inside the tube of interleaving material. The rolled textile items fully covered and secured with extra sheets and tape need to be placed in containers fitted with lids during their movement. For the ease of unpacking, a label indicating the dimensions of textile artifacts is a useful guide for its handlers.

8.6.6 Rolling and Unrolling

Many textiles are stored rolled. Conservation quality rolling materials that do not stick to the fabric surface must be used. Rolling instructions for museum textiles are similar to the packing of museum textiles as written in the earlier subheadings. Unrolling must be done on a large clean platform where the whole surface of the textile can be spread. As the unrolling progresses, the interleaving material on the surface is removed and rolled simultaneously for replacement or reuse.

8.7 Labeling

Methods of labeling of museum textiles differ from museum to museum and objects to objects, but all of these methods generally consist of writing or typing the acceding number on a fabric substrate, allowing the ink to dry if appropriate, and then stitching or binding the label to a stable and easily accessible textile area. "Substrate" is not a technical term used in the industry, but will be used here to refer to the fabric component of the accession mark, whether it is a tape or a nonwoven fabric [26]. As there is no standardized labeling standard, several different substrates and inks are used. Collections trust recommends that labels have the following characteristics [27]:

> Secure—The possibilities of the label or mark being accidentally removed from the item must be relatively lesser.
> Reversible—The label or mark can be removed from the item consciously, with as little trace as possible even after 50 to 100 years.
> Secure for the object—However, neither the materials used on the object nor the process used to apply them is potentially harmful to the artifact.
> Discreet, but visible—The recommended approaches do not detract from the item's appearance or dark information. Nonetheless, the number should be noticeable enough to reduce the item's handling.
> Convenient and safe for staff and voluntary—Resources should be available readily and easily at fair prices, in limited amounts, and if used in conformity with local risk assessment team recommendations, should not pose major health risks.

8.8 Cleaning

Cleaning aged textiles is a constant area of research and development. Cleaning is a method that is irreversible and falls within a curative preservation category. The appearance of textiles is generally affected by the surface dirt and soil. Scientific studies have also demonstrated that dirt on the fabric accelerates the degradation of acidity and can mobilize damage to the fibers. In such situations, the decision on cleaning is complicated and taken after consultations with all stakeholders. The cleaning of historical textiles often involves a sort of consensus; between documentation preservation and improved conservation of textiles over a longer period of time; and between the effective removal of soil and the harm caused by the cleaning process [28].

Surface washing, as the name implies, includes removing dirt, dust, rust, and other foreign substances from the face of the garment. For this reason, only specific materials are used and a wide variety of tools are used, including brushes, erasers, tapes, etc. The choice of the tool depends on the quality of the cloth and the surface to be cleaned. Often surface cleaning serves as a preliminary step before any further cleaning operation is carried out. Also, several times, the conservator can only do surface cleaning for the artifact, depending on the situation and the desired result [29].

Wet cleaning requires the use of water as a method of cleaning textiles. Water has many advantages that make it the best medium for cleaning historic textiles. Textile preservatives add different additives to the water that help to enhance the cleaning of the textile. These may be surface-active agents, sequestering agents, dirt-carrying agents, and even some enzymes [30]. The type and quantity of these agents applied to the washing solution again depends on the type and condition of the textile to be washed and also on the type of dirt that is present on it.

Solvent cleaning refers to the process of cleaning with organic solvents that are used to extract soiling from historic textiles. This method of cleaning is commercially referred to as dry cleaning. Various organic solvents help to dissolve or swell the dirt, making it easier to strip it from the textile. Solvent cleaning is effective because only inorganic solvents are soluble in the soiling present on the artifact. Using this process, soils that are fatty, greasy, or oily in nature can be effectively extracted. Solvent cleaning is often advised where there is a chance of color bleeding or color change due to wet cleaning. If wet cleaning is likely to result in further degradation of the textile, such a process is used. For objects made up of velvet or embossed surfaces, solvent cleaning is desired as water may be detrimental to such structures. If the artifact is a composite piece containing various types of materials that can react differently to water, solvent cleaning is an alternative [31, 32]. However, care should be taken on the selection of the solvent so that the selected solvent should not cause dye degradation for color fading.

8.9 Dealing with Separations

In order to safeguard the artifact, those involved with costumes and textile collections often have to deal with the complete or partial division of the part of the object [33]. The most commonly divided or loose components that cause thread and fabric harm are fastenings and trimmings such as hooks, buttons, leathers, waistbands, beads, lace, and fringes. In these cases, the separation is typical, but sometimes the weak thread is not used to attach it or build it. The second kind of separation consists of clothing, sections, or trims that occur due to the thinness of the textile. A distinction between the two needs to be resolved [34]. When thread damage has led to estrangement, the extent to which it is associated with a local problem or a symptom of widespread weakness is important to be determined. Every item with a widely used textile damage or weakness of thread must be referred to and stored by the conservator for evaluation [35].

8.10 Tools Used for Displaying Museum Textiles

Preserving on-display clothing and textiles necessitates striking a balance between supplying and conserving them. It is also a good idea to seek advice from a textile curator before making decisions concerning textile displays or costumes [36]. The specimen's condition influences its display acceptability and illustrates what is required to keep it in good condition during the exhibition. It is critical to perform a condition evaluation on each object in order to identify its requirements for any scheduled presentation. It is critical to establish the length of the show at the planning stage in order to reach a suitable standard of supporting materials and environmental factors [2].

8.10.1 Showcases and Galleries

The amount of support provided by the textile can differ based on its condition and various other factors. The affected cloth to be vertically placed must be handled with a textile preservative prior to exhibition. Display components should be of a standard of restoration.

The following solutions are available for items in poor condition [8]:

- Laying the fabric flat on a surface, horizontally or at an angle of no more than ten degrees;
- Handling and assembling the object by a textile curator.

For fabrics in secure and satisfactory condition, the conventional procedures can be used [35]:

- Vertical hanging by constant fixation, which evenly distributes the weight.
- Stitching the cloth into a plate that is evenly tensioned and coated with a preservation efficiency. Textiles greater than one square meter may be received from the textile curator if additional sewing is required.
- Drape the cloth over a cushion stand covered with a consistent fabric, such that the majority of the weight is supported by a container stand or the floor.

8.10.2 Frames

Frames serve as mini-cases for objects, protecting them from the environment, physical harm, and theft (Figures 8.4 and 8.5). The materials utilized in construction, particularly

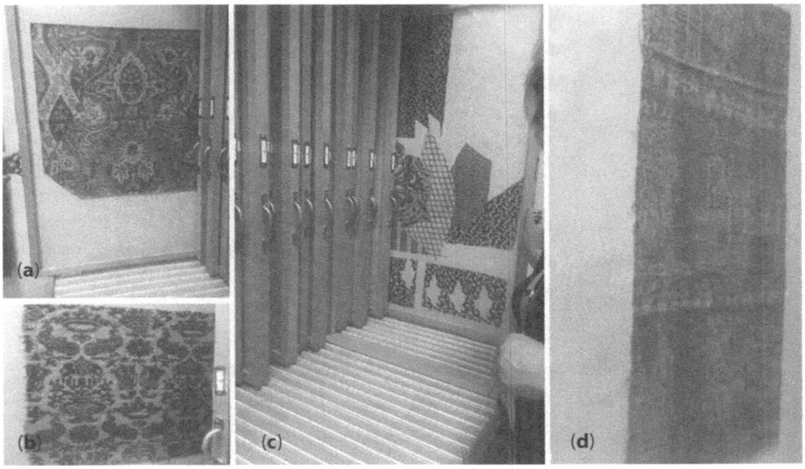

Figure 8.4 Ancient textiles display techniques in Temple Newsam museum (Leeds UK) (a–c) Textiles display on moveable compaction shelves (d) Wall mounting of textile [24].

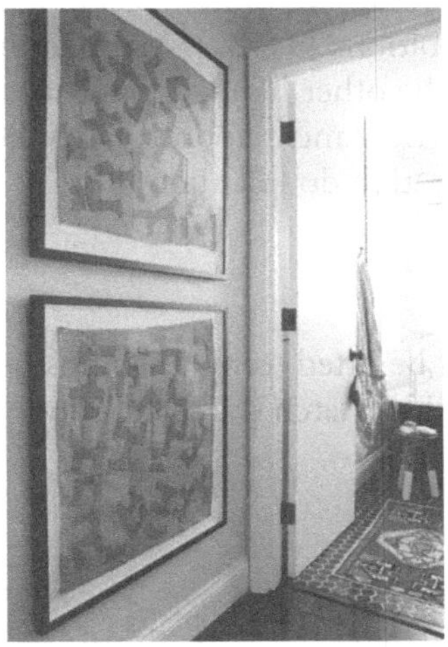

Figure 8.5 Frame [41].

the inside, must be of conservation quality. They should be securely fastened away to keep dust and insects out [2]. Textiles in their original frames should be directed to the textile curator for condition assessment prior to use in exhibitions. Sewing is used to attach textiles to cloth-covered plates. Only tiny durable fabrics in a stable state should be installed for framing without first consulting the textile curator [35].

8.10.3 Mannequins

It is great to observe costumes on mannequins, which fosters a close relationship between humans and clothes, regardless of their universal social culture or special elements of design

Figure 8.6. Mannequins [42].

and wear history. It is critical to meet both aesthetic and conservation standards while purchasing mannequins. There are numerous options accessible [37]. Mannequins are specially designed for historical clothing and exist in a variety of shapes and sizes (Figure 8.6). The mannequins' originality is determined by the exhibition's design. Those with authentic looks and poses are essential to represent individuals in real-life settings, yet they look out of place in historical dress exhibitions [38, 39]. Mannequins built for an ancient costume with toned-down qualities, on the other hand, are not ideal for space. These play a significant part in advertising firms. They must appropriately fit the apparel, ensuring that the mannequin's shape, size, and location do not alter [40].

8.10.4 Hangers

Extremely delicate parts cannot be fitted to mannequins because they must be completely protected from a shape intended to match the cloth interior. These mountings constitute a

Figure 8.7 Hanger [44].

Figure 8.8 Vertical mounting display [41].

distinct task, and the textile curator should be assigned to them (Figures 8.7 and 8.8) [43]. Pins should never be used for costumes since they corrode and discolor [35]. Covering a hanger with cotton batting or acid-free tissue paper and wrapping or laying washed undyed muslin over the cushioning is also recommended. Other issues, such as open boxes for acid leaching, scattering on fabrics and yarn, and spoiling costumes, can be avoided by padding and covering wooden hangers [8].

8.11 Handling During Transportation

The various kinds of transportation used in transit displays will now be discussed. The museum must strike a balance between cost and conservation. The transportation networks covered in this article are those currently used by museums and galleries, depending on the circumstances, by bus, railway, sea, air, and different variations. This is to identify the physical characteristics of movement that affect conservation and condition, rather than to remain entirely in basic documentation, insurance, and cost concerns to registrars and transportation authorities.

8.11.1 By Road

In cities, smaller packages are provided in passenger cars for rapid transport. The wrapping is typical of parchment, plastic corrugated, and polystyrene foam, which implies that there is no adhesive or sticky substance close to the object's surface. Even if the air temperature is too high, there will be no substantial change within the kit for 1 or 2 hours in the microclimate [45]. Car accidents can happen, and the damage can be from a crash or a burn. Many museums, however, ship objects via commercial vehicle for a shorter journey. Each museum has its special vehicles designed for the transport and control of art artifacts with the least amount of load [46]. The exclusive use of a commercial vehicle is only cost-effective if the product is substantial and improved handling is required. In tropical or subtropical climates, vehicles should preferably have cooling mechanisms or be stored in cool environments at least initially until loading [47].

8.11.2 By Rail

Typically, two types of rail transit exist: freight and express transportation. Because the slower methodology is primarily intended for major companies and consumer products, museums rarely adopt it. Typically, freight is the transportation through train from city to city [48]. Items of art that will be transported by freight should be packaged in high-strength, shockproof containers that can withstand large temperature variations and delayed delivery. In extreme cases, express shipments may be supervised directly in the baggage car by a courier, as well as by the management and deployment of such individuals [47]. In exceptional cases, express shipments can be controlled directly by couriers in the luggage wagon.

8.11.3 By Sea

The ship's storage is typically housed in a strong room or in a higher location to protect it from climatic change (i.e. during storms). In such instances, the coating and packaging materials should be designed to prevent the products from shifting or settling and causing damage to them [49]. Vapor barriers are typically necessary for double-box packaging. The customs inspection must be prepared in advance, not in the ports or sheds of the shipping corporations, but the destination museum [50]. Monitoring the handling, storage, and environmental requirements is recommended for case safety. High humidity and salt air infiltration from poorly constructed cases can affect the content [51].

8.11.4 By Air

The most efficient, productive, and faster means of transportation is air travel. According to statistics, the number of injuries per kilometer traveled makes this the safest. Typically, two modes of air transportation are available: air express and air freight [3]. If a nonairtight case is installed on an airplane and the ambient (cabin) pressure drops significantly during flight, air may leak from the inside of the case, depending on the activities of the gases at reduced pressure [52]. The aircraft bags can reach temperatures as low as 4°C during a long voyage at high altitudes [46]. On landing, the cooled cases are removed under normal conditions, and the moisture content will be higher in warm air, and condensation will cause harm. Because air travel is a mode of transportation, the pressure, temperature management, and atmospheric conditions at the landing site must be determined [47]. Airfreight costs are decreased by using lightweight materials and packaging as little as feasible. If a case fails or is deficient in heat isolation, dunnage, shock absorption, and air drainage, the contents are not adequately protected [46, 53].

8.12 Handling Techniques and Conservation Practices of Ancient Textiles in Museums

The earliest methods of textile conservation at Victoria and Albert Museum are sewing threads and adhesive. Professor Sheila Landi first reported about polyvinyl acetate and polyvinyl caprate emulsion as a thermoplastic resin to be applied as backing on supporting nylon tulle. The prepared backing was then attached to ancient textile material by applying warm iron from the backside [54]. Later, natural fabric material and dyes were more commonly used as conservation material in the textile museums.

Few case studies regarding ancient textile conservation techniques adopted in museums are as follows:

8.12.1 Egyptian Shroud

The shroud from one of the excavated mummies was humidified in a humidification chamber to soften the material prior to its cleaning. A softened shroud was kept between two polyester crepe sheets prior to washing in a suction table using ethanol and water. The process cleaned the shroud showing its full details. Silkscreen was used as conservation fabric backing for shroud restoration and display in the museum. Shroud was kept horizontally on a padded nonwoven Teflon coated polyester sheet, facing downward. Silkscreen treated with 10% Lascaux 498 and Mowilith was stretched and placed at the backside of the shroud. A perforated Teflon-coated soft padding impregnated with acetone, a polyethylene covering, and a weight of a wooden board of similar size was pressed over the silkscreen. After 10 minutes, the wooden board and Teflon coated paddings were removed and shroud with silk backing attached to it was left to dry for half an hour. A restored shroud was displayed in a museum, covered in a clear Perspex sheet to prevent any degradation from ultra-violet rays [2].

8.12.2 Jordanian Belt

An ancient historic woollen belt in a Jordanian museum was secured between two layers of stretched nylon mesh using silk sewing thread restoration purpose. Belt was wet cleaned by using neutral soap and gently immersing/pressing in the wash tank. Washed and rinsed belt, free of dirt particles was flat dried with occasional drabbing with a dry cloth. The method of restoring the belt using aqueous emersion was inexpensive. The restored belt was stitched over a framed linen material for preventing damage during further handling and display [2].

8.12.3 Silk Textile

Silk fabric dating the Ottoman period was restored by dyeing the missing design parts of the fabric in natural dye in an Egyptian museum. The restored fabric was supported by new linen fabric stretched on a metal frame for display in exhibitions [33].

8.12.4 Coptic Tapestry

A Coptic textile called "*Kabati*" woven in tapestry style was restored in an Egyptian museum by mounting the tapestry over matching blue linen material. Unravelled weave of the tapestry was engineered using woollen yarns interlacement in tapestry to make it look finished at the edges. The undyed silk sewing thread was used to attach the tapestry over matching linen using blanket stitches at the edges of woven designs. Frayed yarns and spaces of tapestry designs were filled with matching threads and couching stitch [50].

8.13 Conclusions

The useful life of collected artifacts and its accessibility of information are directly influenced by storage and handling methods. Overcrowding, careless, or haphazard storage conditions can all cause damage to archives. Chemically unstable and ill-fitting shelving and storage enclosures hasten the decomposition of the materials they are supposed to protect. Normal use causes wear, but inexperienced or rough handling can cause extensive damage to collections, necessitating expensive repair or replacement. By following the guidelines outlined by the museum authority, the longevity of collections can be significantly increased. Maintenance of museum textiles calls for dust free display conditions with suitable lighting and proper storage. Routine task of collection care must be ensured in a textile museum for maintaining its original brightness. A separate area for display preparation and incoming artifacts inventory formalities must be marked in textile museums. Secure packing, safe transportation, and sterilized display of artifacts play an important role in preserving the pieces of inheritance till eternity. Rules for visitor's interaction with textile artifacts and stored collection will ensure the longevity of ancient textiles and preserve them for the generations to come in a better way.

References

1. Indrie, L., Oana, D., Ilieş, M., Ilieş, D.C., Lincu, A., Ilieş, A., Baias, Ş., Herman, G., Onet, A., Costea, M., Oana, I., Indoor air quality of museums and conservation of textiles art works. *Ind. Textila*, 70, 1, 88–93, 2019.
2. Abdel-Kareem, O.M.A., The long-term effect of selected conservation materials used in the treatment of museum artifacts on some properties of textiles. *Polym. Degrad. Stab.*, 87, 1, 121–130, 2005.
3. Glover, J.M., Conservation and storage: Textiles, in: *Manual of Curatorship: A Guide to Museum Practice*, pp. 302–339, 1992.
4. Kavkler, K., Gunde-Cimerman, N., Zalar, P., Demšar, A., Fungal contamination of textile objects preserved in Slovene museums and religious institutions. *Int. Biodeterior. Biodegrad.*, 97, 51–59, 2015.
5. Kamboj, A., Jose, S., Singh, A., Antimicrobial activity of natural dyes-A comprehensive review. *J. Nat. Fibers*, 1–15, 2021, doi.org/10.1080/15440478.2021.1875378.
6. Gu, J.D., Ford, T.E., Thorp, K., Mitchell, R., Microbial degradation of polymeric materials. Harvard Univ. Boston MA. 439–460, 1994.

7. Gu, J.D., Ford, T.E., Berke, N.S., Mitchell, R., Biodeterioration of concrete by the fungus fusarium. *Int. Biodeterior. Biodegrad.*, 41, 2, 101–109, 1998.
8. Heard, T.A., *Museum storage practices: Evidence in textile and costume collections*, Lowa State University, Iowa, 1992.
9. Gu, J.D., Microbiological deterioration and degradation of synthetic polymeric materials: Recent research advances. *Int. Biodeterior. Biodegrad.*, 52, 2, 69–91, 2003.
10. Fuzek, J.F., Absorption and desorption of water by some common fibers. *Ind. Eng. Chem. Prod. Res. Dev.*, 24, 1, 140–144, 1985.
11. Ponsonby, M., *Faded and threadbare historic textiles and their role in houses open to the public*, Routledge, London, 2016.
12. Summerfield, P.M., Guidelines for curating a costume and textiles collection. *Curator: Mus. J.*, 23, 4, 287–291, 1980.
13. Lobell, J.A., World's oldest dress, Archaeology Magazine, (7 December 1917), USA. Retrieved 5 November 2019.
14. Meier, A.A., 5,000-year-old linen dress is the world's oldest woven garment, Allison Meier on February 29, Hyperallergic, New York, 2016.
15. Jones, J., The enigma of the pleated dress: new insights from Early Dynastic Helwan reliefs. *J. Egypt. Archaeol.*, 100, 1, 209–231, 2014.
16. www.https://aiccm.org.au/wpcontent/uploads/2020/01/Handling_Storage_transport_Display.pdf. Canberra, Retrieved on 15 may 2021.
17. https://en.wikipedia.org/wiki/Tarkhan_dress. London, Retrieved on 15 may 2021.
18. Hicks, C. and Dietmar, R., Improving cleaner production through the application of environmental management tools in China. *J. Cleaner Prod.*, 15, 5, 395–408, 2007.
19. Medha, K., Rajna, S., Devi, L.J., Samant, L., Jose, S., A comprehensive review on moth repellent finishing of woolen textiles. *J. Cult. Heritage*, 49, 260–271, 2021.
20. Sandwith, H. and Stainton, S., *The National Trust manual of housekeeping*, Penguin in association with the National Trust, Hardmondsworth, 1984.
21. Lane, J.B., Oral history and industrial heritage museums. *J. Am. Hist.*, 80, 2, 607–618, 1993.
22. Pinniger, D. and Winsor, P., *Integrated pest management: practical, safe and cost-effective advice on the prevention and control of pests in museums*, Museums & Galleries Commission, 1998.
23. Skals, I., From grave to showcase: modern care for ancient textiles. *Stud. Conserv.*, 41, 162–165, 1996.
24. https://museumsandgalleries.leeds.gov.uk/temple-newsam/. Retrieved on 17 may 2021.
25. Mecklenburg, M.F., Some aspects of the mechanical behavior of fabric supported paintings, in: *The Mechanics of Art Materials and Its Future in Heritage Science*, Washington, 107, 1982.
26. Hinton, M., Professional notes labelling: museum text: Some north american examples of good practice. *Mus. Manage. Curatorship*, 18, 3, 309–321, 2000.
27. Jones, A.R. and Williams, G., Perceptions of fair trade labelling and certification: Three case studies. *J. Fash. Mark. Manage.*, 16, 2, 2012.
28. Brooks, M.M. and Eastop, D., Matter out of place: Paradigms for analyzing textile cleaning. *J. Am. Inst. Conserv.*, 45, 3, 171–181, 2006.
29. Hofenk-De Graaff, J.H., The constitution of detergents in connection with the cleaning of ancient textiles. *Stud. Conserv.*, 13, 3, 122–141, 1968.
30. Museums, C.U., Galleries Commission, *The science for conservators series, cleaning*, p. 2, Routledge, London, 2005.
31. Martin, A.R., Dry cleaning museum textiles. *Bull. Am. Group Int. Inst. Conserv. Hist. Artistic Works*, 7, 2, 9–13, 1967.
32. Rice, J.W., A drycleaning technique for textile conservation. *Stud. Conserv.*, 9, 3, 83–90, 1964.

33. Ahmed, H.E. and Ziddan, Y.E., A new approach for conservation treatment of a silk textile in Islamic Art Museum, Cairo. *J. Cult. Heritage*, 12, 4, 412–419, 2011.
34. Grömer, K., *The art of prehistoric textile making: The development of craft traditions and clothing in Central Europe*, p. 546, Naturhistorisches Museum Wien, Austria, 2016.
35. Robinson, J., Pardoe, T., Council, S.M., *An illustrated guide to the care of costume and textile collections*, Museums & Galleries Commission, 2000.
36. Dancause, R., Wagner, J., Vuori, J., *Caring for textiles and costumes*, Canadian Heritage, Patrimoinecanadien, Ottawa, 2018.
37. Palmer, A., Untouchable: Creating desire and knowledge in museum costume and textile exhibitions. *Fash. Theory*, 12, 1, 31–63, 2008.
38. Niinimaa, G.S., Mounting systems for ethnographic textiles and objects. *J. Am. Inst. Conserv.*, 26, 2, 75–84, 1987.
39. Landi, S. and Hall, R.M., The discovery and conservation of an ancient Egyptian linen tunic. *Stud. Conserv.*, 24, 141–152, 1979.
40. Hjemdahl, A.S., Exhibiting the body, dress, and time in museums: A historical perspective, in: *Fashion Mus.: Theory Pract.*, p. 108, 2014.
41. https://www.montel.com/en/applications/museum-storage-systems. Retrieved on 17 may 2021.
42. https://museummannequins.com/ Retrieved on 17 may 2021.
43. Petrov, J., *Fashion, history, museums: Inventing the display of dress*, Bloomsbury Academic, 2019.
44. https://www.pinterest.com/ARTisticStorage/museum-textile-storage/. Retrieved on 17 may 2021.
45. Morris, J.W., Travelling exhibitions: A team manager's view of the conservator's role. *Conser.*, 11, 1, 18–25, 1987.
46. Westerudd, O., Transportation and handling of art and other exhibits, in: *ICOM Committee for Conservation: 8th Triennial Meeting*, Sydney, vol. 2, pp. 611–615, 1987.
47. Piechota, D., Storage containerization: archaeological textile collections. *J. Am. Inst. Conserv.*, 18, 1, 10–18, 1978.
48. Shelley, M., *The care and handling of art objects: Practices in the Metropolitan Museum of Art*, Metropolitan Museum of Art, New York, 2019.
49. Couper, A., *The geography of sea transport*, Routledge, London, 2015.
50. Amin, E.A., Technical investigation and conservation of a tapestry textile from the Egyptian textile museum. *Cairo Sci. Cult.*, 4, 3, 35–46, 2018.
51. Fung, W. and Hardcastle, J.M., *Textiles in automotive engineering*, p. 13, Woodhead Publishing, London, 2000.
52. Stolow, N., *Procedures and conservation standards for museum collections in transit and on exhibition*, Unesco, 1981.
53. Vom Lehn, D., Heath, C., Hindmarsh, J., Exhibiting interaction: Conduct and collaboration in museums and galleries. *Symb. Interact.*, 24, 2, 189–216, 2001.
54. Landi, S., *The textile conservator's manual*, Routledge, London, 1998.

9

Roles and Responsibilities of Museum Professionals

Kanika Sachdeva

Conservation Scientist, London, UK

Abstract

Museums are becoming increasingly impactful vehicles of community education. As living become more layered and complex due to factors like economic patterns, migrations, and now pandemics; there are increased expectations from museums and other modes of community education. To be able to meet these expectations, museums have constantly upgraded there offer utilizing expertise from science, technology, and visual arts. In contemporary heritage management, there is an army of specializations that come together to manage a museum. Museum professionals in any museum today are a team of numerous superspecialized individuals that complement various aspects of museum management with their skills and knowledge. It is interesting to note here that there is not a single heritage management activity that can be completed by any one of these professionals in isolation. In fact, every single task performed by museums requires collaboration of multiple experts, fully or partially. This chapter presents an exhaustive analysis of various roles involved in running a museum. It also, reflects the interconnection between various roles and how each activity requires collaboration of different roles at some level.

***Keywords*:** Museum professionals, conservator, curator, conservation scientist, museum registrar, documentalist, collections manager, exhibit designer

9.1 Introduction

The quintessence of a museum is guided by the expertise and day-to-day decision making of its various professionals. The tasks and responsibilities of museum professionals evolved with evolution of the concept of museums, which dates back to ancient civilizations. The shift of museums from private collections to state owned institutes of heritage preservation and education has been gradual and so has been the role of professionals involved in its day-to-day management. One of the essential ethical obligations of the present-day museum professionals is to ensure that the collections are passed on to future generations in as good and safe a condition as possible. This is possible only with the collaborative efforts of all concerned, whether a director, curator, scientist, conservator, or restorer by ensuring adequate conservation technology for preventive and curative conservation, including the provision of suitable environmental protection against the known natural or artificial

Email: Dr.Kanikabhatiasachdeva@Gmail.Com

Seiko Jose, Sabu Thomas, Pintu Pandit, Ritu Pandey and Vandana Gupta (eds.) Handbook of Museum Textiles: Volume I Conservation and Cultural Research, (181–192) © 2023 Scrivener Publishing LLC

causes of deterioration of art objects and museum specimens [1, 2]. The role of conservators and curators has been central to any museum's existence, while curators play a vital role in acquiring, cataloging and exhibiting the artifacts, whereas a conservator works tirelessly behind the scenes in scientific analysis, repairs and restorations, slowing the rate of degeneration and maintenance and monitoring of storage and exhibition setups; these two are continuously guided and aided by other professionals to achieve the end purposes of a museum. Where on one hand it is imperative for all these professionals to be in constant communication and work in total coordination, their expertise lies in the realms as different as chalk and cheese. Similarly, there are numerous other museum roles like historians, information technologists, exhibition officers, and many more, who are integral part in building, exhibiting, and maintaining museum collections.

9.2 History of Museums Professionals Training in India

The oldest known publication concerning museum ethics dates from 1898 and concerns the field of entomology [3]. Available literature shows that research in conservation in India started in the early 1930s due to the scientific vision and foresight of Dr. F.H. Gravely, Superintendent of the Madras Government Museum (today Government Museum, Chennai) and Dr. S. Paramasivan, after his appointment as the chemist and the first curator of the laboratory on paintings and metallic antiquities at the same museum [4]. Conservation of cultural property had its beginnings in India in later part of nineteenth century when historic structures were repaired using civil engineering materials and techniques. The methodology underwent a change with the establishment of Archaeological Survey of India in 1861, placing greater emphasis on maintaining the historic authenticity of structures. Subsequently, a conservation policy was formulated by the Government of India and a conservation laboratory was established at Calcutta by the Archaeological Survey of India in the second decade of the 20th century. With the establishment of *International Centre for the Study of the Preservation & Restoration of Cultural Property* (ICCROM) in Rome [5], of which India was a member state, training opportunities in the field of conservation increased substantially. This led to the establishment *of National Research Laboratory for Conservation of Cultural Property* (NRLC). Thus, conservation training picked up momentum in the country in the latter half of the 20th century with the establishment of NRLC. Many other institutions, such as National Museum-New Delhi, The Government Museum, Chennai, The Indian Museum, and The Victoria Memorial Hall-Kolkatta, followed suit and started conducting different types of training courses from time to time [6].

9.3 Roles in a Textile Museum

A study of the professional hierarchy structure of various museums across the world, reveals interesting roles and job profiles. Where there is no standard guidance on the roles and hierarchy, these differ in different countries, cultures, and organizations, depending on the size and objectives of a particular organization. It is interesting to note here that these varied roles do not work as water-tight compartments, rather their responsibilities are overlapping. This ensures that each individual function/activity of a museum involves more than one of these professionals, adding layers to decision making and implementation. This process ensures that every decision has been adequately weighed from different point of views before being finally implemented.

While various roles performed in a museum can be categorized into the following:

1. Administrative
2. Conservation
3. Curation
4. Research & Education
5. Design and Exhibition
6. Public Outreach and Communication
7. Digitization
8. Security

These roles categorize constantly work in coordination with each other to perform the four primary functions of a museum, i.e.,

1. Accession/assessment,
2. Conservation,
3. Exhibition/education,
4. Storage.

The diagram in Figure 9.1 explains various tasks associated with each of these functions. We will make use of this graphic representation to understand the overlapping, layered structure of museum roles, and how they coordinate toward specific end purposes.

As mentioned earlier, museum professional roles differ in different organizations. A quick Internet survey of job vacancies offered by various museums across the world will provide you with a variety of roles. However, a detailed survey of the job description tells us

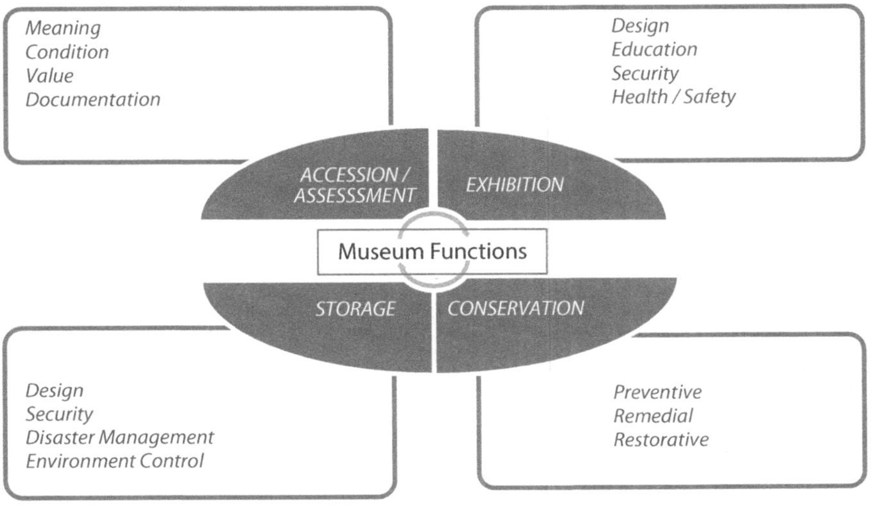

Figure 9.1 Museum functions.

that where museums with larger collections and high footfalls, look for super-specialized roles in specific disciplines, smaller museums often bundle the parallel roles with similar expertise for functional purposes. For the sake of clarity and uniformity, the chapter discussion will be based on the museum roles specified on *International Council of Museums-Committee for Conservation* (ICOM-CC) website [7].

The detailed function of each of these roles is being tabulated in the accompanying chart (Figure 9.2) where *partial* & ***full*** involvement has been marked.

	ACCESSION/ ASSESSMENT	EXHIBITION	STORAGE	CONSERVATION
CONSERVATOR	• **Condition** • *Meaning* • *Value*	• *Design*	• *Environment Control*	• **Preventive** • **Remedial** • **Restorative**
CONSERVATION SCIENTIST	• **Condition** • *Meaning* • *Value*	• *Design*	• *Environment Control*	• **Research** • **Preventive** • **Remedial** • **Restorative**
CURATOR	• **Meaning** • **Value** • *Condition* • *Documentation*	• *Design* • *Education*	XXX	• *Preventive* • *Remedial* • *Restorative*
COLLECTIONS MANAGER	• Meaning • Condition • Value • Documentation	• *Security*	• **Disaster Management** • Design • Security • Environment Control	XXX
REGISTRAR / DOCUMENTALIST	• **Meaning** • **Condition** • **Value** • **Documentation**	XXX	• *Disaster Management*	• *Remedial*
HISTORIAN	• **Meaning** • **Value** • **Documentation**	• *Design* • *Education*	XXX	• *Remedial*
EXHIBITION COORDINATOR	• Meaning	• **Design** • **Education** • **Security** • **Health / Safety**	XXX	XXX
PHOTOGRAPHER	• **Meaning** • **Condition** • **Documentation**	• **Design** • **Education**	XXX	• **Preventive** • **Remedial** • **Restorative**
MUSEUM EDUCATION OFFICER	• **Meaning** • **Value** • *Condition* • *Documentation*	• **Design** • **Education**	XXX	XXX
INFORMATION TECHNOLOGIST	• **Meaning** • **Value**	• **Design** • **Education**	XXX	XXX
HEALTH & SAFETY OFFICER	• **Condition**	• **Health / Safety**	• *Design*	• *Preventive* • *Remedial* • *Restorative*
SECURITY OFFICER	XXX	• **Security**	• **Security**	XXX

Figure 9.2 Museum roles and their involvement in various.

9.3.1 Conservator

As per the ICOM-CC website, a conservator is a person educated in conservation-restoration techniques and ethics, who is responsible for maintaining objects in as stable condition as possible, so that they remain accessible as long as possible and still represent the meaning attributed to the objects. According to American Institute of Conservation a Conservator is a professional whose primary occupation is the practice of conservation and who, through specialized education, knowledge, training, and experience, formulates and implements all the activities of conservation in accordance with an ethical code, such as the AIC Code of Ethics and Guidelines (American Institute for Conservation of Historic & Artistic Works) for Practice (Figure 9.2). It has been noted in various museum epilogs that while the profession originates from highly skilled craftsmanship and the arts, it was developed academically during the latter half of the 20th century to include humanistic and natural sciences, applied chemistry and physics, as well as analytical, organizational, and philosophical elements. Conservation specialists and technicians need to be highly skilled in scientific research [8]. It is important to note here that conservators are expected to treat, cure, and protect aged artifacts, which have different levels of susceptibility. In doing so, they take important decisions about chemical treatments, environmental and climate control, pest management, storing, and packaging along with the conservation scientists. It is a profession that requires the highest degree of precision and patience. Conservation procedures on textile artifacts can often be painstakingly long and tedious, moving by just about few inches after 6 to 8 hours of hard labor in a day. Along with preservation, conservators also look to reverse damages, prevent long-term deterioration, and retain the aesthetic appeal of certain items. All of these functions have direct long-term impact on the useful lifetime of an artifact. As per ICOM, 'Conservator-restorer work in museums, in official heritage protection services, in private conservation enterprises or independently. Their task is to comprehend the material aspect of objects of historic and artistic significance in order to prevent their decay and to enhance our understanding of them so as further the distinction between what is original and what is spurious.'

9.3.2 Conservation Scientist

A conservation scientist needs to be a person trained and educated in the natural sciences and engineering, who conducts research and advises conservators and other collection professionals on technical and scientific issues and problems related to the active and passive conservation of objects and collections. She/He is responsible for the identification and analysis of incoming museum artifacts, advise conservators on repairs, restoration procedures and support archivists with documentation information. All procedures related to decontamination, cleaning, and reconsolidation need to be finalized in consultation with conservation scientists (Figure 9.2).

In most small and medium-scale museums, often the role of conservator and conservation scientist is played by the same professional. However, as the scale of a museum increases, the nature of its collections becomes more complicated. Since larger museums might house artifacts that are rare and carry huge national or cultural significance, their associated conservation decisions might need detailed study and experimentation. This often demands the expertise and precision of a scientist trained in material sciences.

Thus, the role of conservation scientists, although hugely overlapping with conservators; retains a position of niche affordable only for larger organizations.

Most importantly, a conservation scientist carries out largely research based projects within a museum laboratory. Most textile artifacts are complex objects, each unique due to its specific combination of fibers, varied construction parameters, finishes, dyes, and most importantly, their life journey. Therefore, they have a potential to behave differently as compared to their similar looking counterparts. This implies that detailed research needs to be undertaken before the conservation pathway for each of these is finalized. Thus, a conservation scientist might not be directly involved in application of conservation procedures, however, plays a decisive role in charting the way an artifact should be treated.

9.3.3 Curator

Curators hold one of the highest positions among the museum hierarchy of employment. A curator could be a person educated, trained, and/or with experience in the appropriate cultural heritage field, who has responsibility for the content of a museum or collection, including selection and acquisition, research and planning of exhibitions, and the associated research (Figure 9.2). The specific responsibilities, however, could vary from museum to museum. At a small independent museum or gallery, a curator might be the museum manager as well. On the other hand, at a large national museum or gallery, a curator might be responsible for one specific area of the collection, for research in a specific field of knowledge and for the management of a small team of assistants and volunteers. In general, a curator has responsibility for a collection of artifacts or work of art and typically works with the artifacts right from the time of acquisition to final deaccession many-many years later. In between these two ends of the spectrum, the curator works with collections performing diverse functions like background research, records/catalog management, exhibitions and display, write publications, internal and external collaborations to name a few. It is customary in bigger museums to have specific curators for each collection's category. This is admissible as curators are expected to have in-depth knowledge about the material, cultural, and historical aspects of a certain category of artifacts. Any enquiries from public, researchers, and other stakeholders might need active engagement from the curators. They also need to actively liaise with voluntary groups, local communities, industries, local schools, management boards, governors, trustees, and local councils to manage the functioning of the museum, including planning, forecasting, funding, and budgeting. The role has evolved from being interpretational to one of maintaining balance between various museum functions so that final museum objectives are delivered.

Curators play most active role in organization and interpretation of exhibitions, and this role has hugely transformed in the past few decades. Where originally, the exhibitions were solely based on the curator's point of view, recent trends suggest a more liasionary role where curators actively communicate with relevant public and communities to curate exhibitions that provide audience an open-ended, unprejudiced view. Some contemporary researchers have compared the role of curators to that of activists, in respects that curators have huge influence over the direct and indirect messaging to public by the vehicle of exhibitions [9]. As discussed earlier, museums play a significant role in public education and can often be seen as reflecting thought-provoking messaging through exhibitions on subjects that are not openly explored in formal and informal education. The role of curators has thus evolved from being storytellers about the past to harbingers of change in modern society. It is important to note here that this aspect of curatorship is one of the most challenging of its roles across the spectrum. Where the curator's work becomes a vehicle of education about the past and the changes to be seen in future, the conveyance of this education can be extremely complicated. A curator has to constantly avoid falling for personal judgment while expressing the exhibitions in an objective manner. A modern-day museum exhibition is not expected to pass a judgment or lead a viewer's thought process in a specific direction. Whereas a contemporary museum curation has to be a balanced piece of research, thoughtfully expressed in sensitive manner that allows audience an open-ended environment to make their own inferences. Thus, curatorship is transitioning to a riskier model where a curator is expected to engage museum visitors in topics that could be historical, environmental, social, or even controversial in nature. The curator can no longer assert one truth and make effort to convince the audience of its importance. Instead of providing ready interpretations of exhibits, the curators now encourage audience to discover exhibitions from their own perspective, making sense from what they see, hear, touch, or experience during the exhibition.

Thus, a curator can be aptly called the captain of the museum ship where her/his work involves constantly engaging diverse audience with utmost sensitivity, while keeping relevant to the demands of cultural ethos and voices of change. Managing these complicated educational aspects, along with the pile loads of administrative work, makes curatorship one of the pivotal roles in the museum environment; the one which demands a well-informed and objective mind, dynamic personality and loads of energy to deal with everything simultaneously.

9.3.4 Collections Manager

This is essentially a person overseeing for collecting and managing the entire logistics of collection ownership, including registration, tracking and tracing, storage, and transport (Figure 9.2). The role is completely administrative and can be often confused with that of an archivist/registrar/documentalist. It is important to note here that a collection manager takes physical ownership of the collection and is directly responsible with its current storage, movement, and security. While she/he is also partially involved in the accession process, the role is largely administrative. At the time of accession or deaccession, a collections manager might be involved in determining the physical condition of the artifact and the storage logistics associated with it while an archivist would focus on all other aspects of accession/deaccession. As outlined for some other roles, in small- and medium-scale museums, the role might be played by the same person. However, the complexities of larger museums would prefer different people in these roles [10].

9.3.5 Registrar/Documentalist

Also known as archivist, this is one of the most prominent position in a museum. An archivist is responsible for maintaining the collection of artifacts, art, objects, documents, and historical or otherwise relevant items to the museum (Figure 9.2). He is the person responsible for identifying, describing, and cataloging objects in a collection; knowing where they are and keeping track of their movements (tracking and tracing). This role is essentially meant for people who are extremely detail-oriented as the job as archiving requires great detail and organizational skills. The role of an archivist differs from that of a collections manager in respects that an archivist plays a more interpretational role to the management of collections in the museum. The archivist might not play active role in the storage, upkeep, and security of the artifacts like a collections manager. However, he would take decisions about the cataloging & categorization of the artifacts, depending on their characteristics, condition, meaning, culture, etc. This would decide the way an artifact is identified, defined, and marked in museum storage and how it is used at the time of exhibitions. She/he is deeply involved in the interpretational tasks while managing the artifact. This means that remedial conservation process requires his inputs to ensure that the original history of the artifacts is not tempered with during restoration process. A job listing posted by ICOM in January 2020, also records digitization support as part of the expected responsibilities from an Archivist [11]. Therefore, an archivist, like a curator is expected to be a multitasker, holding expertise in various aspects of heritage management.

9.3.6 Historian

As per the description on the ICOM CC website, A historian is a person educated, trained, and/or with experience in the appropriate cultural heritage field, responsible for all the information related to the story behind a project, its development and fabrication, and for placing this information in context, context meaning, among others, time, location, use, meaning to people and society, political environment, etc. (Figure 9.2). The role of a historian in a museum starts right from the inception of the artifact. She/he would play an important role in identifying and verifying the importance and position of a proposed acquisition in the realms of history. Once this task is completed, the documentation, categorization of the artifact, and its curation for various exhibitions would be largely done in consultation with the historians. They would also aid in the process of restoration to a certain extent, where they would inform conservators restorers about the characteristics of the artifact that reflect its historic importance and thus should not be fiddled with. The historian needs to accurately validate the story a curator wants to tell through artifacts so that museum education is true and unbiased [12]. It is important here to understand that this might or might not be a full-time role in a museum job. In most museums, a registrar/documentalist would be expected to be adequately trained in history, to be able to perform these functions. However, museums might need to consult historians specialized in certain time spans and geographies to be able to curate accurate stories for their exhibitions.

9.3.7 Exhibition Coordinator/Designer

These are the people responsible for logistics of presenting objects, collections or parts of collections in a given venue. Exhibit designers are the creative minds behind an exhibition [13]. They give expression to the museum's most treasured items by creating meaningful installations for exhibitions. Exhibit designers work closely with curators to create setups that meet the exhibition objectives by displaying objects or collections in a way that makes them accessible and engaging to the general public (Figure 9.2). An exhibit designer needs to be well versed with the discipline of material design. They work with a range of exhibition materials like paper, cardboard, etc. to recreate the original culture and timeline of the exhibit. An exhibit designer's creativity can go a long way in transforming exhibition space into a time capsule, thus enriching visitor experience. While curator decides and conceptualizes the theme and mood of the exhibition, it is the exhibition designer who enables this concept into physical reality. The success of any exhibition hugely depends on the creativity and versatility of the exhibit designer.

9.3.8 Museum Education Officer

This is one of the most visible roles in museum environment. Museum education officers can be safely called the public interface of the museum environment. From the moment one steps inside a museum and approaches the reception for advance information, during visits across various galleries, right till the feedback collection and future event information at the exit door; a museum education officer and his team, tirelessly works toward communicating the museum content to audience (Figure 9.2). It is their task to engage visitors in a meaningful and interesting manner while communicating all aspects of exhibitions on display. A museum education officer is the person responsible for explaining information about objects and collections to increase public enjoyment and appreciation. Primary task of museum educators is to help visitors understand the history, and backgrounds of exhibits [14]. One can find museum educators providing guided tours, sitting at information booths, or publishing information on the museum's website to expand online educational resources. Museum educators are also expected to assist curators in conducting research for the museum. They would also be actively involved in writing reference material for exhibits or galleries. The profession is best suited for people who have a passion for history, culture, and learning.

9.3.9 Photographer

Person responsible for proper visual documentation of objects and documents using photo reproduction techniques such as film, video, photography (digital imaging) (Figure 9.2). In large museums, photographer can expect a fulltime engagement. Unlike general perception, the work of a photographer does not limit to documentation of the artifacts. The role of the photographer can be defined as assisting in maintaining a picture journal of each artifact. This file/journal could be stored online or offline in various formats. However, it captures each and every stage of artifact right from inception to restoration, sometimes storage and most importantly the way it has been displayed in various exhibitions. Museum photography is a specialized area as museum artifacts cannot be photographed under the same protocols as ordinary objects. Museum artifacts count light as their biggest enemy. A simple camera flash can prove lethal for an aged artifact. At the same time, a healthy artifact might not show immediate signs of deterioration after photography attempts; however, a series of camera flashes might silently trigger or accelerate the process of deterioration, which might shorten the life of artifacts in the long term. Therefore, a museum photographer needs to be aware of all the aspects related to vulnerability of aged museum textiles and choose correct lighting and lenses to be able to work with these limitations. The role has become more important than before as a lot of public communication is now being done via websites a digital medium. Also, with growing trend of virtual exhibitions, the dependency of a museum on its photographers is more than before. The role has effectively evolved from being just documentary, behind the scenes to being a decisive force in exhibitions and public communication.

9.3.10 Information Technologist

Person educated and trained in software technology, including social media and Internet, responsible for all aspects of gathering and communicating digital information about objects, collections, and museums (Figure 9.2). An information technologist is one of the most important roles in modern museum. The museums are increasingly engaging existing and potential visitors via digital mediums like social media channels, online streaming of galleries, workshops discussions, etc. It is imperative for a vibrant public organization to have a fully functional website, social media handles and a strategic outflow of public engagement events that keep visitors hooked on to the museum. The tasks of an information technologist in museums would involve designing and maintaining the museum website, actively engage with audience on the museum's social media channels, design a creative stream of online events that effectively communicate events like new exhibitions, addition of important artifacts and other events that would attract footfall for the museum. It is also the task of information technologists to gather responses toward queries posted online and help visitors maximize benefit from their museum visits by providing required information. With advent of virtual museums and digital exhibitions, role of information technologists in museums has widened than ever before [15]. In effect museums these days might require a whole team of Information technology experts to be able to communicate effectively via various digital mediums like websites, social media channels to name the few.

9.3.11 Health and Safety Officer

Health and safety officer is one of the most important museum professionals working tirelessly behind the scene to ensure safety of all the people who keep the museum running (Figure 9.2). He is the person responsible for ensuring that museum professionals have proper conditions for their work, and that visitors have the proper conditions for moving about in collections and exhibitions. This includes adequate protection from hazardous chemicals, proper ventilation and lighting, and protection from physical danger. It is important here to note that museums can often be pockets of increased bio-chemical activity because of the nature of materials it stores and exhibits. Since most artifacts stored in museums are aged, there is a greater possibility of harmful emissions, as a result of constant deterioration. Most of these emissions might go unnoticed if proper monitoring and analyzing systems are not in place. These fumes and gases could be toxic and impact both workers and museum visitors. Although the risk to museum visitors is much higher as they remain in vicinity of these through the day and all working days of the week. Similar amount of risk can be posed by the restoration & conservation processes which sometime require use of strong chemicals and reagents for specific end purposes. Restoration-conservation procedures can a lot of time include uncommon practices like use of certain toxic gases for neutralizing pests to working at extremely low temperatures for stabilizing artifacts. It is important to have adequate policies in place to ensure that the artifacts are not being stabilized at the cost of health and safety of the museum workers, as well as visitors. Therefore, a museum officer has to conduct routine audits and checks of every part of the museum, i.e., exhibition, storage, restoration-conservation laboratories, etc. to ensure that they are safe places for anyone who walks in. To be able to deliver this, museum health and safety officers need to be well versed with the nature of chemicals, reagents, and gases; their identification etc. apart from a thorough knowledge of the equipment required to monitor these circumstances.

9.3.12 Security Officer

Person responsible for measures to prevent damage or loss of objects due to events other than natural ageing and use, such as fire, flood, theft, or vandalism (Figure 9.2). The role of a museum security officer looks routine like manning and safeguarding any other establishment like a corporate office or a public garden. However, it is more likely to be compared with that of securing a huge bank as museums are storehouses of priceless artifacts that carry unparalleled value. The role of security officer is both preventive an anticipatory. A museum security officer has to think through all potential security threats to a museum, its workers, visitors and of course the artifacts and then take appropriate measures to mitigate these potential risks. The task becomes more complicated because of the susceptibility of aged museum artifacts as lot of commercially available disaster management processes might not be appropriate for museum artifacts. Thus, ensuring security of a museum would involve detailed planning and assessment at all levels.

9.4 Conclusion

It is interesting to note here that to be able to fulfill its objective of public education efficiently, a museum needs seamless coordination between people from different areas of expertise. A museum as a workplace is a perfect amalgamation of various disciplines and roles that often overlap each-other's scope of work. In a way, it reflects the multidimensional nature of life in general, where an artist cannot perform without the support of a chemist while being braced by a historian, under the detailed supervision of an archivist. Not to mention the relentless work of the photographers, security officers and information technologists, and many more who complete the cycle of communication and education under the umbrella of a creative and safe learning environment.

References

1. Bisht, A.S., Ethics in conservation-problems and prospects, in: *Conservation of Cultural Property in India*, vol. XXIII, pp. 17–22, 1990.
2. Bisht, A.S., Documentation in a conservation laboratory through computers-problems and prospects, in: *Conservation of Cultural Property in India*, vol. XXIII, pp. 63–66, 1990.

3. Mensch, P., *Towards a methodology of museology*, University of Zagreb, Croatia, 1992.
4. Nair, M.V., *Status of conservation research in India', heritage conservation and research in india-60 years of indo-austrian collaboration*, pp. 121–127, National Museum, New Delhi, 439–454, 2009.
5. https://www.iccrom.org/
6. Jain, K.K. and Dalela, N., Academic Conservation Training Programs in India and International Networking, in: *Heritage Conservation and Research in India-60 years of Indo-Austrian Collaboration*, pp. 139–146, National Museum, New Delhi, 2009.
7. http://www.icom-cc.org/330/about-icom-cc/what-is-conservation/conservation:-who,-what-amp;-why/#.YQKIR45KhPZ
8. Hutchings, J. and Corr, S., A framework for access to the conservation–restoration profession via the mapping of its specialist competencies. *High. Educ.*, 63, 4, 439–454, 2011.
9. Cachia, A., 'Disabling' the museum: Curator as infrastructural activist. *J. Vis. Art Pract.*, 12, 3, 257–289, 2013.
10. https://www.britishmuseum.org/our-work/departments/collection-care/collection-management.
11. https://icom.museum/wp-content/uploads/2019/09/Job_Offer_Archivist-EN_110919.pdf.
12. https://www.historians.org/jobs-and-professional-development/career-resources/careers-for-students-of-history/historians-in-museums.
13. https://www.theartnewspaper.com/feature/exhibition-design-the-art-of-seeking-attention.
14. Cooper, Y., A day in the life: The qualifications and responsibilities of an art museum educator, in: *From Periphery to Center: Art Museum Education in the 21st Century*, P. Villeneuve (Ed.), pp. 68–73, National Art Education Association, Reston, VA, 2008.
15. https://www.museumnext.com/article/how-technology-is-bringing-museums-back-to-life/

10

Ancient Weaving and Dyeing Techniques

Hannah Dewey[1†], Meghan Lord[1†], Seonyoung Youn[2], Januka Budhathoki-Uprety[1*] and Kavita Mathur[2]

[1]*Department of Textile Engineering, Chemistry and Science, Wilson College of Textiles, North Carolina State University, Raleigh, NC, United States*
[2]*Department of Textile and Apparel, Technology and Management, North Carolina State University, Raleigh, NC, United States*

Abstract
Dyeing and weaving techniques have existed since ancient times with some of the same materials and techniques still used today. The main objective of weaving is to interlace warp and weft threads at right angles to produce a fabric. However, the technique and meaning behind weaves produced in various regions may differ. Various techniques included plain weaves of warp and weft-faced patterns, ikat, brocade, gauze, and leno fabrics woven on warp-weighted, back-strap, and drawlooms. By studying weaving techniques based on the chronological period from the Neolithic Age through the Iron Age and utilizing geographic location, ancestral, societal, and religious information can be inferred. Thereby, giving life to the story ancient civilizations tell through their clothing. Ancient textile dyes were natural dyes as opposed to most of the synthetically produced dyes used in modern times. Dyeing techniques in ancient times followed what we now refer to as a vat process and a mordant process. In this chapter, various weaving and dyeing techniques will be identified and discussed based on their region, time period, and social importance. Examples of looms, fabric types, dye classes, and dyeing techniques are provided.

***Keywords*:** Natural dyes, weaving, pigments, dyeing techniques, loom, ikat, brocade, gauze weaving

10.1 Introduction to Weaving

Weaving can be defined as interlacing two sets of threads (warp and weft) at right angles to form a fabric. Technically, the warp is the longitudinal thread of a textile, "a pre-arranged and more or less fixed set," while the weft is the "transverse thread of the textile as a second set" [33]. Archeologists believe that the oldest fragment of fabric was originated at the Turkish site of Çatalhöyük. Based on the radiocarbon analysis of neighboring artifacts, this fabric is dated at the beginning of the 6th millennium B.C. in the Neolithic period [33].

**Corresponding author*: jbudhat@ncsu.edu
†These authors contributed equally.

Seiko Jose, Sabu Thomas, Pintu Pandit, Ritu Pandey and Vandana Gupta (eds.) Handbook of Museum Textiles: Volume I Conservation and Cultural Research, (193–208)

In Egypt, there has been evidence of weaving through the discovery of tomb of Chemhòtpe at Beni Hasan where horizontal and vertical looms were reported dating back to the XII dynasty (1976–1784 B.C.) [2]. In contrast, textiles from Greece and Italy were woven via warp-weighted looms [1, 3, 4]. While, weavers in ancient China utilized backstrap looms to weave a pattern by the use of the supplementary threads on warp or weft to produce ikat [5, 6], brocades [7], warp-patterned, and weft-patterned fabrics [8, 9]. Ikat is the woven fabric widely used to apply resist-dyed yarn either on warp or weft; if the dyeing is only applied on the warp, then it is called the warp ikat, whereas if on weft alone, it is referred to as weft ikat. If both the warp and weft are dyed, then it is called double ikat [34]. While weaving originated in the eastern hemisphere, there is evidence of weaving in the Americas dating back to 1000 BC [10]. For example, backstrap, frame, and continuous warp looms were used for a warp-faced fabric, the side of a textile on which the warp dominated or gauze, and a weave in which binding is achieved by twisting the warp respectively [1, 8, 10, 11].

Although many different types of looms were used in ancient civilizations, the fabrics produced were of a similar technique. Each loom aided in the interlacing of warp and weft threads to create different patterns and designs on the fabric. These designs would signify social status, ancestry, and culture [4, 6, 7, 9, 10]. Therefore, studying weaving techniques in ancient civilizations may reveal the types of technologies available and ultimately lead to a better understanding of ancient culture. This section will discuss and identify various weaving techniques in Europe, Asia, and the Americas as well as give detailed descriptions of the types of looms used and how they functioned.

10.2 Ancient Weaving by Geographical Region

10.2.1 In the Middle East and Central Eurasia

It is believed that weaving originated in Mesopotamia and eventually migrated throughout Europe and Asia. Evidence of textiles in Jordan and central Eurasia have revealed that warp-weighted looms were used to produce plain weaves of warp/weft-faced patterns [12, 13]. The warp-weighted loom, a simple ancient type of loom, uses a system of holding the warp threads "hanging freely from a bar supported by two uprights placed on a slight slant," as seen in Figure 10.1 [34]. Numerous weights made of stone were used to hold the warp threads parallel under tension. During the Bronze Age (3000 BC – 1200 BC), looms were most likely small, lightweight, and portable for migration purposes [13]. At least 68 small loom weights were discovered from Begash, a seasonal camp in Kazakhastan no later than 2450 BC [13]. Given the archeologists' assumptions about the military function of Mudaybi fort, the loom weight seemed to be preserved and survived by being enclosed with sturdy basalt [12]. In this fort, the loom weights were hand-molded from local clay ranging from 32 to 61 mm in height, 48 to 68 mm in width, and weighed 70 to 437 g [12]. The loom weights would be used to provide tension in the warp threads in a vertical warp loom to allow the weaver to integrate warp and weft yarns that hung from the ceiling (Figure 10.1) [12]. Textiles produced from a vertical warp loom could be of any length only limited by how much could be rolled at the top of the loom [12]. However, textiles could also be produced via tension-loom, like the warp-weighted looms or hand twining techniques. Twining is weaving with "two or more warp ends twisted around each other" as they interlace with the

Figure 10.1 A warp-weighted loom (top image) [35] and a loom weight (bottom image) dating to 1295–1070 B.C. molded with clay/mud by hand [29]. Reproduced under open access policy.

weft [34]. The vertical warp loom was not only used in the Middle East and Central Eurasia but also used in Egypt [2, 15, 36], and Greece [1, 3].

10.2.2 In Egypt

Little evidence was preserved in the prehistoric era, especially in the Neolithic period, because most textiles were made of vegetable fibers, such as flax and linen, which decomposed easily [33, 36]. If it had not been for the expert's effort, it would not have been possible to discern Egypt's ancient replace sources with artifacts weaving history. Egypt is unique in the wealth of different types of historical artifacts such as different types of looms offered [36]. For example, the tombs of Thot-nefer at Thebes XVIII Dynasty (1425 BC) [37] and Meket-Re had murals of weavers using horizontal and vertical looms dating back to 2000 BC [1]. According to Anderson, Egyptian weaving comprises three stages, as shown in Figure 10.2 [15]. The first stage uses horizontal looms where two transverse bars are pegged to the ground with the warp extending between these two bars [15]. Weft yarns are then inserted by passing over and under alternate sections while the weaver sits upon the cloth as it is being woven (Figure 10.2a) [15]. The second stage is the use of the vertical warp loom, which was depicted in Beni Hasan's tomb during the XII Dynasty (1976 - 1784 BC) as shown in Figure 10.2b [2, 15]. Beams used in the horizontal frame were raised to the vertical position where warp threads were suspended from the fixed beam at the top of the loom [2, 15]. Once the warp is created, a weaver threads a shuttle containing yarn through the warp to create the weft [2]. However, before the weft is inserted, half of the warp is lifted by a shaft tool. The shaft can raise and lower the warp ends to form a shed so that the shuttle can be inserted [2, 15, 34]. Once the weft yarn is inserted, it is beaten into position by a reed,

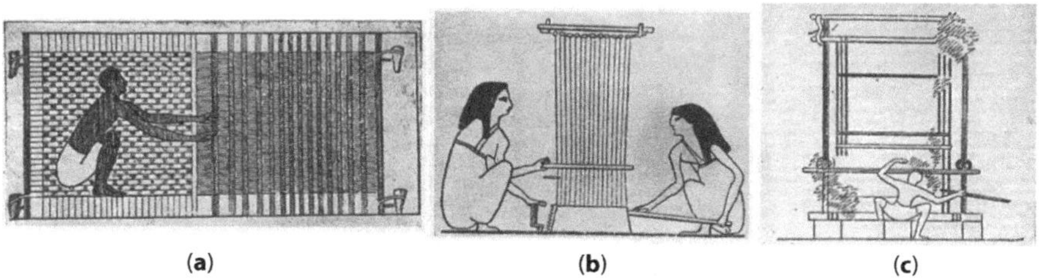

Figure 10.2 Three stages of weaving in Egypt: (a) depicts the first stage of weaving via a horizontal loom; (b) depicts stage two using a vertical loom; (c) depicts stage three where the vertical loom is only operated by one weaver [15].

and the weave is rolled [2]. Weaving with the vertical-warp loom led to low production times due to the amount of time required for lifting the warp yarns for weft insertion [2].

Lastly, the third stage allowed one weaver to operate all three steps of the previously described second stage. This stage is represented in Figure 10.2c by the weaver opening the shaft and receiving the shuttle from the right-hand side while the horizontal bars act as heddle shafts in conjunction with the top bar to control the warp [15]. This stage marked the development of heddles which were metal wires linked to rings to aid in alternately lifting the warp yarns by threading them through the rings, and the frame alternately lifts the warp up and down with foot pedals [2]. These looms produced plain weaves (also called tabby weaves, taffeta, and linen weaves) [14, 15, 44]. A plain weave is the "simplest possible form" of warp-weft weaving [44]. Many different terms are used for a plain weave to identify a specific fiber or weave structure itself. For example, the term linen weave reflects the commonness of plain weave, which is made out of linen at some particular period [44]. In other instances, plain weave structures also have been used to characterize the weave itself, such as "tabby, canvas-, gingham-, and taffeta-weave," which are examples of this kind of transfer of terms [44].

10.2.3 In Greece, Italy, and Romania

In Greece and Italy, plain weave fabrics were produced by maids who wove fabrics in gold, silver, or bright colors [3, 15]. Looms were of the vertical warp-weighted variety where open tabby, twill, and tapestry weaves could be produced [1, 3, 4]. Paintings on Etruscan vases depict Grecian looms as having a frame with a roller at the top, seen in other Greek vases of the period. Sometimes, a border of the roller shows a design of "winged creatures" in other Greek vases [3]. The frame consisted of two poles set at a distance a little farther than the desired width of the fabric and was connected by three horizontal supports: top castle (held loose yarn, spools, and bobbins via upward projecting pegs), warp beam, and a brace at the bottom [1]. Loom weights attached to each warp thread or group of threads would create the tension needed for inserting the weft, and the warp was kept from becoming entangled by two rods (a reed) [3, 4]. In the absence of heddles, the warp was manipulated by hand to make the shed for weft insertion, which was then beaten from below to create a tight weave [1, 3]. Grecian looms containing heddles would have the warp passing over and under the heddle sword and rod, respectively [1]. Women were often depicted as operating the Grecian warp-weighted loom shown in Figure 10.3 [3]. In the 1st millennium BC in Greece, there are descriptions about a girl using the heddle and the heddle bar in a passage of the Iliad (Hom. Il.

Figure 10.3 Vase painting dated back to 550-530 B.C. illustrating Greek women weaving on a warp-weighted loom [30]. Reproduced under open access policy.

23) [38]. The heddle rooms were originally invented as a simple shedding device for narrow bands, straps, or trim. The heddle sword (or shuttle) and rod were operated via pulling cords or pegs attaching these devices to the top castle to form the shed, as shown in Figure 10.4 [1, 33]. The action of beating the weft from below was unlike Egyptian weaving where beating occurred above the warp and pushed downward with a comb [1, 3]. Beating the weft upward was commonly used by weavers. The weft was loosely inserted to cover the front and back of all warp threads to create elaborate patterns in ornamental and tapestry textiles [1, 3].

In a small settlement in Luciana, Italy, loom weights of pyramidal and rectangular shapes were discovered dating back to the archaic period [4]. Given the weight and shape of the loom weights found in these buildings, they would have produced weaves ranging from a very open tabby of three to six threads per centimeter to fine twill weaves, resulting in a thread count of 13 to 31 threads per centimeter [4]. These clothes would have been made for ceremonial or gift purposes and worn as a symbol of wealth and status for people in the community [4].

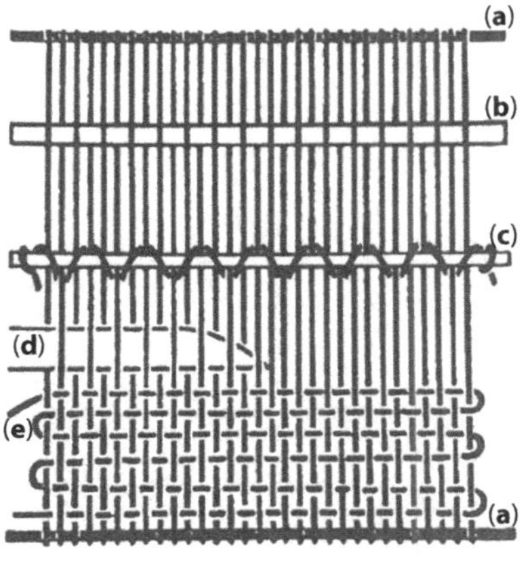

Figure 10.4 Diagram of the simplest loom: (a) device for fixing the warp, (b) shed rod, (c) heddle rod with heddles, (d) sword, (e) weft stick for insertion with the weft thread wound around it or a shuttle [33].

On the other hand, the local soil conditions in Transylvania, a plateau in northwest central Romania, preserved a larger unit of textile imprints during the Neolithic and Copper Age [41]. In Romania, textile imprints were significantly important for different fabrication techniques of tabby weaves and the "evolutionary shift" in textile production [41]. According to Mazare, several technical improvements were implied from a much more flattened and denser tabby appearance (about 12 threads per cm) during the Copper Age than those of the Neolithic period. Such morphological differences and adjustments would be possible because of differentiated loom-weights in the Romanian sites [41]. The developed technique of twinning and tabby was also used in later times, as discovered in Ukraine in the Bronze Age [42].

10.2.4 In India

Indian weaving techniques consisted of creating bright and colorful patterns via backstrap looms and handloom weaving [1, 16]. The backstrap loom was popular in India, Tibet, China, Korea, Japan, Peru, and Mexico [1]. Backstrap looms contained six sticks held together by the warp: front cloth bar, string heddle rod, shed roll, two release sticks, and a back-warp bar [1]. A strap wrapping around the back of the weaver was used for support during the weaving process [1]. As weaving progressed, the weaver would move closer to the back bar, and the cloth was rolled up onto the front cloth bar [1]. The shed was created by lifting the string heddle rod to allow for weft insertion; for pattern weaving, pattern string heddle bars were placed in front of the regular string heddle rods [1]. It has been reported that the oldest looms with heddles were found in India, where heddles were moved by strings attached to the weaver's feet [2]. Handloom weaving in India took priority in many households, evidenced by their homes, which were constructed to allow light to shine on the loom [16]. The weaver would sit in a pit dug into the ground in their living room with a wooden frame in front of them which held several carved sticks that functioned as the reed

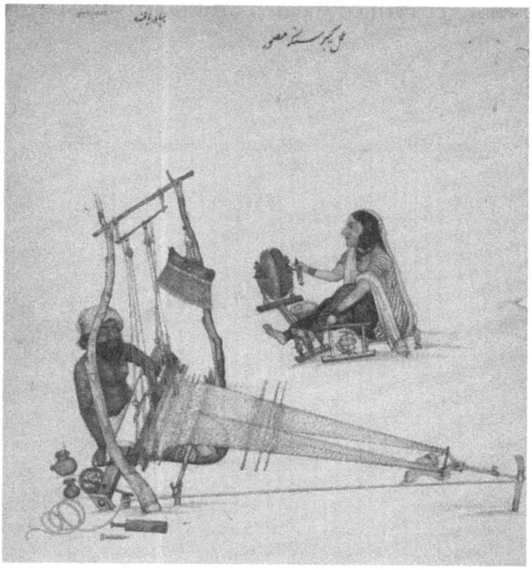

Figure 10.5 Painting of handloom weaving and spinning [16, 31]. Permission to reproduce under Creative Commons.

and heddles [16]. While the warp was stretched and attached to a horizontal stick by tying or hanging, a shuttle would be passed manually by the weaver for weft insertion, and the warp was manipulated by hand to weave complex designs [16]. Typically, the designs produced were created during the weaving process, which required dyeing the yarns beforehand to achieve the desired pattern [1, 16]. An illustration of this weaving technique can be seen in Figure 10.5.

10.2.5 In Southeast Asia and China

In Southeast Asia and China, several weaving techniques, such as ikat, brocade, warp-faced (patterned warp threads are prominent in finished fabric), and weft-faced patterns (weft threads are prominent) were produced on backstrap (body-tensioned) and draw looms [1, 5–9, 14, 17]. Ikat weaving, meaning to bind, tie, or wind around, was produced mainly on body-tensioned looms where the warp or weft (depending on a warp or weft patterning) threads were decorated with dyed designs before weaving [5, 6]. Repetitive geometric and floral designs were common in weft ikat fabrics, while simple shapes, such as dashes, hooks, and curls were used to make shapes in warp ikat fabrics [5, 6]. Other motifs, such as crocodiles/reptiles and ancestral symbols, were found on ikat fabrics across Southeast Asia, indicating the migration patterns and passing down of traditions in Austronesian civilizations [6]. The ikat weaving technique is as follows: i) binding bundles of warp/weft threads with a resist material such as elastic string for the resist pattern, ii) dyeing the warp/weft threads, iii) removing the resist material, and iv) mounting the patterned warp onto the loom for warp ikat or winding the weft onto bobbins to be inserted through a plain colored warp for weft ikat fabrics [5, 6]. The warp is stretched between two parallel wooden beams that are approximately the width of the finished cloth which are called the breast beam (closest to weaver) and the warp beam (furthest from weaver). Tension was provided by a strap spanning from the breast beam and passing behind the weavers back [6]. Methods of bracing the warp beam varied from bracing directly by the weaver's feet or tied to a tree/stake depending on the region in Southeast Asia [6]. Fabrics produced were used to make tube skirts of asymmetrical shape, as seen in Figure 10.6.

Brocade weaving can be dated back to the Zhou Dynasty (1046–56 BC) in China. Brocade fabrics were colorful, had high warp density, and were considered a symbol of wealth and prosperity in the community [7]. Gold or silver threads were often used during the production of brocades as well as colored warp/weft to create floral and raised surface patterns via jacquard or draw loom [7]. Furthermore, the unique Huose technique was employed by weavers where weft shuttles were used to create fabric surfaces that changed color from different viewing angles [7]. With silk as the main fiber, four different types of brocade can be identified: double, silk, box, and tiny brocade [7]. Further details on each type of brocade can be found in Table 10.1.

Prior to and during the Han Dynasty (206 BC–220 AD), warp-patterned fabrics such as silk fabrics, plain tabby weaves, twill weaves with repeating motifs, and gauze weaves were predominant in China [9, 14, 17]. Gauze weaves can be described as open lacelike weaves where pairs or groups of warp cross over and under each other (one yarn is below and the other is above the weft) [1, 11]. Weft-faced textiles were used as belts or decorative trim for robes [17]. Warp-patterning can be defined as colored warp yarns being brought to prominence on the fabric surface through selectively raising warp yarns for weft insertion.

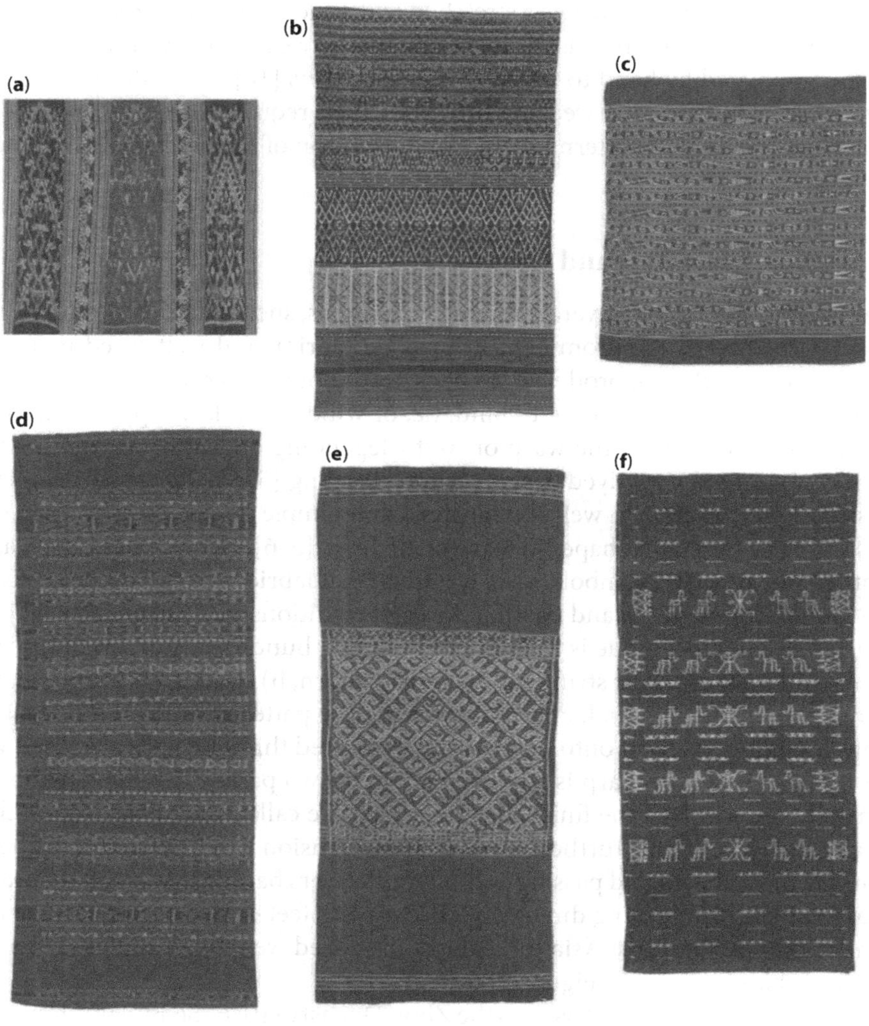

Figure 10.6 Various tubeskirts (a–f) created via ikat weaving technique throughout Southeast Asia [6]. Reproduced under Creative Commons policy.

In contrast, weft-patterning uses colored weft yarns that are brought to the fabric surface in the same manner as weft-patterning [17]. As weft-patterned textiles became more common, the use of the drawloom was adopted to aid in the production of weft-patterned fabrics with complex designs such as linked-pearl pattern brocades [9]. As shown in Figure 10.7, a drawloom can be considered a loom with a flexible patterning device that allows warp yarns to be lifted independently in any combination [17]. It provided the ability to weave valuable fabrics, such as double and silk brocades, as shown in Figure 10.8.

10.2.6 In The Americas

In Peru and Central America, weaving was depicted in a ceramic figurine as women using two-bar backstrap looms and can be dated back to 1000 BC [1, 10]. The main weaves

Table 10.1 Descriptions of four brocade styles [7].

Type	Rank	Description
Double	Most valuable	Two groups of warp yarns used: plied silk yarns with sericin and dyed silk as ground warp. Single stranded silk yarns or plied light colored yarns as face warp. Refined and dyed silk or golden silk used as weft.
Silk	Valuable/Most common	Close to double brocade; short-throwing shuttle used for weaving main floral motifs and long-throwing shuttles used for weaving geometric motifs, flower branches, leaves, stems, etc.
Box	Middle-grade	Plied yarns of degummed and dyed silk used as warp while dyed silk plied yarns as ground weft. Dyed cotton or dyed silk plied yarns used as pattern making weft.
Tiny	Low-grade	Plied yarns of dyed silk used as warp and plied raw silk used as weft. Satin and rearranged twill weaves are common structures.

Figure 10.7 Draw loom (lesser draw loom), China National Silk Museum [39].

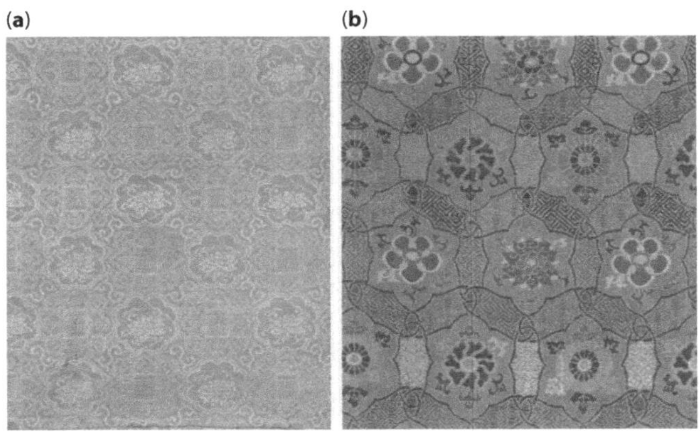

Figure 10.8 An image of Song brocade with lotus motifs (a) and brocade with multiple contrasting colors (b) using a drawloom. Ming Dynasty (1368–1634 AD). Source: The Palace Museum, Beijing 100000, China [7].

produced included gauze and twining techniques [1, 10, 11]. This was confirmed through the discovery of weaving tools such as ceramic spindle whorls (spinning thread), bone needles (for sewing garments, adding fringe to selvages, and embroidering cloth), and bone picks (for brocading) in elite households [10]. Twinning can be done in the warp or weft threads of the fabric. Warp twining consists of warp yarns held intact by courses of straight, regularly woven weft yarns, and pairs of twisted yarns to form columns in the warp of the fabric [1]. Whereas weft twining has pairs of weft yarns that pass over and under the warp forming horizontal twisted weft columns across the warp [1]. These weft yarns can be twisted, looped into a pile, braided, crocheted, or ornamented with beads, seeds, or sequins [1]. The ancient Peruvians, Amazon Indians, and North American tribes used this weaving method to create different patterns via different colors or variations of yarn and varying twist directions [1].

In North and South America, free warp, vertical-warp weighted, and backstrap looms were used to produce the weaves previously described [1, 8, 10, 11]. The free warp loom did not require heddles or a frame but consisted of heading cords stretched between two trees or supports. Warp yarns were then attached to the heading cords by slip knots, and the weft was inserted via twining [1]. The Ojibway Native Americans frequently used this loom to make handbags in North America [1]. Unlike the Grecian looms, the vertical-warp weighted looms used by North and South Native Americans were the same as the free warp looms (without heddles and weft insertion by twining), but with weights at the ends of the warp to provide tension [1]. These two-stick variety looms could be used for continuous warps where rugs and serapes were reported to be made in Peru and Ecuador [1]. Furthermore, ancient Peruvians produced double cloth weaves on these looms to make heavy and durable fabrics with two sets of warps [1]. Lastly, backstrap looms were used to create labor-intensive gauze weaves for elite Mayan women and Aztec commoners for tribute cloth [10]. A detailed description of the backstrap loom can be referred to in the discussions of Chinese and Indian weaving techniques. This type of loom could use as many as 20 different heddles to create different gauze structures and plain weave combinations in Peru and Central America, as seen in Figure 10.9 [11].

Figure 10.9 Peruvian tunic with double headed serpents ca. 400-200 B.C. made by gauze weaving. Double headed serpents may have had cultural significance [32]. Reproduced under Creative Commons policy.

10.3 Conclusion on Weaving Techniques

Throughout the discussion of weaving techniques in varying regions, the foundation of each technique has a common thread: interlacing warp and weft threads at right angles. However, the type of weave, patterns created, and what the fabric represents varies by geographical region. Furthermore, the patterns, motifs, and fiber types represented in the weaves signified ancestry, social status, and cultural practices.

By comparing the loom types of each region, one can distinguish technological advancements of weaving over time, exemplified by the transition from manipulating warp yarns by hand to the use of heddles to create the shed for weft insertion. This innovation led to the production of more complex designs in fabrics and faster output. For example, the draw-loom used by weavers in China allowed for the lifting of warp threads in any combination, which resulted in highly valuable brocades made of silk. Whether these looms were of the two-stick variety with hand manipulation or more advanced, such as the backstrap/draw-loom, the fabrics woven on these devices illustrate the stories of ancient civilizations and provide insight into ancient practices.

10.4 Introduction to Dyes and Dyeing Technologies

Dyes and pigments are molecules that have a chromophore, a chemical structure that produces color, and are used to impart color to another object. The major difference between a dye and a pigment is that dyes are soluble or can be made soluble, while pigments are always insoluble. Dyes and pigments can be classified based on their origin - naturally occurring or synthetic, chemical structures (e.g., chromophore groups) and uses to dye) [40].

Naturally occurring dyes and pigments have been used since ancient times to impart color to textiles. Although less commonly used in the modern textiles industry, natural dyes are depicted as sustainable and eco-friendly.

10.5 Ancient Dyes, Pigments, and Dyeing Technologies

10.5.1 Indigoids (Indigo and Tyrian Purple)

Indigoids, which contain indigo, and Tyrian Purple, are one of the most historically important dye classes. Indigo, a dark blue dye that is typically obtained from *Indigofera* species and woad, is considered a global ancient dye. Indigo was traded throughout ancient Mediterranean civilizations [18]. The dye was obtained by soaking the leaves of *Indigofera* sp. plants in water, then aerating the liquid, which allowed the indigo dyestuff to precipitate [19, 43]. The precipitate was collected and boiled for purification before being filtered, washed, and dried [19]. Another common source of indigo is woad, a yellow-flowered European plant of the cabbage family [19]. The indigo dye is also obtained from the woad plant by crushing the leaves into a pulp and forming a ball which was left to dry, then crushed into a powder and watered on and off for about nine weeks before being dried and stored for future use [19]. Textile fibers dyed with woad dating back to the Bronze Age have been discovered in the rock salt mines at Dürrnberg near Hallein in Austria [43].

Tyrian Purple was obtained from Murex snails that live in the Mediterranean Sea [20]. The ancient extraction method of the dye for textile use was lost around the 15th century AD and only recently rediscovered by John Edmonds in 2000 [20]. The snails are opened with a hammer or stone, and the hypobranchial gland is removed and added to water. Potash is added, and the liquid is heated in a closed container for several days before use [20].

Indigo and Tyrian Purple are dyed in the same manner, known as the vat dyeing process. The fabric is dipped into the dyebath for a period of time, then removed from the dyebath to the air. This process is repeated several times to obtain darker shades. The concentration, pH and amount of time in the dyebath affect the shade and depth of color of the textile. For Tyrian Purple, the best purple is obtained by letting the fiber oxidize in the dark, otherwise a bluer shade is produced [20].

Due to limited resources and expense of creating Tyrian Purple, it was considered a luxury dye only the richest could afford. Tyrian Purple was mentioned in the Old Testament as an example of royalty and was used in the Holy Tabernacle [20]. Later Tyrian Purple became rebranded as Imperial Purple by the Romans. In modern times, Tyrian Purple remains one of the most expensive dyes while synthetic indigo is used to dye denim, a ubiquitous textile [18].

10.5.2 Quinonoids (Madder)

Quinonoids dyes are typically used to produce shades in the red to yellow region and have a large diversity of chromophores, containing benzoquinones, anthraquinones, and α-naphthoquinones [21]. A well-known example of a quinonoid is Madder. The use of Madder in ancient times can be found globally in Phoenicia, Egypt, Mesopotamia, Greece, Rome, India, China and even an analog in the Americas [22]. It is mentioned in the Old Testament in Genesis, Chronicles, and Judges [22]. Madder produces a variety of red dyes from scarlet to a bluish red (purpleish) and pink [22]. The dye was extracted from the roots of the Madder plant. The roots were steamed, then dried and crushed to produce a powdered unpurified form of the dye [22]. Two other well-known quinonoids are cochineal and kermes from the Americas.

10.5.3 Carotenoids (Saffron)

Carotenoids are natural pigments found in the chloroplast of almost all plants, and are red, orange, and yellow in color [21]. Carotenoids are characterized by their simple arrangement of conjugated double bonds [21]. There are many possible examples of plants that produce carotenoids, but one common plant used for dyeing due to its carotenoids is Saffron. Saffron utilization can be found in many ancient civilizations including Mesopotamia, Egypt, Crete, Greece, Iran, Rome, and India [23]. It was used not only as a textile colorant but also as a spice, medicine for a variety of ailments, as cosmetics, and in religious ceremonies [23]. Saffron flower stigmas and pollen produce a brilliant shade of yellow [24]. However, it takes a large amount of raw material and a long time produce enough Saffron dye for textiles. In ancient Greece, it was regarded as a luxury textile dye which was traded throughout the Mediterranean Sea [24].

10.5.4 Flavonoids

Flavonoids are plant dyes that produce yellow, orange, red and blue color [21]. Although a wide variety of flavonoid dyes can be produced, most of the yellow flavonoid colorants have been lost to antiquity due to lack of historic documentation [25]. Additionally, yellow flavonoid dyes are less resistant to fading as compared to most of the blue and red dyes used. Therefore, the dyes that are not resistant to environmental exposure may pose challenges in characterization of ancient textiles dyes [25].

10.5.5 Dihydropyran (Brazilwood and Logwood)

There are two sources for dihydropyran dyes, Brazilwood and Logwood [21]. In ancient times, the bark of the Redwood trees, also known as Brazilwood, was used to produce a red dye [26]. Logwood trees, found in the Americas, were used to produce violet and black dyes to obtain colored textiles. The dyes were extracted from the bark of the trees, after the bark was boiled and the solution was oxidized [27].

10.5.6 Tannins

Tannins are chemical compounds found in vegetables and other plants [21]. When mixed with other natural dyes, these compounds can produce yellow, brown, grey, and black shades [21]. Tannins are typically used as a mordant to help attract the dye to the fiber and are not used as a colorant [28].

10.6 Conclusion

In ancient times, weaving was not merely a way to produce cloth, but also to demonstrate social status and celebrate the wearer's culture. Although the same basic premise of interlacing threads at right angles to produce cloth is universal, each region had a unique method or style of weaving. The cultural significance of weavers, patterns, and fibers also varied from region to region. Dyeing techniques also followed the same trend where different materials were used to produce varying shades based on the region where the dyes were produced. However, some dyes, such as Tyrian Purple were universally acknowledged as a rare luxury.

Studying how textiles were woven and dyed gives insights to the culture where the textiles originated and the cultures around it. For example, if textiles with dyes and weaving techniques have been found in one geographical region that the dyes and techniques are not native to, the civilizations had some contact with each other.

References

1. Birrell, V.L., *The textile arts, a handbook of fabric structure and design processes: ancient and modern weaving, braiding, printing, and other textile techniques*, Harper, New York, 1959.
2. Rossi, C. and Russo, F., Spinning and weaving, in: *Ancient Engineers' Inventions*, 2016, Available: http://link.springer.com/10.1007/978-3-319-44476-5_19.

3. Hooper, L., The technique of greek and roman weaving. *Burlingt. Mag. Connoisseurs*, 18, 95, 276–284, 1911.
4. Quercia, A., Textile production and technological changes in the archaic societies of Magna Graecia: The case of Torre di Satriano (Lucania, Italy). *Origini*, 40, 243–258, 2017.
5. Wang, H., Exploratory study of the origin of the ancient Ikat dyeing technique and its spread in China. *J. Text. Inst.*, 96, 2, 105–108, 2005.
6. Buckley, C.D., Investigating cultural evolution using phylogenetic analysis: The origins and descent of the southeast Asian tradition of warp ikat weaving. *PLoS One*, 7, 12, e52064, 2012.
7. Chen, K. *et al.*, Song brocade in the ming and qing dynasties. *Clothing Text. Res. J.*, 38, 4, 285–297, 2020, Available: https://journals.sagepub.com/doi/full/10.1177/0887302X20932657.
8. Buckley, C., Looms, weaving, and the austronesian expansion, in: *Spirits and Ships: Cultural Transfers in Early Monsoon Asia*, A. Acri, R. Blench, A. Landmann (Eds.), pp. 273–324, ISEAS - Yusof Ishak Institute, Singapore, 2017.
9. Li, Q. and Li, B., Study on the relation between the Linked-pearl Pattern and Weaving Technology in Ancient China. *Adv. Mat. Res.*, 3==32-334, 2093–2098, 2011, Available: https://www.scientific.net/AMR.332-334.2093.
10. Brumfiel, E.M., Cloth, gender, continuity, and change: Fabricating unity in anthropology. *Am. Anthropol.*, 108, 4, 862–877, 2006.
11. Rowe, A.P. and Bird, J.B., Three ancient peruvian gauze looms. *Text. Mus. J.*, 19, 20, 27, 1980.
12. Wade, J.M. and Mattingly, G.L., Ancient Weavers at Iron Age Mudaybi. *Near East. Archaeol.*, 66, 1, 73–75, 2003.
13. Doumani, P.N. and Frachetti, M.D., Bronze age textile evidence in ceramic impressions: Weaving and pottery technology among mobile pastoralists of central Eurasia. *Antiquity*, 86, 332, 368–382, 2012, Available: https://dx.doi.org/10.1017/S0003598X00062827.
14. Zhao, F. *et al.*, The earliest evidence of pattern looms: Han Dynasty tomb models from Chengdu, China. *Antiquity*, 91, 356, 360–374, 2017, Available: https://dx.doi.org/10.15184/aqy.2016.267.
15. Anderson, F., Some aspects of linen weaving, ancient and modern. *J. Text. Inst.*, 15, 8, 348–373, 1924, Available: https://doi.org/10.1080/19447012408660973.
16. Roy, T., *The crafts and capitalism : Handloom weaving industry in colonial India*, Routledge, London, 2020.
17. Boudot, E. and Buckley, C., Chapter 2: Textiles and looms in ancient Chinese culture, in: *The Roots of Asian Weaving*, E. Boudot and C. Buckley (Eds.), pp. 20–45, Oxbow Books, Limited, Philadelphia, 2015.
18. Melo, M.J., History of natural dyes in the ancient Mediterranean world, in: *Handbook of Natural Colorants*, T. Bechtold and R. Mussak (Eds.), pp. 3–20, John Wiley & Sons, Incorporated, United Kingdom, 2009.
19. John, P., Indigo - extraction, in: *Handbook of Natural Colorants*, J. Bechtold and R. Mussak (Eds.), pp. 105–134, John Wiley & Sons, Incorporated, United Kingdom, 2009.
20. Landenius-Enegren, H. and De Meo, F., *Treasures from the sea: Purple dye and sea silk*, Oxbow Books, Oxford, 201730.
21. Yusuf, M., Shabbir, M., Mohammad, F., Natural colorants: Historical, processing and sustainable prospects. *Nat. Prod. Bioprospect.*, 7, 1, 123–145, 2017, Available: https://search.datacite.org/works/10.1007/s13659-017-0119-9.
22. Chenciner, R., *Madder red: A history of luxury and trade: Plant dyes and pigments in world commerce and art*, London, 2000, Available: https://www.taylorfrancis.com/books/9781135796136.
23. Mousavi, S.Z. and Bathaie, S.Z., Historical uses of saffron: Identifying potential new avenues for modern research. *Avicenna J. Phytomed.*, 1, 2, 57–66, 2011, Available: https://doaj.org/article/7e183dcd8c1f48099039595365c401d9.

24. Dewan, R., Bronze age flower power the minoan use and social significance of saffron and crocus flowers, *Chronica,* 5, 42–55, 2015.
25. Biertümpfel, A. and Wurl, G., Dye plants in Europe, in: *Handbook of Natural Colorants*, T. Bechtold and R. Mussak (Eds.), pp. 39–52, John Wiley & Sons, Incorporated, United Kingdom, 2009.
26. Ahmed, H.E., History of natural dyes in North Africa 'Egypt', in: *Handbook of Natural Colorants*, T. Bechtold and R. Mussak (Eds.), pp. 27–36, John Wiley & Sons, Incorporated, United Kingdom, 2009.
27. Ortiz-Hidalgo, C. and Pina-Oviedo, S., Hematoxylin: Mesoamerica's gift to histopathology. Palo de Campeche (Logwood Tree), pirates' most desired treasure, and irreplaceable tissue stain. *Int. J. Surg. Pathol.*, 27, 1, 4–14, 2019, Available: https://journals.sagepub.com/doi/full/10.1177/1066896918787652.
28. Chandrika, U.G., Carotenoid dyes - production, in: *Handbook of Natural Colorants*, T. Bechtold and R. Mussak (Eds.), pp. 237–242, John Wiley & Sons, Incorporated, United Kingdom, 2009.
29. Rogers Fund, *Loom weight*, vol. Mud, 1050. Available: https://www.metmuseum.org/art/collection/search/568484?searchField=All&sortBy=Relevance&ft=loom+weight&offset=0&rpp=20&pos=5.
30. Painter, A., *Terracotta lekythos (oil flask)*, vol. Terracotta; black-figure, ca. 550-530 B.C. Available: https://www.metmuseum.org/art/collection/search/253348?searchField=All&sortBy=Relevance&ft=terracotta+lekythos&offset=0&rpp=20&pos=1.
31. Singh, K., *Weaving and spinning*, vol. Ink and color on paper, 1860. Available: https://clevelandart.org/art/1987.162|author=KeharSingh. Accessed online on Feb 2021.
32. Dudley, P. and Allen Fund, *Tunic with double-headed serpents*, vol. Cotton;gauze, ca. 400-200 B.C. Available: https://clevelandart.org/art/2005.17. Accessed online on Feb 2021.
33. Barber, E.J.W., *Prehistoric textiles: the development of cloth in the Neolithic and Bronze Ages with special reference to the Aegean*, p. 9, Princeton University Press, Princeton, NJ, 1991.
34. Burnham, D.K., *A textile terminology: Warp & Weft*, Routledge & Kegan Paul, London, 1981.
35. Nils Oppervoll of Manndalen, Nord Troms, Norway, the Royal Ontario Museum, Historisk museum, Bergen, Norway, Norsk Folkemuseum, Norway, https://sallymclean1.wordpress.com/2015/04/04/the-warp-weighted-loom/. Accessed 16 September 2021.
36. Mossakowska-Gaubert, M., *Egyptian textiles and their production: Word and object*, p. 157, Lulu. com, Lincoln, Nebraska, 2020.
37. Broudy, E., *The book of looms: A history of the handloom from ancient times to the present*, University Press of New England, Lebanon, New Hampshire, p. 46, 1993.
38. Nosch, M.L., The wool age. In *Tradition and Innovation in the Mycenaean Palatilal Polities*, 167. 1st ed. Verlag Der österreichischen Akademie Der Wissenschaften, 2016.
39. *Lesser draw loom*, China National Silk Museum, https://artsandculture.google.com/asset/%AD-lesser-draw-loom/UQEu_2D_MLLiZg. Accessed 22 September 2021.
40. Chequer, F.M.D. *et al.*, Textile dyes: Dyeing process and environmental impact, in: *Eco-Friendly Textile Dyeing and Finishing*, vol. 6.6, pp. 151–176, 2013.
41. Mazăre, P., Investigating neolithic and copper age textile production in Transylvania (Romania). Applied methods and results, in: *Prehistoric, Ancient Near Eastern and Aegean Textiles and Dress: An Interdisciplinary Anthology*, Oxbow Books, Oxford, vol. 18, p. 1, 2014.
42. Gleba, M. and Nikolova, A., Early twined textiles from Sugokleya (Ukraine). *Archaeol. Text. Newsl.*, 48, 7–95, 2009.
43. Ballard, M.W., Natural dyes: Sources, tradition, technology and science. *Stud. Conserv.*, 335–408, 2007.
44. Emery, I., The primary structures of fabrics, an illustrated classification, p. 76, Watson-Guptill Publications, New York, NY, 1980.

11

Armours: Ancient Metallic Textiles

Ritu Pandey[1], Ragini Dubey[1*], Pintu Pandit[2], Suruchi Pandey[3], Mukesh Kumar Sinha[4] and Amarish Dubey[5]

[1]*Chandra Shekhar Azad University of Agriculture & Technology, Kanpur, India*
[2]*National Institute of Fashion Technology, Department of Textile Design, Patna, India*
[3]*Symbiosis Institute of Management, SIU, Pune, India*
[4]*Technical Textile Division, Ministry of Textiles, New Delhi, India*
[5]*Rajiv Gandhi Institute of Petroleum Technology, Jais, India*

Abstract
Ancient metallic armours were used by cavalry during the war, melee combat, and jousting. Kings and noblemen of the royal court donned metallic armours embellished with inlay designs and polychrome glass for ceremonial purposes. Different parts of the armours were used to protect different parts of the body such as head, breast, back, torso, shoulder, hand, legs, and foot. Armours were embossed and engraved with geometric or figurative decorations. Armours also conveyed the social status, reputation, and lifestyle of the wearers in the ancient era through heraldry. Plate armours covered the soldiers from head to toe. The chapter is to connect with ancient metallic armours. The chapter also describes materials, designs, significance, uses, and armours' parts. A case study of the ancient Egyptian king; Tutankhamen's ceremonial apron is also presented.

Keywords: Armours, metallic textiles, mail armour, armour parts, ancient textiles

11.1 Introduction

Metal armour is one of the highly polished contour-defining costumes of the middle ages, exhibiting knights' remarkable personage [1]. No loose folds, only statuesque figures, and great strength were the impression of an armour wearer. Padded garments were worn under the armours to protect the body parts [2]. Metallic textiles in fibers, threads for embellishments, and finished costumes are classic ancient textiles. The art continues today in decorations, accessories, breast plates, and costumes. Currently metallic textiles are part of specially designed essential wear for shape retention. Knight armours are additional clothing worn over regular textiles as part of the defence, cavalry, adornment, and sports costumes [3]. Metallic costumes are reusable and can be worn by several generations. A prized heirloom, today knight armours are the shining jewels of any museum. Several museums like Royal armouries at Leeds, archeological and Haldighati museum at Chittorgarh are

**Corresponding author*: raginidubey91979@gmail.com

Seiko Jose, Sabu Thomas, Pintu Pandit, Ritu Pandey and Vandana Gupta (eds.) Handbook of Museum Textiles: Volume I Conservation and Cultural Research, (209–218) © 2023 Scrivener Publishing LLC

dedicated to kings armours and related accessories like swords, helmets, shields, and boots. Metallic textiles are the most serviceable and eco-friendly natural textiles that do not require laundering and thus save water and detergents. Periodical wax polishing is necessary to maintain the sparkle of the costume for a long life span. Precision is required to manufacture armours to support a perfect fit with ease of movement during military games and war. Ancient knights, kings, and noblemen of the court were the primary users of armours. Armours were often inscribed with a family seal indicating higher social status. Armours are metallic dresses hammered to mimic human anatomy to fit the chest and back perfectly. Different parts of armours covering the whole body are individually made, undergoing additional moulding and processing techniques. The sculptural metallic armours are engraved with stylized, animal, mythological, and geometric patterns besides heraldic designs. Heraldry is blazoning or describing armourial bearings where every symbol and line of the design convey a mark of the distinction having specific meaning [4]. This way the lady recognized her knight during tournaments and medieval jousting. Besides heraldry, hieroglyphics and the hierarchical display were also evident in military armours and shields [5]. Accessories worn along with armours to protect the body were helmets, shields, gloves, and boots. Ancient armours (400–1000 B.C.) were wood with leather lining. Monster and vulture-shaped helmets with wings were also used in jousting armours [6]. Materials for manufacturing armours in the medieval era were bronze, iron, steel, and gild engraved and embossed in decorative patterns. Gold, silver, and rock solid animal shells/scales were also used to make decorative armours [7].

11.2 Parts of Armour and Accessories

Armours existed in different parts of the world from Europe to Asia and Egypt. Different parts of the armours protected every part of the body. European armour's breast plates depicted gothic styles [8, 9]. Armour parts and their uses are presented in Table 11.1. Helmet, mail armour, greave, gauntlet, pauldron, and sabatons are depicted in Figures 11.1 to 11.8.

11.2.1 Helmet

Sallet (celata), barbuta, bascinet, helm, spangenhelm, and pollaxe were medieval helmets. Helmets were designed with visor (eye and nasal opening). Metal plates of the helm headwear were hinged to cover cheeks, nose, and top of the shoulders. The head-wears padded with quilted material and tied with strings were fastened at the chin [8]. Spider, monster-shaped, winged, tower helmets, and masks (Figures 11.1 a and c, 11.2, 11.6 a) were also prevalent for face and head protection. Conical helmets with distinctive top crests were worn over mail coif to protect the head, face, and neck (Figures 11.3 a–c).

11.2.2 Coif

Coif was made up of iron wire in interconnected rings/loops [8]. It was worn with a band, attached to ventail and wrapped around and above the lower edge of the helmet. Cervelliere, a curved skullcap, was worn under the coif (Figures 11.3 a–c).

Table 11.1 Parts of ancient armours and its significant purpose.

Armour parts	Description/use
Helmet	Head protection
Shield boss	Hand cover
Mail armour	Looped metal chain mail shirt
Cuirass	Backplate
Cubitiere	Elbow cap
Tasses	Thigh covering
Mail hood	Coif (to hold the helmet)
Mail mitten gauntlet	Muffler
Greave	Leg protecting armour
Pauldron	Shoulder armour
Sabatons	Armour for feet

Figure 11.1 (a) Close helmet with visor (face mask with sight and breaths opening). (b) Tournament ring at Royal Armouries, Leeds; Ancient tournaments recreation where armour laden knights are performing with lances. (c) Helmet, visor, gorget, gauntlet, greave, tasses for whole body protection (Source: Royal Armouries, Leeds, UK).

Figure 11.2 Armours covering for head to toe protection (a) helmet, gorget, breastplate, and plackart (torso armour). (b) Tasset (thigh protection), gauntlet (arm and palm protection), cuisse (upper leg covering), Poleyn (knee defence), greave (lower leg armour), sabatons (metal plates shoes). (c) Pauldron (shoulders, sleeves, and chest armour), Guard of Vambrace (elbow armour). (Photo source: Royal Armouries, Leeds, UK).

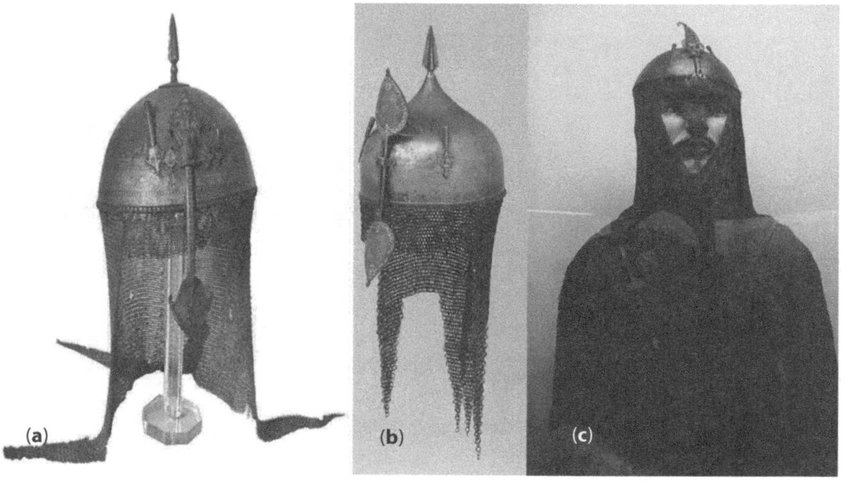

Figure 11.3 (a–c) 19th C.E. helmet designs with top crest and mail-coif. (c) Helmet with mail standard collar, face, and movable nose guard.

11.2.3 Ventail

Flap of mail shirt, used for face protection, was known as ventail. Ventail was used to cover the face and wrapped around the neck, mouth, and lower face for protection.

11.2.4 Mail

Mail armour shirts were made of iron wires interlocked to form continuous rings joined together by rivets or weld techniques (Figure 11.4). Interlocked loops allowed the flexibility to armour shapes and size. Extended mail to cover the sleeve's back were known as muffler/gauntlet. Chain Mail shirts were in use from 300 B.C. to 1600 A.D. [10].

11.2.5 Hauberk

Armour for sleeves was known as hauberk or mail sleeve. Mail sleeves were extended to mufflers and mail mittens.

11.2.6 Gauntlet and Pauldron

Gauntlets were metal plates to protect sleeves and palm. Small disc shaped metal plates provided elbows, palm, fingers, and shoulders' protection. Metal plates were joined using sliding rivets for ease in moving. Pauldrons were armours for arms extended to the chest and back. Metal plates were rounded to cover chest and back (Figures 11.2, 11.5 a–c).

Figure 11.4 Mail (interconnected rings/loops) armour shirt made up of steel, iron, gold, leather.

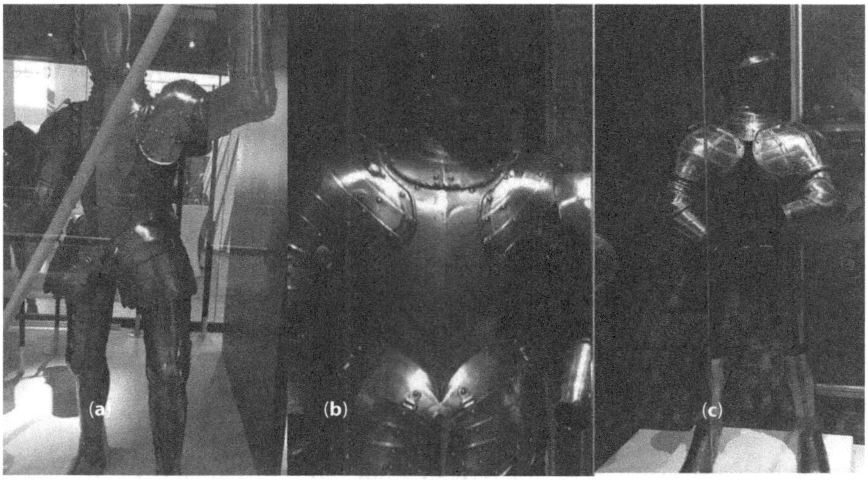

Figure 11.5 (a, c) Front of knight armours made of metal composite (pauldron, gauntlet, tasset, greaves, cuisse, sabatons) (b) back of knight armours (pauldron, hinges, tasses, and culet of lames plates).

11.2.7 Sabatons and Greaves

Both were used for leg protection. Mail leggings known as chausses were used in the ancient medieval period. Padded chausses known as gamboised cuisses (upper leg covering) protected from blunt objects. Mail flaps hanging and fitted on either sides of legs using clasps and mail hose with fitted garter for support were used. Greaves were tubular plates strapped around lower legs and sabatons. Sabatons were overlapping metal plates strapped around lower legs. Lamellar plates were riveted with leather lining and attached with the shoes through laces (Figures 11.5, 11.6 d-e).

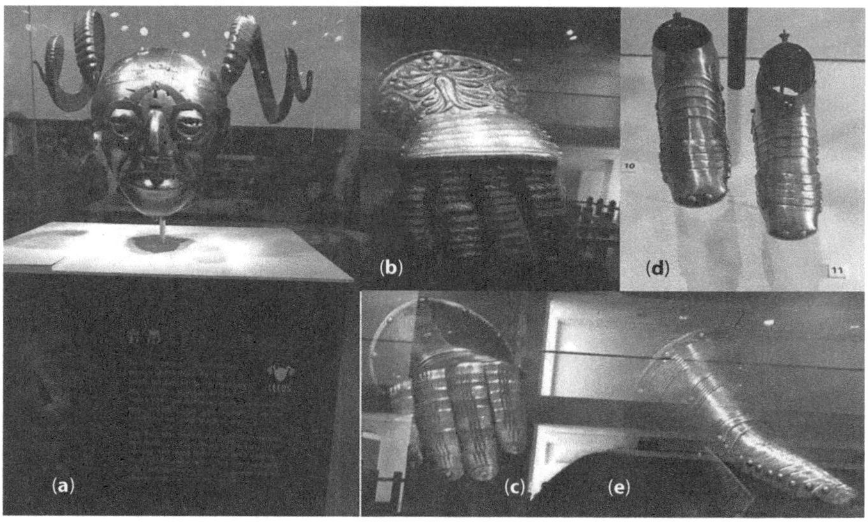

Figure 11.6 (a) Embossed face mask with winged ears. (b) Anthemion design in knuckle armour plate. (b, c, e) Aesthetics with practicality—Gauntlet with small metal discs for individual fingers provided ease of movement. (d) Sabatons (photo source: Royal Armouries, Leeds, UK).

11.3 Armour Designs

Floral, geometrical, mythological, and animal motifs were common in ancient armours (Figure 11.6b). Helmets were also decorated with figs, feathers, and horse hairs. Heraldic symbols were used in the armours specially during military games to identify the team members from their family [11]. Helmets were decorated with wood, feathers, molded leather, papier-mâché, and textiles also as part of heraldry and hieroglyphics symbols. The metal breastplates were decorated with latten, gilded copper with engraved, etching, and embossing floral designs. German armours was decorated with ridges, grooves, and various cut decorations. The Japanese introduced the art of stitching together the laminated plates of iron [12].

11.4 Armour Materials

Materials used to manufacture different parts of ancient metallic armours are described in Table 11.2. Gold, silver, wood, leather, iron, cast iron, wrought iron, steel, blister steel, shear steel, wootz, or Damascus steel were common materials used to manufacture medieval armours [12, 13]. Japanese steel, bronze were engraved, embossed, etched, pointille, gilded in figurative and geometric patterns (Figure 11.9). Blueing and Murcury gilding highlighted the designs (Figure 11.7c). Use of copper alloy (latten) was done to decorate

Table 11.2 Material and pattern used to manufacture metallic armours in different eras [12].

Period	Armours	Material/pattern
900 B.C.	Helmet, breastplate, backplate, greave, shield	Wood, leather, Bronze/painted, engraved and embossed geometrical, fauna, mythological patterns
500 B.C.	Helmet, greave, gauntlet, shield	Iron, bronze/heraldic display, human face, animal head, decorative pattern
400 C.E.	Helmet, belt, clasp, shield boss	Iron, copper, silver, leather lining/ embossed, inlay, gilding, engraving, figurative and mythological scenes
1000 C.E.	Mail armour, scale armour, mail hood, shield	Metallic, gold, hardened leather/ extremely decorated—gilded, painted, lined with velvet, brocade
1400 C.E.	Mail armour, scale plates, steel plate from head to toe, Helmets, breast plates lined with textiles	Metallic armours lined textiles/heat-patination, embossing, etching, engraving, pointillé, gilding, cut work, and inlay of figurative and gothic designs
1600 C.E.	Plate armour	Steel plate/heat-patination

Figure 11.7 Ceremonial armours of the royal family of Britain on display at Royal Armouries, Leeds UK. (a–b) Small size half armours (without leg defence) for boys mainly for joust and military training, open helmets (visor opening for eyes, nose) were padded. (c) Pollaxe—used for foot combat in tournaments. The hammer head has three weapon components (three knife claws, the rear fluke, and top spike).

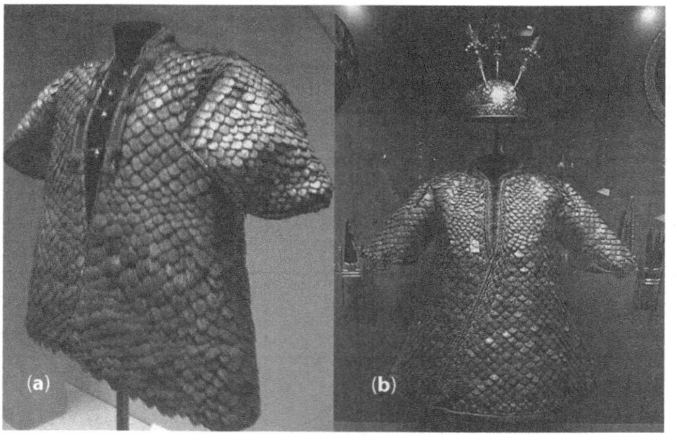

Figure 11.8 Pangolin (Manis crassicaudata) scale armour, splendid craft from India; Both armours were presented to British royal members. (a) King George III in 1820. (b) Albert Edward, Prince of Wales (1875–1876).

the edges of metallic armours [14, 15]. Armours made of animal scales (Figure 11.8) provided protection with aestheticism, flexibility, and comfort to the wearer. Blacksmiths artistically shaped the iron and other metals into effective armours using tools, such as hammer, swages, punches, moulds, bits, drifts, fullers, etc (Figure 11.9). Large crucible, bloomery, and blast furnace-shaped steel and iron [16].

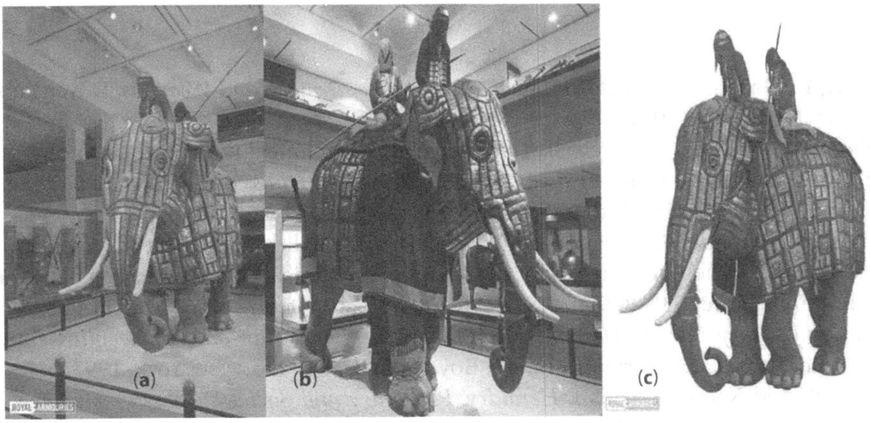

Figure 11.9 Elephant armour made of metal, textiles, leather. War-craft acquired from India between 1798 and 1801, now in Royal Armouries, Leeds.

11.5 Metallic Costume of King Tutankhamen

The mummy-case of King Tutankhamen, who ruled Egypt as pharaoh for 10 years till his death in 1324 BC, opened the rare treasures of ancient armours, textiles, and lavish lifestyle. British archeologist Howard Carter discovered the tomb of Tutankhamen on November 26, 1922. Tutankhamen's ceremonial apron consisted of several gold plates graduated in various sizes and rhythmic patterns. Each gold plate was inlaid with opaque polychrome glass. Edges of the royal apron were lined with a row of asps signifying royal insignia. A golden girdle with tucked in dagger at one side, encircled the king's waist. Girdles were richly embellished with colored leathers, feathers, and opaque polychrome glass. Two separate girdles kept beside king were composed of ornamental faïence and gold beads [17]. "Treasures of Tutankhamen," now displayed at Egyptian Museum in Cairo, were once travelled around the world and exhibited in 1960s and 1970s.

11.6 Conclusion

Ancient metallic armours were functional, fashionable, and serviceable costumes. Armours protected during war and military games involve use of weaponry. Adults and young ones alike donned metal armours. Designer armours were also available for women warriors, queens, and princesses. New innovative materials and designs were used to manufacture armours in the past. Armours may still fascinate the present generation, but the history of metallic armours took a halt by 1900 CE. Later, the advent of nylon and kevlar led to the development of lightweight combat body armours taking over the place of metallic armours. Some accessories, such as helmets, mittens, and boots, are still used for body protection during war. However, the materials employed for manufacturing modern combat armours and accessories are lightweight and different from the ancient one. Thus, traditional with futurism, avant garde with elements rooted in the past, and leveled as assets of the men-at-arms summarizes metallic armours existence in the present scenario.

References

1. Wagner, E., Drobná, Z., Durdik, J., *Medieval Costume, Armour and Weapons*, Dover Publications, Mineola, New York, 2014.
2. Williams, A.R., Edge, D., Capwell, T., Tschegg, T., A technical note on the armour and equipment for jousting, in: *Gladius: Estudios sobre armas antiguas, armamento, arte militar y vida cultural en Oriente y Occidente*, vol. 32, pp. 139–184, 2012.
3. Bérard, E., Dillmann, P., Renaudeau, O., Verna, C., Toureille, V., Fabrication of a suit of armour at the end of Middle Ages: An extensive archaeometallurgical characterization of the armour of Laval. *J. Cult. Herit.*, 53, 88–99, 2022. https://doi.org/10.1016/j.culher.2021.11.008.
4. Brault, G.J., *Early Blazon: Heraldic Terminology in the Twelfth and Thirteenth Centuries with Special Reference to Arthurian Heraldry*, Boydell & Brewer Ltd, New York, 1997.
5. Coss, P.R. and Keen, M. (Eds.), *Heraldry, Pageantry and Social Display in Medieval England*, Boydell Press, New York, 2003.
6. Zyguslki Jr., Z., Armour as a symbolic form. *Waffen Kostümkd.*, 26, 77–96, 1984.
7. Bruet, B.J., Song, J., Boyce, M.C., Ortiz, C., Materials design principles of ancient fish armour. *Nat. Mater.*, 7, 9, 748–756, 2008.
8. Waddell, J.N. and Palermo, B.S., *Medieval Arms, Armor, and Tactics*, Worcester Polytechnic Institute, 2002.
9. Heller, J., *Medieval Armor and Armored Combat*, Doctoral Dissertation, Worcester Polytechnic Institute, 2005.
10. Richardson, T., The introduction of plate armour in medieval Europe, in: *Medieval Warfare 1300–1450*, pp. 441–446, Routledge, UK, 2017.
11. Wise, T., *Medieval Heraldry*, Bloomsbury Publishing, UK, 2012.
12. Allen, M.W., *Designing successful e-learning: Forget what you know about instructional design and do something interesting (Vol. 2)*, John Wiley & Sons, San Fransisco, 2011.
13. Vervhoeven, J.D., Pendray, A.H., Gibson, E.D., Wootz Damascus steel blades. *Mater. Charact.*, 37, 9–20, 1996.
14. Fox-Davies, A.C., *The Art of Heraldry: An Encyclopaedia of Armory*, TC & EC Jack, London, 1904.
15. Hazell, P.J., *Armour: Materials, Theory, and Design*, CRC Press, Boca Raton, 2015.
16. Williams, A.R., *The Knight and the Blast Furnace: A History of the Metallurgy of Armour in the Middle Ages & the Early Modern Period (Vol. 12)*, Brill, London, 2003.
17. Allen, S.H., *Tutankhamun's Tomb: The Thrill of Discovery*, Metropolitan Museum of Art, New York, 2006.

12

Textile Conservation in India: A Case Series

Deepshikha Kalsi[1]*, Elizabeth-Anne Haldane[2] and Lynda Hillyer[2]

[1]Textile Conservation Studio, Green Park Main, New Delhi, India
[2]Textile Conservation Studio, Victoria and Albert Museum, London, United Kingdom

Abstract
In this chapter, the focus is on how an established paper and paintings conservator trained in India developed her training in textile conservation through internships and courses in Europe, including working in the Textile Conservation studio of the Victoria and Albert Museum (V&A), UK and the Abegg-Stiftung Foundation, Switzerland, and how she has transferred the skills she learnt in Europe to conserving textile in India. Themes discussed include sourcing, adapting, and applying the materials and techniques employed at the V&A to set up her private Textile Conservation studio in New Delhi and finding alternative solutions for some of the specialist equipment used in Europe. These issues will be discussed through case studies of objects primarily treated in India allowing for an in-depth discussion of conservation methods and materials and equipment used. Methods for displaying costume and textiles will also be discussed, and the contribution of textile conservators to the technical analysis of textiles. Strategies to create awareness of the need to preserve India's textile heritage and to share skills and knowledge through training are discussed. The importance of continuing professional development and maintaining a professional network to draw on for continued advice and mentorship is emphasized.

Keywords: India, textile conservation, treatment, mounting, training, internships

12.1 Introduction

Conservation as a profession is well established in India; however, the focus has until now been on the preservation of buildings, archaeological artifacts, decorative art, paintings, paper and archives. For the conservation of textiles, India relies heavily on her esteemed master darners even in leading cultural institutions. Although the origins of textile conservation lie in the traditional practices of embroidery and darning, it has evolved as a scientific branch in the field of conservation. In conservation practice the function of stitching has developed, and its aim is to stabilize and support without being invasive. However, stitching is only one aspect of textile conservation, the role of the textile conservator lies far

**Corresponding author*: deepshikha.kalsi@gmail.com

Seiko Jose, Sabu Thomas, Pintu Pandit, Ritu Pandey and Vandana Gupta (eds.) Handbook of Museum Textiles: Volume I Conservation and Cultural Research, (219–238) © 2023 Scrivener Publishing LLC

beyond as discussed in the accompanying chapter 'The conservation and display of Indian textiles at the Victoria and Albert Museum' by the same authors.

In this chapter, the focus is on how Deepshikha Kalsi has transferred the skills she learnt during internships and courses in Europe to working in India, including sourcing, adapting and applying the materials and techniques employed at the V&A to set up her Delhi conservation studio, and finding alternative solutions for some of the specialist equipment used in Europe [1]. These themes will be discussed through case studies of objects primarily treated in India including a 19th century jama belonging to the Maharawals of Dungarpur and a frock coat belonging to Maharajah Sawai Man Singh II and Maharajah Sawai Bhawani Singh of Jaipur.

The importance of continuing professional development will be emphasized, both to keep up to date with treatment options and to extend the range of services that can be offered by a private conservation studio. A second internship at the V&A gave Deepshikha the opportunity to expand her treatment skills and gain more experience in the display and mounting of textiles allowing her to develop her skills in exhibition display as shown through the case study of the display of flat textiles and costumes for the special exhibition PRA-KASHI Silk, Gold and Silver from the City of Lights at the National Museum, New Delhi. Similarly, an internship at the Abegg-Stiftung enabled the development of skills in the technical analysis of textiles as illustrated by the technical study of textiles from the Aurel Stein collection at the National Museum in India.

Maintaining a professional network to draw on for continued advice and mentorship is crucial to success. It is also equally important to reach out and create awareness of the need to preserve India's textile heritage and to share skills and knowledge through training. The final section of this chapter will discuss workshops on conservation and preservation of textiles developed for the Indian Museum, Kolkata and The Registry of Sarees, Bangalore.

12.2 Internship Training in Textile Conservation at the V&A

In 2011 Deepshikha Kalsi, already an established paper and paintings conservator, came to the Victoria and Albert Museum for the first of two six-month internships in textile conservation. Although conservation training is available in India there are no specialist courses focusing on textiles. The authors originally met when Deepshikha was attending a short textile conservation workshop in Kolkata, organized by SUTRA, and taught by members of the V&A Conservation Department [2]. The aim of the internship was to build upon her existing conservation knowledge and develop specific skills in the treatment of textiles, which could then be developed for use in an Indian context. The topics covered have been addressed in the first chapter by the authors and included object investigation, assessing condition, documentation, preventive conservation methods, and treatment methodologies including cleaning and humidification, stabilization and support, and object display. The projects chosen were all Indian textiles requested for display in the South Asia gallery at the V&A. The complexity of the projects increased as skills were developed, culminating in the treatment of a mid-17th century painted and dyed cotton chintz panel from the Coromandel Coast (IS.16-1956), discussed below.

12.2.1 Condition Assessment

A vital part of the conservator's skill is expertise in the study of deterioration in an artifact. Objects undergo continual change and the conservator's job is to try and stabilize that change. An object may have been physically or chemically altered. It could be due to fading, acidity, staining, dye bleeding, accretions, loss of pile, abrasions, cellulose degradation, fungal growth, bleaching, sweat, body fluids, dust and dirt, insect activity, adhesives, salt migration, charring, tarnishing, corrosion, religious customs and many more. For large textiles or textiles with complex surface decoration the weight of the textile needs to be assessed. The textile conservation studio at the V&A has identified more than eighty parameters that can be used to evaluate the condition of an artifact. These are mapped on documentation sheets for visual reference and include causes of deterioration where they are known. The condition assessment and end purpose of conservation have a direct influence on the approach undertaken. The object may be going into storage, on display in a permanent gallery, on display in a temporary exhibition or on loan to another museum in another country. For some objects a well-constructed mount can minimize the amount of interventive conservation needed. Textiles which can be handled safely may not need any conservation, but their condition still needs to be assessed in order to display them safely.

Understanding the condition of an object and making a precise and accurate assessment is one of the most important aspects of a conservator's work. It forms a vital part of the process of decision making regarding inventive conservation, display, storage or handling and movement of an object.

12.2.2 Case Study: Conservation of a Painted and Dyed Cotton Chintz Appliqué Panel

The textile panel in this case study, (Figure 12.1a, IS.16-1956), is thought to have been made in Europe in the 18th century from fragments of hand painted, mordant and resist-dyed cotton derived from a 17th-century Indian hanging [3]. The irregular shape of the panel suggests that it was made to decorate the back of a daybed. The design features figures in Iranian-style dress of the type seen in chintzes made for the Golconda court. The figures are set within an architectural framework surrounded by floral motifs, all composed from pieces of appliqué textile outlined with couched silk (Figure 12.1b). Indian Chheent was both hand painted and printed with mordants and dyes. Chheent over a period of time became known as Chintz in the West, much like Palanposh (bedcover) became Palampore. Kalamkari was originally only hand painted with mordants and dyes and never printed. Although the appliqué fragments in this object appear to be all hand drawn like a kalamkari, the entire original object is not present, so it is safer to classify it as a Chintz rather than a Kalamkari. The ground fabric upon which the applique is stitched is itself made from ten separate plain cotton components stitched together and supported by a thick cotton fabric at the back. Paper has been inserted beneath the appliquéd patches to achieve an embossed effect, and the chintz fragments are outlined with couched silk in pale blue and cream.

12.2.3 Condition Assessment

The panel had gathered surface soiling, and the cotton fabrics have acquired a yellow cast due to cellulose degradation. There are multiple stains primarily caused by interaction with moisture which has caused tide lines to form at the wet/dry interface on the textile. Some stains are red in colour, possibly caused by iron. The object had become distorted with extensive fraying along the top and bottom edges. The top ground fabric had numerous holes, tears, splits, losses and weak areas. The paper underneath the appliquéd areas had become a solid brittle mass in many areas and was causing strain on the appliquéd chintz patches as it was bursting out, resulting in splits, tears and losses (Figures 12.1c, d). The couched silk thread holding these chintz patches was also fraying in places. Some holes had also resulted from apparent burn marks. Remnants of a textile edge binding are present along the left and right of the object; multiple small pieces of white and red threads at regular intervals around the perimeter suggest that they were once used to hold the fabric in place at the bed head. The textile was cleaner and less damaged around the edges, consistent with its use as a bed head. The thicker support textile on the reverse had similar patterns of damage in the form of stains and holes as to the front of the textile panel. Overall, it was found to be suitably robust and was performing its function to support the decorative panel.

12.2.4 Conservation Treatment and Mounting

Once the condition was assessed, the conservation treatment was discussed in consultation with the curator. As the components of the object were of a diverse nature, it was decided that the object would be surface cleaned only. The front and the back were both vacuum cleaned using controlled suction through a nylon mesh and then surface cleaned using a latex smoke sponge with light uplifting movements followed by further vacuum cleaning. The textile was then humidified using water vapour applied through Sympatex, a semi permeable membrane. Once the textile fibres had relaxed, they could be gently manipulated to reduce the distortions in the panel.

The weak areas were assessed, and a strategy was devised wherein both adhesion and stitching would be used to support the holes, splits, tears and weak areas. A key part of this project was to develop skills in dyeing support fabric that is colour-matched to the original object. The support/infill fabric was dyed with Novacron reactive dyes for cellulose fabrics. These dyes have been tested to ensure that they are fast to both water and light. For this project 27 different coloured fabrics had to be dyed to match the original object. Adhesive was chosen as the main method of attaching support patches under the appliquéd chintz in order to avoid stitching through the layers of brittle paper underneath (Figures 12.1e, f). The losses in the appliquéd chintz were in-filled with dyed cotton fabric adhered using the modified cellulose adhesive Klucel G (low molecular weight hydroxypropyl cellulose). For practical purposes Klucel G, in a 3% w/v solution of 75% IMS: 25% water (Industrial Methylated Spirits (IMS) is a mix of 95%Ethanol/5%Methanol) was cast as free film [4]. Once dried it was carefully lifted and was then transferred onto one side of the requisite dyed fabrics for in-fills (Figures 12.1g, h). The lost areas were traced, and the prepared in-fill textile was cut as needed and inserted under the lost areas with the adhesive film facing down. The adhesive on the in-fills was reactivated with IMS vapour applied through Sympatex. Once the infill was secured the lifting edges around the loss were held down with approximately 2 mm wide strips of the free film of Klucel G. These were also reactivated with IMS vapour applied through Sympatex, followed by some pressure with a spatula.

As the original support fabric was stable, it was decided to provide localised support to give strength to the weak areas, and so, the rest of the tears, splits, weak areas, losses and

Figure 12.1 (a) IS.16-1956 Indian Chintz appliqué panel before treatment, (b) detail of appliqué panel, (c) detail of annotated image to document condition of the panel, (d) detail of paper under appliqué fabric, (e) detail before treatment, (f) detail after treatment, (g) (top) Klucel G free film being peeled from polythene, (h) applying Klucel G free film to infill fabric, (i) stitched repair to support hole in ground fabric, (j) conservation complete, laid on Hexalite board and measurements being taken, (k) stitching the panel to the padded board using a curved needle and polyester threads to match.

holes in both layers of background fabric were supported with dyed fabric patches inserted from the back, and wherever necessary with patches inserted through damage on the front (Figure 12.1i). Zigzag stitches were used to secure the edges of the support patches at the back and laid thread couching stitch was used over the front to support and secure the loose and fraying ends/edges.

The textile was mounted on a fabric covered board and displayed vertically within a case. A rigid lightweight Hexalite board (aluminium honeycomb between a fiberglass and resin outer skin) was chosen and custom cut to suit the shape of the bed head (Figure 12.1j). It was then covered with an even layer of polyester wadding to provide cushioning, followed by a layer of thick cotton display fabric stapled to the reverse. Fixings to attach the board to the case wall were attached to the reverse prior to attaching the object. The object was then pinned onto the board and gently tensioned to ensure it would not sag when vertical. It was then secured in place around the perimeter using fine polyester thread and long and short stitches (Figure 12.1k).

12.3 Setting Up a Textile Conservation Studio in India

In 2012, following her return to India, Deepshikha planned to establish a private practice in textile conservation. The primary reason being India was still not ready for textile conservation as a profession, and all the significant museums still employed master darners. The private practice allowed for a focus on practical work and an opportunity to build awareness of textile conservation and the skills offered by the studio. There was no funding support, and the studio was set up from scratch (Figure 12.2a). The materials, equipment and instruments were acquired in stages as per the needs of the object being conserved.

India is a global manufacturer and supplier in almost every field related to textiles. Thus, local manufacturers and suppliers of Gütermann threads (Skala/Mara), Lanaset and Novacron dyes (Huntsman), Dehyphon LS 54 surfactant (BASF) were located within India. Fine darning needles (Pony no. 11/12/13) were identified as the best for curving by hand at various angles for conservation stitching. Surgical sutures employed in various surgeries were also procured for the same purpose. The finest support fabrics in different materials, weight and weave structure were sourced from local export markets. To address mounting concerns, archival aluminium foam board used for photographs is being used as an alternative to Hexalite. Fibre glass mannequins of grades 400 and 600 are customized locally as per the object dimensions. Museum grade and conservation grade acid free mount boards, acid free blotting paper, Japanese tissue paper and acid free tissue paper is sourced locally from CTS Restoration Products India Pvt Ltd. Polyester film (brand names Melinex/Mylar) is available in rolls and sold by weight in different microns. Dupont has local vendors in Delhi who sell Tyvek per square meter in various grades and width. Material not available in India was sourced from abroad so as not to compromise the quality of work and includes silk crepeline, polyester Stabiltex, nylon net, smoke sponges and latex free sponges for cleaning, conservation grade adhesives, gels etc.

Getting equipment custom made in India is easier and more economical than acquiring from overseas. The engineers and manufacturers are willing to customize and fabricate if the end function and purpose is clearly communicated. This has been a rewarding experience. A dye hotplate, bulk dyeing bath machine with temperature controls and a steel

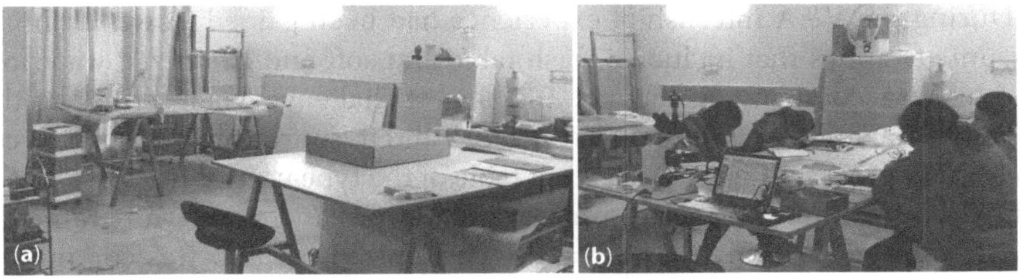

Figure 12.2 (a) The Textile Conservation Studio, New Delhi, (b) compound microscope and Dinolite for analysis.

wash table were all locally manufactured in Delhi and Deepshikha is presently working on getting a suction table made. Adapting available tools and devices happened simultaneously. A soldering iron with fine tip, used for repairing mobile phones became a useful instrument to cut and seal polyester Stabiltex. A soft water machine has been installed and de-ionized water is bought in cans. The services of a jeweller were employed for fine weighing before resources were available to purchase a digital weighing scale. A Dinolite, digital microscope, and a compound microscope were bought to help study and identify the weave structures and fibres (Figure 12.2b).

Storage boxes are made from acid free mount boards. But practicing in India has also made one realize that acid free mount board boxes are not practical for every collection unless appropriate environmental parameters can be maintained. Thus, aluminium and steel boxes customized to the dimensions of the objects are now being used. The objects are packed in layers of acid free tissue, cotton fabric and Tyvek during storage and for transportation. This approach of using knowledge with creativity helped in establishing the studio. Slowly but steadily, the studio has been equipped to undertake challenging treatments, many of which are being conducted professionally for the first time in India.

12.4 Conservation of an 19th Century Jama

An early project carried out by the Textile Conservation Studio was the conservation of a 19th century jama. This case study discusses the conservation methodology employed to establish a suitable approach for cleaning and stabilizing the garment.

This style of garment was introduced to India in the early 16th century by the Persians. Over time the jama became an acceptable form of attire for Indian royalty and members of the court, though with a subtle variation, the Muslims tied the jama on the right and the Hindus on the left [5]. This jama is part of the royal collection at Dungarpur in Rajasthan, northwestern India (Figure 12.3a). It is made of extremely fine translucent cotton muslin and measures 55 metres in circumference at the hem and the skirt is composed of 201 kalis (triangular panels). Each panel was documented in detail along with the bodice. It was heavily soiled, stained and discoloured due to cellulose degradation, indicating that the cotton would also be very acidic (Figure 12.3b). There were remains of case bearing moths, insect fracas, accretions, open seams, splits, tears and holes and creases (Figure 12.3c).

The jama was initially cleaned using low suction vacuum to extract the insect fracas and the dust from the surface, wet cleaning would be required to address the rest of the soiling. During the V&A internship, experience had been gained in wet cleaning using the Museum's facilities that include a wash table and soft and deionized water on tap as discussed in the accompanying chapter. The biggest challenge when setting up the Delhi studio was to simulate the equipment at the V&A and to source the materials to carry out the wet cleaning process. The cost of a wash table of the type used at the V&A was prohibitive, but without compromising on conservation ethics and standard parameters some innovations were made along the way and a customized bathtub was commissioned to wash the jama. It was made from stainless steel, measured 8' x 4' (approximately 2.44 m x 1.22 m) and 10" (25 cm) deep with a tap to drain water at one end. As chemicals for conservation can be expensive to source Frances Hartog, Senior V&A Textile Conservator advised on the purchase of a non-ionic surfactant suitable for treating all types of textile fibres and suggested a new version Dehypon LS 54 that could be used at a higher temperature, particularly important as the treatment was to be undertaken during the summer in India [6]. Information was also conveyed about the wet cleaning at the V&A of a similar jama and the approach taken [7].

As the jama was heavily soiled it was decided to give three subsequent baths, reducing the working strength of the surfactant to half of the previous concentration with each successive bath. The initial soak was done in deionized water and the main washing process was carried out in soft water due to budget considerations.

The Jama was placed on a sheet of Melinex (clear polyester film) in the steel tank and the Dehypon solution gently added. Each panel was sponged twice and then it was left for 10 mins and sponged all over again (Figures 12.3d, e). Thereafter it was thoroughly rinsed, and the same process repeated for the next baths. When the jama was turned over it was sandwiched between two layers of Melinex. The jama was thoroughly rinsed in soft water to remove all traces of the surfactant and given a final soak then running rinse with deionized water to remove all unwanted ions.

The Jama was taken out of the bath and excess water was soaked by absorbent cotton towels, followed by blotting with acid free paper. The bodice was then supported on a thin roller passed through the sleeves to hold the bodice out of the way of the heavy skirts, which were spread out flat on polythene covered soft boards (Figure 12.3f). Hair dryers, on cold and warm settings, and fans were further used to help in the drying process. The use of hair dryers at this stage prevented the layers of muslin becoming stuck together and further ensured even and crease free drying. The wet cleaning began at 6 am and it took 19 hours for a team of 5 to wash and dry the Jama, it was a great team effort!

Tears were supported with patches cut of sheer polyester Stabiltex as even the finest muslin fabric available would be visible. The Stabiltex was heat cut with a fine soldering iron to create a clean, discreet edge that does not fray (Figure 12.3g). The tears and weak areas were supported with Stabiltex patches using a combination of laid thread couching using fine threads pulled from Stabiltex, and zig zag stiches to secure the edge of the patch to the muslin.

A young boy's mannequin was chosen as the jama was too small to fit an adult sized form. The Jama was displayed seated, in a royal darbar scenario, so that there was no stress on the delicate bodice from the weight of the voluminous skirts. The mannequin was covered in washed stretch jersey fabric. and perched up on a stable support underneath to raise the height, allowing the skirts to spread out around the form (Figure 12.3h).

Figure 12.3 (a) Jama before conservation, (b) stains on jama hem, (c) detail of insect fracas and holes, (d, e) washing with Dehypon LS 54, gently pressing and releasing the sponge to push solution through the jama, (f) Jama spread out to dry after washing, (g) cutting Stabiltex with soldering iron, Jama displayed after conservation at Udai Bilas Palace, Dungarpur.

12.5 Case Study—Conservation of a Military Frock Coat

As the work of the studio became known, the scale of projects commissioned increased. Clients were still primarily private individuals or privately run museums. A collection of military uniforms and accoutrements of HHMS Man Singh II and HHMS Bhawani Singh were sent by the Jaigarh Fort Museum in Jaipur for treatment at the Textile Conservation Studio [8]. The uniforms were in poor condition, in need of urgent conservation intervention to stabilize them for study and display at the Jaigarh Fort Museum. This case study discusses the approach undertaken to conserve a black frock coat and highlights the costume mounting techniques employed.

12.5.1 Historical Context

Maharaja Sawai Man Singh II reorganized the Jaipur State Forces, including the Kachhawa Horse, in 1939 and set up the Rajendra Hazari guards in 1942. He trained at the Royal Military Academy at Woolwich, England, and his selection as the first and only Indian Prince to serve in the King's Life Guards exposed him to the pomp and ceremony of British military life. Maharaja Sawai Man Singh II held the rank of a Lieutenant General in the Indian Army. The reorganizing of the Jaipur State Forces and the raising of regiments meant that several new uniforms had to be created. These uniforms reflected the Maharaja's innate sense of style and aesthetics with the splendour of the Raj, and the comfort and practicality of traditional Indian garments. The costumes include Rajput styled Jodhpur breeches and turbans, Mughal influenced patkas (waist bands), European fashioned caps, boots and accessories. These elements were worn in various permutations and combinations for both day and evening wear and as ceremonial and combat dress.

In contrast, the uniforms of Maharaja Sawai Bhawani Singh of Jaipur reflect a military career spent largely at the warfront for which he was awarded the Mahavir Chakra. In a distinguished military career, he first served as a young subaltern (1954) in the Presidential Bodyguard under the first President of India Dr Rajendra Prasad (1950-62). He was subsequently posted to the headquarters of the 50 Para Brigade in 1963 from where he was selected and posted as Adjutant at the Indian Military Academy, Dehradun (1964-67). After his retirement from active service, Maharaja Sawai Bhawani Singh continued with his public role as titular head of the Kachwaha Rajputs and was appointed India's High Commissioner to Brunei (1994-97). All of these appointments included great ceremony and protocol, and this is seen in his formal clothes of the time.

12.5.2 Documentation of Construction and Condition Assessment

Interestingly photographic evidence shows both Maharaja Sawai Man Singh II and Maharaja Sawai Bhawani Singh of Jaipur wearing the frock coat. It was thought to be tailored in London as it was clearly inspired by the uniform of the Coldstream Guards of the British Army, but as the label was missing the origin was uncertain. The epaulette with a Major's rank insignia; the paratroopers wings and the colors indicates that this was probably last worn by Maharaja Sawai Bhawani Singh as a Major in the Indian Army.

The coat is constructed from multiple layers and parts. The outer fabric is a fine, plain weave black wool, it is lined on the inside with twill weave black silk fabric. The upper bodice of the coat lining is partially quilted in a grid pattern and stiffened with a cotton interlining; the sleeves are lined with twill weave silk. The collar is adorned with thin black cord braid, it has been made to stand upright by using three layers, an outer black wool layer, a stiff cotton fabric layer in the middle and a black silk lining. The coat is front opening, with concealed metal hook and metal eye closures at the top and all along the front edge with metal hooks and metal loops on alternating sides to the waist. It has six flat black mohair lace braid loops with pointed ends on each side; with drops across the chest, fastened with two vertical olivettes (toggles) on each loop; flat braid on the back seams of the arms and back extending down from the waist with tassels on the end; shoulder strap epaulettes of the same black wool material edged with black lace supporting rank insignia of a Major. The gold embroidered cuffs show sun and moon motifs ending with a heraldic crest flanked by lions.

The military frock coat could not be exhibited due to its poor condition. There were visible signs of use, it appeared worn, there was pilling, bruising, abrasion, discolouration, fading, detached and loose embellishments (Figures 12.4a, b). It was creased and distorted from a long period of poor storage. There was dust, dirt and soiling, which was particularly prominent on the fold and crease lines. The goldwork/zari had tarnished over a period and dirt was embedded in the accoutrements. There were the remains of crushed neem leaves (traditionally used to deter pests) and insect debris but no signs of an active infestation.

There was a rip on the outside right shoulder and the sleeve and epaulette were dislocated. Insects had caused substantial damage to several areas of the coat, leaving holes and in some areas causing loss of parts. The collar was the most severely affected, there was partial loss of pile, discolouration, accretions, holes, and associated fraying (Figure 12.4c). The local arid conditions had led to embrittlement of the lace on the collar, which was fragmentary and further loss was likely if not treated. The inner layer of silk lining on the collar appeared to be a later repair as it was secured with uneven herringbone stitches, and it extended into the torso lining rather than retaining the shape of the collar on the inside.

The bodice silk lining was also damaged due to insect infestation, but more disturbing was the patched up unsightly repair and indiscriminate use of herringbone stitches to hold it all together, which was causing more damage. The silk was still flexible, but the holes caused by insects had led to further splits and tears, a large hole, and a detached seam. The lining in the underarms was noticeably discoloured from sweat.

12.5.3 Conservation Treatment

The curators wanted to display this coat in an upcoming exhibition as this was most significant object in the collection having been worn by both HHMS ManSingh II and HHMS Bhawani Singh. It was not possible to display this without remedial conservation. The intervention was aimed at structural stabilization so it may be displayed for six months every year. In consultation with the curators, it was decided to remove the old repairs to the lining that were causing damage.

The cleaning process was strategized based on the condition of each area and a customized approach was adapted. The collar being the most vulnerable was not subjected to any vacuuming, as it could not even be brushed without causing further damage. It was stabilized first, then cleaned. The rest of the areas were lightly vacuumed, cleaned with a latex smoke sponge, vacuumed again and then swabbed with IMS, with the released soiling absorbed onto blotting paper. Humidification was carried out locally using Sympatex to control application of water vapor to gently ease out folds and creases, always after cleaning to ensure that the ingrained dust and dirt did not get embedded in the object when moisture was introduced.

To support the losses and weak areas one had to first source and prepare the support fabrics. The areas that needed the support were identified along with the type/approach to support. Certain areas needed full support; some full support and overlay; some partial overlay; some localized support backing; some localized support and overlay. The support materials were sourced accordingly as follows, for the inner lining twill weave silk fabric; for the outer plain weave wool; for overlay plain weave silk crepeline was chosen. Lanaset dyes were used to dye these support fabrics in the required colors, shades and quantity (Figures 12.4d–f) [9]. Lanaset is a range of 1:2 metal complex and reactive dyes designed to be used on protein fibers. A chromium molecule is pre attached to the dye complex making them 'pre-metallized', making it the modern equivalent of mordant dyeing.

One must decide upon the choice of thread depending upon the material of the object, the nature of deterioration, the required strength, purpose, colour/shade and thickness respectively. After taking all factors in consideration, fine threads pulled from polyester

Stabiltex were used for laid thread couching, and monofilament polyester Skala and polyester Mara (spun from staple) were used as required. No silk or cotton threads were used as neither is as stable as polyester. Only appropriate conservation stitches were used: back stitch, running stitch, long and short stitch, herringbone stitch, zig-zag stitch, slip stitch, laid thread couching and tailors' tacks, no darning was carried out. Melinex and very fine marker/drawing pen/pencil were employed to make stitching plans for the damaged areas. Planning helps to ensure that the support stitches begin and end in strong areas and helps when deciding on the balance of support stitching and laid thread couching. It also helps with estimating the approximate time required to conserve the object.

When the inner silk lining of the collar was removed, what awaited underneath was the original label of the tailor from England along with a hand-written inscription on the manufacturer's label that reads 'H.H. Maharaja of Jaipur'. Once the original inner lining was discovered, it was noticed that it too had been badly damaged in the past and thus prompted the later addition to cover up the visible losses.

The original collar lining was also separated from the tunic for treatment. The tailor's label was kept aside to be replaced last. The inner collar lining was cleaned with applications of IMS, then humidified using Sympatex allowing the weave to be realigned. It was supported onto a new piece of Lanaset dyed twill weave black silk using conservation stitching. The corresponding outer layer of black wool collar was not detached. It had many holes and missing parts. It was supported *in-situ* with Lanaset dyed plain weave black wool. Utmost care was taken to not take any stitches through the brittle silk braid, only the wool base, as it would have resulted in shattered and fragmented parts of the braid. It was treated by encapsulating it with an overlay of sheer plain weave silk crepeline, dyed to match, secured by stitching to the wool (Figure 12.4g). The lapel was also undone, humidified and straightened, reinforced and stitched back in place. As the inner lining overall was crudely repaired with indiscriminate use of herringbone stitches that were causing further strain and damage, it was decided to undo the herringbone stitches. Often historic repairs are left in place as they are part of the history of the object, but in this case the type of stitch and the thread used were far too strong and thus inflicting more damage to the original. The unsightly stitches were removed. The damaged silk was humidified, and the weave re-aligned allowing it to be supported with new twill weave custom dyed silk. The majority of the work was carried out in situ. However, due to its deteriorated condition the decision was made to remove the central upper back panel of the lining. This allowed for more extensive conservation treatment to be undertaken that was required to be able to mount the coat onto a mannequin illustrating how a lot of decisions in conservation also depend on the nature of proposed display. Once the back panel was conserved it and the collar were both stitched back in place and the tailor's label was reattached in its original place on the collar lining. Most of the supported areas in the inner lining were further stabilized by providing an overlay of silk crepeline. This was to protect it from further abrasion during mounting on the mannequin. The holes and tears on the sleeves were supported. The distortions along the inner sleeve lining and the outer layer were aligned and support stitched. The silk braids where fragile and damaged, were given an over lay of silk crepeline. The goldwork/zari embellishments and epaulettes were further spot cleaned using mechanical cleaning methods. The coat was given a final all over humidification treatment, including realigning the mohair braid loops. The loose olivettes were secured and two were given an overlay of silk crepeline.

Figure 12.4 (a) Frock Coat before conservation, (b) documenting coat measurements with digital photography, (c) detail of damage to collar, (d–f) measuring dyestuffs, using hotplate for dyeing, record of dyeing with samples, (g) collar after conservation, (i) fibre-glass torso on left, on right pinning cotton jersey cover to shape on the torso prior to machine stitching, (j, k) on left, padding up torso to shape, on right with jersey top cover and attaching silk covered soft arms, (l) coat after conservation and mounting.

12.5.4 Customizing the Mannequin Mount

Measurements were taken from the coat in order to customize the mannequin. A fiberglass torso of an appropriate size was selected and covered with a base layer of cotton jersey (Figures 12.4h–i). Then the form was built up using layers of polyester wadding secured to the jersey with herringbone stiches, until the correct shape and profile to support the structure of the Frock Coat was achieved (Figure 12.4j) [10]. A double layer of jersey was used to cover the polyester wadding to ensure a smooth surface next to the object, similarly soft arms were made from silk, stuffed with polyester wadding, to enable them to slip easily into the sleeves of the coat (Figure 12.4k). The arms were then stitched to the shoulders of the torso. The treatment was successful, the coat was stabilized, it could be safely handled, and it now has a supportive mount for display at the Jaigarh Fort, Jaipur (Figure 12.4l).

12.6 Developing Display and Mounting Solutions for Flat Textiles and Costumes for the Special Exhibition PRA-KASHI Silk, Gold and Silver from the City of Lights at the National Museum, New Delhi

Every object has unique needs when it comes to mounting for an exhibition display. The vision of the curator and the exhibition designer is taken into consideration along with the physical condition and the construction of the object to present it in an aesthetically pleasing way to the viewer. The experience of working on the 2015 V&A exhibition 'The Fabric of India' during her second internship gave Deepshikha a valuable insight into the process of mounting an exhibition and lots of hands-on experience.

Textiles in the "PRA-KASHI" exhibition were primarily from the ASHA workshop in Varanasi, supplemented with historic examples from the National Museum collection. They represented the art of silk-weaving from the beginning of the first millennium. Technical skills that were lost over the last few centuries were brought alive by the art historian Rahul Jain. Taquete, Samite, Lampas, Velvet, Voided Velvet, Double-Weave Brocades, Damasks and Extended Samite breathe a new life at the drawlooms in his workshop ASHA.

Costumes, fabric lengths, floor coverings, Altar piece, Shawls, Patkas, textile panels, jewellery, and borders from the National Museum and ASHA workshop were to be displayed. Textiles which can be handled safely may not need any conservation to display them as a well-constructed mount can minimize the amount of interventive conservation needed as in this project. Heavier textile panels were supported on slanting boards, some were suspended on rollers following the vision of the exhibition designer (Figure 12.5a). Shorter textiles were attached to the roller with a fabric extension, whereas longer textiles did not need that as there was enough original fabric to be rolled on the roller to support the rest of the hanging fabric without causing stress.

Two 17th century Patkas from the National Museum were to be displayed. These had previously been lined on paper, one had been stored flat and another on wide roller. The design plan was for them to be displayed together in a glass box. Upon examination it was discovered that the flat Patka had some fungal growth, the solution was to display it separately in a glass frame away from other objects (Figure 12.5b). The display roller for the

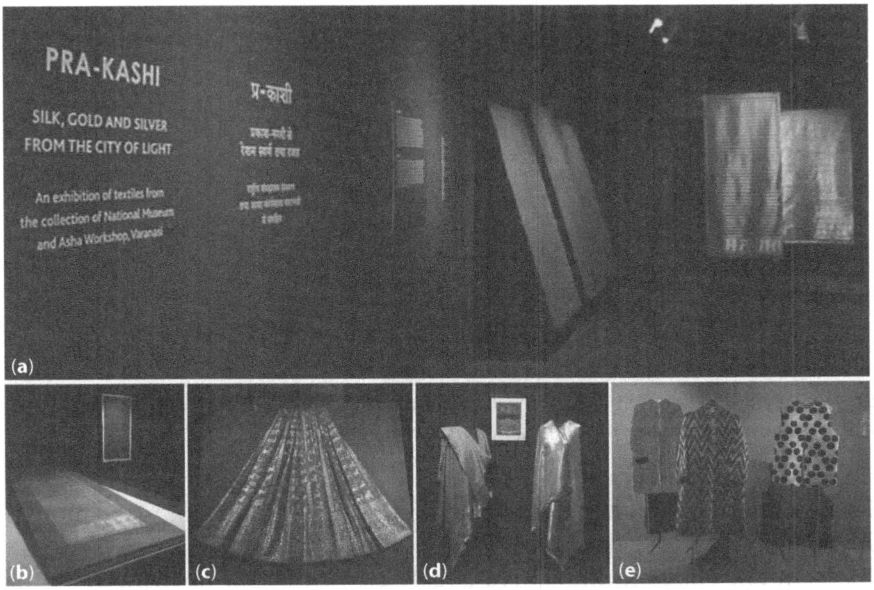

Figure 12.5 (a) Pra-Kashi exhibition at National Museum, Delhi, (b) cased textile, (c) shikargah lehenga supported with magnets, (d) shawls draped on mannequins, (e) velvet jackets supported on Pi bars.

rolled Patka was made of the same diameter as the one on which it was stored, to avoid undue stress to the object and to maintain its equilibrium in the rolled state.

Early 20th century Shikargah brocaded lehenga panels were supported for display with rare earth magnets [11]. The magnets, covered with dyed silk, were first stitched to the display board at strategic points following the planned drape of the Lehenga. The Lehenga was placed on the board and corresponding fabric covered magnets were placed on top to hold the Lehenga in place (Figure 12.5c). The board was placed at an angle instead of vertically for better support. A 17th century Altarpiece was also displayed on a slanted board as it couldn't be displayed vertically. On the reverse side of the Altarpiece, there was a provision for a drawstring, which was wide enough to pass a flat cotton band through that was fixed to the slanted board to hold the Altarpiece in place. A 17th century velvet floor covering that was very weak, was displayed flat in a raised box to provide a better viewing experience.

Shawls were draped on mannequins; costumes were displayed on specially constructed collapsible Pi bar rather than T bars (Figures 12.5d, e). This helped provide a more balanced support and enhanced the visual experience because of shifting the central vertical bar to either side under the shoulder. Magnets covered with Lanaset dyed fabrics were attached on the inside of the garments to hold them in place on the mount to avoid slippage [11].

All the covered boards and rollers were lined with polyester film as a precaution and the fabric used to cover them was scoured to get rid of any surface finishing. The light levels were always maintained under 50 lux.

12.7 Technical Analysis and Documentation

Technical analysis and associated documentation are vital components of the conservation process. They can reveal valuable information for researchers, historians, and curators, and

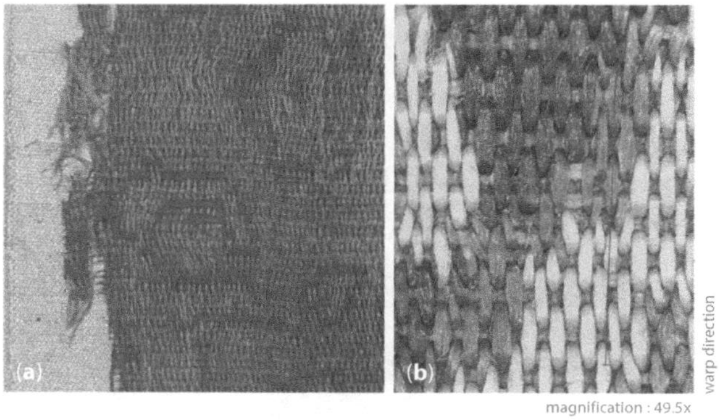

Figure 12.6 (a) Detail of a fragment of figured silk, pasted on silk, lined on Japanese paper 6th - 9th century CE, National Museum, New Delhi, (b) silk fragment under magnification.

with greater technical and material information the conservator can make more informed decisions about object treatments.

Textiles from ancient civilizations are largely created by the weaving of natural fibres that occur locally. Differences in cultivation, processing, spinning, weaving and dyeing are often unique to the place of origin [12]. Ancient cotton textiles woven with a z twist thread [13], for example, are usually associated with India. Weave structures evolve over time due to trade, socio political and cultural influences. Plain weave, tapestry weave, warp faced compound tabby, weft faced compound tabby, damask, leno weave, twill, samite, lampas, velvet, and voided velvet are some examples of woven textiles that define the history of textiles. Examining weave structures, loom widths, dyes and patterns can help in establishing provenance.

The Cieta course (Centre International d'etude des Textiles Anciens) provides intensive training in weave analysis. Deepshikha Kalsi undertook this training after being introduced to it during her internship at the Abegg-Stiftung Foundation in Switzerland. It proved to be an invaluable tool in the study of ancient textile fragments from the Stein collection at the National Museum of India, New Delhi [14], which was carried out in collaboration with textile historians Rahul Jain and Vinay Singh and curator Binoy Sahay, see Appendix I Technical study of NM: 2003/10/717 [15] (Figures 12.6a, b). The study of fragments from the Stein Collection at the National Museum, New Delhi, also revealed information on the early development of textile conservation. Most of the silks fragments from this collection had been pasted and supported on Japanese paper. Some fragments were pasted on silk and then lined with Japanese paper [16]. Painting and paper conservation were established as professions much earlier than textile conservation which eventually evolved out of them.

12.8 Training and Outreach

Over the last 10 years, there has been a growing awareness in India of textile conservation as a profession. Significant advances have been made but huge challenges remain. There are academic qualifications in textile conservation, but the teaching of advanced textile practical techniques is still lacking. Many museums in India employ the rafugars to repair historic textiles but the difference between their skills and those of a professional textile conservator is still not fully understood. Individual textile conservators have been coming to India to work on important Indian textiles since the 1980s [17]. More recently textile conservators from the West have conducted workshops to raise awareness of the profession in India.

12.8.1 Case Study—Indian Museum, Kolkata

SUTRA (www.sutratextilestudies.com) was founded in 2002 in Kolkata with the aim of helping to raise awareness of the need to preserve living textile traditions, to disseminate knowledge about textiles and to promote the preservation and conservation of India's unique textile heritage. SUTRA organized a series of workshops on textile conservation in conjunction with the Victoria and Albert Museum beginning in 2010 [2]. In 2018, Deepshikha Kalsi was invited by Lynda Hillyer to participate in the teaching of a third practical workshop organized by SUTRA and the Indian Museum in Kolkata, following workshops taught by V&A textile conservators in 2010 and 2011 (Figure 12.7a).

The purpose of this workshop was to teach basic handling and safe ways of storing the museum's textile collections. A course book based on handouts prepared by textile conservators at the V&A was created for the participants. Two pieces of equipment, an ultrasonic humidifier and a Dino-lite (a digital microscope which enables the viewer to see magnified images on a computer) were purchased for the museum. There were about 16 participants the majority of whom were museum staff. Two days of lectures were followed by intensive demonstration and practice of basic techniques. Handling and movement of objects, cleaning, humidification, and basic conservation stitching were covered, as well as the key components of condition reporting (Figure 12.7b). The participants worked in teams to practice making covered rollers, padded hangers and support cushions, all necessary for the safe storage of textiles.

12.8.2 Case Study—The Registry of Sarees, Bangalore

Workshops are a useful tool in raising awareness, but continuity of good practice can be at risk because of a shortage of trained textile conservators to continue the process. Some private museums and collectors are actively engaged in contributing to good practice.

Figure 12.7 (a) Lynda Hillyer teaching workshop participants at the Indian Museum, Kolkata 2018, (b) practising conservation stiches at the 2018.

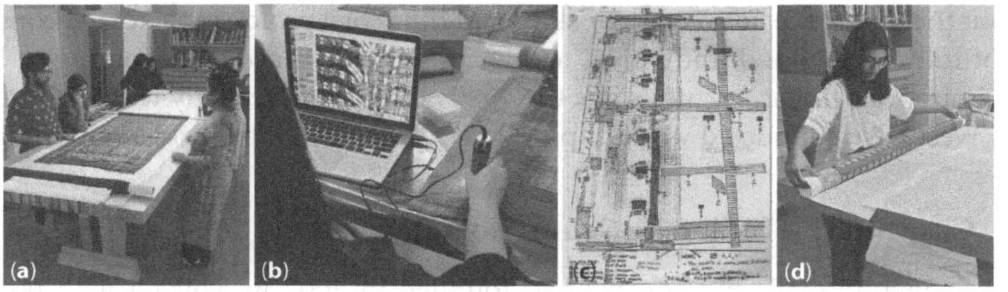

Figure 12.8 (a) Examining sarees at the Registry of Sarees, Bangalore, (b) analysing saree weave structure with Dinolite microscope, (c) documenting saree condition, (d) practising rolling a saree during workshop.

The Registry of Sarees in Bangalore, for example, has initiated staff training so that processes that have been shared with them can be continued. Deepshikha Kalsi conducted a week-long workshop for this collection on documentation (Figure 12.8a–d). A form was created to document every aspect of a saree from weave structure to its present condition. The team became very skilled in this task and were able to document a significant collection of Kalakshetra sarees. The team standardized the condition assessment form and the clarity of its terminology facilitates communication within the team. It can also be used more widely in the field. Sharing of knowledge is absolutely essential to the growth of the textile conservation profession in India.

12.9 Conclusion

The practical experience gained through the internships at the V&A and the Abegg Stiftung provided a solid ground for Deepshikha Kalsi to build up her own Textile Conservation Studio in India. However, one needs to constantly keep up to date with developments in the field across the globe and this is only possible through the support of colleagues and sharing of information. For example, the introduction of a new surfactant Dehyphon LS 54 and advice on managing the wet cleaning set up for the Jama was crucial for the treatment undertaken at the studio. Treatments developed in the West are often reevaluated as conservators gain more insights into material knowledge and new products are developed.

It is important to share skills and knowledge locally to raise awareness of the profession. The opportunity to undertake in person training courses cannot be easily replaced when it comes to teaching the practical elements of conservation practice but thanks to advances in modern information technology, it is easier than ever to keep up to date with developments in the field and to nurture and maintain links with colleagues across the world. Following the COVID-19 pandemic many conservation organizations have hastened the shift to providing digital content, streaming talks, and conferences via online platforms and by doing so democratizing access to information. There have never been so many opportunities to share information across the globe, all that is required is a shared vision to serve the field!

Acknowledgments

Deepshikha Kalsi would like to thank: my Gurus, Frances Hartog and Rahul Jain. My mentors Girikumar Sekharakurup and Elizabeth-Anne Haldane. Lynda Hillyer, Marion Kite and Joanne Hackett former Heads of Textile Conservation, V&A, and the team of textile conservators at the V&A. I am deeply grateful to Dr. Regula Schorta (Director), Bettina Niekamp (Head of Textile Conservation) and Corinna Kienzler (Deputy Head of Textile Conservation), from the Abegg-Stiftung to take me deeper in the field and give exposure to research and methodology. Manuela Scheurer and the team of Abegg-Stiftung Foundation for sharing their knowledge and special thanks to Andrea Meuer, Head of Library.

We are deeply indebted to Parasol Foundation (formerly The Bonita Trust), Nehru Trust, Charles Wallace India Trust, Simon Digby Trust, Abegg-Stiftung Foundation, Rosemary Crill and Richard Alford for helping secure the fellowships. Eka Archiving Services Ltd, Pramod Kumar K.G. and Deepthi Sassidharan. Jaigarh Fort Charitable trust, Princess Diya Kumari, Giles Tilotson, Mrinalini Venkateswaran, Udai Bilas Palace, Maharaj K Harshvardhan Singh Dungarpur, Maharaj K Priyadarshini, Sutra Textile Studies, The Registry of Sarees. Special thanks to team at National Museum, B.R. Mani (Former Director General), P.K. Nagta(former Head of Conservation) and Binoy Sahay(Curator, Central Asian Galleries) for inspiring to undertake the technical studies of the Stein Collection at the National Museum. Vinay Singh for his invaluable insights.

Heartfelt thanks to Elizabeth-Anne Haldane and Lynda Hillyer for help in editing this chapter.

References

1. Kalsi, D., Creative Pathways: Developing a Career in Textile Conservation in India, Lessons learned Textile Conservation-Then and Now. *12th North American Textile Conservation Conference*, Ottawa-Gatineau, Canada, pp. 341–355, September 23-27, 2019, http://www.natc-conference.com accessed 24/08/2021.
2. Haldane, E.A., Glenn, S., Hunter, S., Hillyer, L., RAKSHA: Raising Awareness of Textile Conservation in India, The Textile Specialty Group of the American Institute for Conservation (AIC). *40th Annual Meeting Post-Prints*, vol. 22, pp. 19–31, 2012.
3. Crill, R., *Chintz: Indian Textiles for the West*, p. 69, V&A Publishing, London, 2008.
4. Haldane, E.A., Encounters with paper conservation: the treatment of a Chinese painted silk dress. V&A *Conserv. J.*, 49, 14–16, Spring 2005.
5. Goswami, B.N., Men's Garments-Jama, Angarkha, in: *Indian Costumes in the Collection of the Calico Museum of Textiles*, pp. 25–86, Sarabhai Foundation, Ahmedabad, 2010.
6. Sato, M. and Quye, A., Detergency evaluation of non-ionic Dehypon® LS54 for textile conservation wet cleaning. *J. Inst. Conserv.* (ICON), 42, 1, 3–17, 2019.
7. Hillyer, L., RAKSHA–Raising Awareness of Textile Conservation in India Elizabeth-Haldane, Sarah Glenn, Susana Hunter, and Linda Hillyer. In Textile Specialty Postprints Presented at Textile Subgroup Sessions. 22, 19–31, 2012.
8. Venkateswaran, M. and Tillotson, G. (Eds.), *Masterpieces at the Jaipur Court*, pp. 188–189, Niyogi Books, New Delhi, 2021.

9. http://www.vam.ac.uk/blog/conservation-blog/textiles-dyeing-support-fabric. Accessed 22/10/2021.
10. Flecker, L., *A Practical Guide to Costume Mounting*, Routledge, UK, 2007.
11. Spicer, G., Types of Magnetic Systems, in: *Magnetic Mounting Systems for Museums & Cultural Institution*, pp. 69–96, Spicer Art Books, New York, 2019.
12. Böhmer, H., *KOEKBOYA Natural Dyes and Textiles-A Colour Journey from Turkey to India and Beyond*, The University of Chicago, Chicago, 2002.
13. Barnes, R., *Indian Block Printed Textiles in Egypt, 2 vols*, p. 49, Oxford University Press, Oxford, 1997.
14. Sahay, B.K., Technical Studies of Central Asian Textiles(with emphasis on Astana, Kara-Khoja and Lou-Lan Sites), in: *Textiles: Binding threads Between Cultures from National Museum Collections*, National Museum, New Delhi, 2019.
15. Jain, R., Kalsi, D., Singh, V., *Ancient Textile Fragments From The Central Asian Antiques Collection National Museum*, INTACH, New Delhi, 2021.
16. Joshua, J., Restoring of Ancient Textiles. *Embroidery*, 2, 15–18, September 1933.
17. Hillyer, L., *Textile Conservation in India–a View from the West, Eastern Review*, Press, Poland, 2022.

13

Symbolism and Conservation of Indigenous African Textiles for Museums

Raphael Kanyire Seidu[1]*, Ebenezer Kofi Howard[2], Edward Apau[3] and Benjamin Eghan[2]

[1]*School of Fashion and Textiles, The Hong Kong Polytechnic University, Hung Hom, Kowloon, HongKong*
[2]*Department of Industrial Art, College of Art and Built Environment, Kwame Nkrumah University of Science and Technology, Kumasi, Ghana*
[3]*Department of Visual and Industrial Art, Faculty of Built Environment and Applied Art, Sunyani Technical University, Sunyani, Ghana*

Abstract
Indigenous textiles from Africa are iconic products that eulogies the culture and history of the people. These textiles, made of different techniques, are collected and exhibited in museums across the globe to promote and conserve the symbolism, craftsmanship, uniqueness, and identity of African tradition for sustainability. This chapter identifies and discusses indigenous African textiles with specific reference to aesthetics, techniques, and symbolism for museum exhibition. African ethnic patterns come in unique varieties and the chapter would specifically look at the Kuba raffia cloth (D.R. Congo), Akwete cloth (Igbo, Nigeria), Ukara cloth (Igbo, Nigeria), Aso Oke fabric (Yoruba, Nigeria), Adire cloth (Yoruba, Nigeria), Fugu/Smock weaves (Northern Ghana), Kente cloth (Asante, Ghana), Adinkra cloth (Asante, Ghana), Kete cloth (Ewe, Ghana), Bark cloth (Ganda/Buganda, Uganda), Berber cloth (Morocco), Mud cloth (Bambara, Mali), Kanga (Tanzania), Shweshwe (South Africa), Shuka cloth (Kenya and Tanzania) and Ankara or African Wax Prints (West Africa) or Kitenge (East Africa). Further details on museums in Africa especially, and not limited to their challenges, organization and prospects, would also be dealt with.

Keywords: Indigenous African textiles, museum textiles, symbolism, aesthetics, traditions

13.1 Introduction

Indigenous African Textiles are iconic crafts that are produced and cherished by the indigenes. These textiles collectively define and represent the philosophy, history, concepts, culture, ethics, political activities and philosophies of the people. Acquaye *et al.* [1] positioned that indigenous African textiles express the ethnicity, religion and social status which aids in knowledge acquisition and understanding of the multitude of human issues [2]. Essentially defined as dress and cloth, these textiles reflect the practices, philosophy, beliefs,

**Corresponding author*: seiduraphael@gmail.com

Seiko Jose, Sabu Thomas, Pintu Pandit, Ritu Pandey and Vandana Gupta (eds.) Handbook of Museum Textiles: Volume I Conservation and Cultural Research, (239–266) © 2023 Scrivener Publishing LLC

and culture of the people [3]. As a contributor to the sustenance and preservation of the heritage of the people, indigenous African textiles are produced using different materials, such as raffia, cotton, wool, silk, bark fibers coupled with natural dye pastes, and techniques, such as embroidery, applique, weaving, dyeing, hand painting and printing. These textiles carry motifs, symbols, images, colors, and texts, which highlight the symbolic connotations about the political, social, beliefs, events, and religious complexities among others of the indigenes. A clear path to significantly connect and associate their past with the present era. The use of colors, motifs, and symbols produces a metaphoric ideology which according to [4] communities and connects with the people. For example, an *akwete* cloth with tortoise motif is used by a royal to symbolize his status [5] and a *berber* cloth woven in white colors which symbolizes purity is generally used during Friday prayers.

As a major craft form that communicates the cultural heritage of Africans to the World, these indigenous textiles has remained a vital museum collection and source of knowledge by anthropologists, students and the very indigenes. Due to the composition of these textiles, which are normally from natural sources, its conservation is treated with the appropriate conditions to prevent damages. In such situation, indigenous African textiles are preserved under controlled humidity and temperate conditions, stored in pest-free areas to ensure its structures are kept. Exhibited in museums, these textiles provide the history and philosophy about the culture of the indigenes hence an in-depth understanding and appreciation about their craft.

13.2 Types of Indigenous African Textiles

With an inherent symbolic connotation, these textiles are made of different motifs, colors, shapes, and symbols. Indigenous African textiles have a metaphoric import that connects the symbolism and characteristics of the cloth with the people [4]. They come in different features, techniques, origin, and symbolism from the parts of Africa; North, South, West, East, and Central. Table 13.1 identifies the popular indigenous African textiles that define the craftsmanship and identity of the people in the continent. Indigenous African textiles are categorized generally according to their mode or tradition of production.

13.2.1 African Weave Traditions

Simple structured traditional looms are used to produce plain or design weaves that are either predominantly warp-faced, weft-faced with geometric patterns and figurative images. Popular weave traditions in Africa are smock/fugu, aso-oke, kente, kete, akwete, berber, shuka, and kuba raffia.

13.2.1.1 Smock Weaves/Fugu

Smock is the name given to a narrow strip woven fabric normally 3 to 5 inches [6] originally produced for the royals and chiefs by the indigenes in Northern Ghana. These strip fabrics are woven on narrow double heddle looms [7] with plain weaves, which have existed for centuries. The smock is a warp-faced fabric where the warp arranged in different colored stripes are woven to cover completely the weft yarns. Traditionally, cotton yarns obtained

Table 13.1 Summary of the indigenous African textiles and techniques.

Location	Country	Indigenous African textiles	Techniques
West Africa	Ghana	Smock/Fugu	Weaving and embroidery (to create embellishing effects on the neckline of fugu
		Kente	Weaving (hand picking)
		Kete	Weaving (hand picking)
		Adinkra cloth	Dyeing and block-printing
	Nigeria	Akwete Ukara	Weaving Resist-dyeing
		Oso-oke	Weaving
		Adire	Resist-dyeing
	Mali	Mud cloth/Bogolanfini	Weaving and hand painting
		Ankara/African Wax Prints	Printing
East Africa	Uganda	Bark cloth	Peeling and Beating
	Tanzania	Kanga	Printing
	Kenya/Tanzania	Shuka cloth	Weaving
		Kitenge/African wax prints	Printing
Southern Africa	South Africa	Shweshwe	Printing
Central Africa	DR Congo	Kuba raffia cloth	Weaving
North Africa	Morocco	Berber cloth	Weaving

from hand spinning are used or further dyed with natural indigo dyes (popular in Daboya). Even though, weavers still adopt cotton dyed yarns, majority of the strip fabrics are woven with synthetic yarns. The smock has symbolic significance where the naming is created from the arrangement of the colored warp, the end user and the type of colors used [6]. For the smock weaves, a double heddle loom is used by weavers to create plain weaves and warp-faced structures. Designs are created by the dyeing effects on the warp or the arrangement of the colored warps. The use of different weft colors adds no effects to the fabric woven since the warp covers the weft entirely. These strip woven *smock* are fashioned either

Figure 13.1 *Fugu* of Northern Ghana [9].

by hand and machine stitches to produce a garment *fugu* (Figure 13.1) for both males and females. This *fugu* was placed on the spotlight when Dr Kwame Nkrumah and his associates worn it to declare the independence of Ghana on March 6, 1957 [8]. The *fugu* comes in different types; sleeves, short sleeveless, long sleeves and the regalia for the traditional priest in the society.

13.2.1.2 Aso-Oke

It is a woven fabric of the Yoruba people in the Southwest of Nigeria. *Aso-Oke* (Figure 13.2) is sometimes known as *Aso-Ofi* [10] and translates as *Aso awon ara oke*, meaning cloth of the people from the hinterland [11]. The traditional fabric is woven with elaborate strands of dyed yarns which features a predominantly warp-faced structure [12] even though weavers sometimes employ supplementary wefts to produce pronounced effects in the strip. Furthermore, weavers in Yoruba employ two distinct loom types: upright single heddle loom and double heddle loom. The upright single looms are commonly used by women to produce 30 to 90 cm width of fabric on a fixed vertical loom with tensioned warp, whereas the double heddle loom is used by men to produce 14 to 15 cm width of fabric on a horizontal loom with tensioned warp stretched in front of the weaver [12]. With the double heddle loom, depending on how many of the warp threads are lifted, the fabric

Figure 13.2 *Aso-oke* of the Yorubas in Nigeria [2].

can be designed either in warp or weft-faced patterns. The symbolic values or significance of the *Aso-Oke* cloth by the Yoruba people are embedded among three major types: *sanyan* (woven in pale brown color), *etu* (woven in dark blue with light thin blue stripes) and *alaari* (woven in crimson color). Olajide *et al.* [12] further opines that variant factors are considered during designing and weaving of *Aso-Oke,* and they include the nature and color of fibers employed, relationship between warp and the weft, which may be affected by the loom, and the possible methods of embellishing a fabric after manufacture. According to [12], *Aso-Oke* comes in three major types; *Etu, Alaari* and *Sanyan* with many variations achieved with the use of extra weft brocading technique which are identifiable by their patterns and color to inform their uses at a designated traditional ceremony. *Etu* (fowl) features strips woven with local wild silk fiber (dyed repeatedly in indigo blue dye). *Alaari* is traditionally woven with locally spurned silk yarns dyed in red cam wood solution severally to achieve permanence in color fastness. *Sanyan* is traditionally produced from fibers made from the cocoons of the anaphe silkworm. Historically, the silk fibers are hand spun into silk threads, washed, and soaked in corn starch to strengthen the yarn. *Sanyan* is regarded as *baba-aso* (father of fabrics) and is classified as the most expensive of all Yoruba woven fabrics. The use of these fabrics to events and ceremonies symbolizes productivity, protection, wealth, prosperity, and blessings [13, 14]. The choice of a color type in the fabric has a significant connotation. For example, white symbolizes purity; red, productivity; and black, sadness or strength [15].

13.2.1.3 Kente

Kente is known as *kenten* (basket) in the Akan word. *Kente* (Figure 13.3) is a traditional woven fabric produced by the Asantes in the Ashanti Region of Ghana. This craft weaving is popular in the Bonwire and Adanwomase communities. *Kente* was originally woven with silk yarns unravelled from imported European fabrics [16], later with cotton yarns and currently with synthetic yarns. The fabric features geometric shapes with recent introduction of the art of lettering inspired from the Ewe weavers [17]. Weaving was restricted to men in the quest to prevent infertility in women should they weave. With varying debate on its origin, *kente* was traditionally woven for the chiefs and royals in the society. Designs are produced on *kente* with different shaft looms to produce simple to complex figurative images on the woven fabric. These looms are used alongside its accessories and materials to produce designs dependent on its complexities. Designs or motifs are created by hand-picking to produce geometric patterns that are weft-faced structures. *Kente* is produced with four different types of looms: two-shaft loom, two-shaft looms with supplementary healds, four-shaft looms and six-shaft looms, depending on the simplicity or complexity of the design and motifs (either single, double and triple weaves). These fabrics in narrow strips about 4 cm [18] come with very bright colors, interesting designs and motifs that carry symbolic meanings. Names are derived from proverbs, religious beliefs, political ideologies, customs, historical events, wise sayings and important personalities in the society. For example, *akyempim* cloth symbolizes the military prowess of the Asante military in the war era [19]. Colors carry certain meanings (e.g., yellow for royalty, wealth, prosperity; blue for love and happiness; red and black for sadness, death; green for fertility, growth; and white for purity) and as such weavers employ particular color type for specific events and people (e.g., red,

Figure 13.3 *Kente* of the Asantes in Ghana [3].

black, and dark hues for funerals; blue and silver for Queenmothers; brown for war and white *kente* for funerals for the very aged individuals).

13.2.1.4 Kete

The etymology of *kete* (Figure 13.4) as *ke* "to open" and *te* "to press" [17], is a traditional fabric produced by Ewe weavers in the Volta Region of Ghana. *Kete* weaving is widely centered in Agotime-Kpetoe and Agbozume. *Kete* is a hand-woven textiles produced on both fixed and movable narrow looms which features figurative and nonfigurative images. Unlike the Asante *kente, which* is made of bright colored yarns, the Ewe *kete* utilizes calm and muted yarn colors. In fact, [20] identified three main categories of Ewe *kete*: cloths with figurative images, plain weave cloths and cloths woven with supplementary warps. The weaving of figurative motifs in *kete* cloth is a distinguishing feature of Ewe weaving which sets them apart from their Asante *kente* counterpart that focus exclusively on geometric patterns. Just like the Asante *kente*, weavers hand-pick the weft to create both nonfigurative and figurative images in *kete* cloth. According to the nature of design, supplementary healds in the form of cords through which selected yarns are threaded are not controlled by the feet as the others, but are lifted in turn by the hands to create the desired effect. The art of intersecting letters in the woven structure originated from the Ewe weavers [17] which was later adopted by the Asante weavers to produce strip fabrics commonly used at graduation ceremonies. Symbolic significance is derived based on color used, arrangement of warp colors and how

Figure 13.4 *Kete* of the Ewes in Ghana [4].

the strips are joined to form a cloth, and the structure of the strips [20]. Design or pattern symbolism draws on the daily activities of life, e.g., tree cloth *Ativo* symbolizes the attention and care required for adults to train children to be responsible individuals. The meanings and names further has an association with confidence, moral alertness and good governance [21]. *Kete* is generally worn with no restrictions to the colors used for events and occasions in the society. Generally, some colors are commonly used for certain events e.g. black, brown, red for funerals, and black, wine for enstoolment of new chief.

13.2.1.5 Akwete

Mkpuru Akwete meaning Akwete fabric [22] or *Akwa Miri* (cloth of the water) meaning towel [23] is a traditional woven fabric produced in the *Akwete* community of the Abia State in Nigeria. The name of the town influenced the naming of the cloth, which remain a cultural pride and source of livelihood. It is woven by the Igbo women on a wide vertical wooden loom using coarse raffia, spun cotton and sisal-hemp. *Akwete* cloth (Figures 13.5, 13.6) is characterized by abstract, geometric and figurative patterns, produced by utilizing supplementary weft patterning and tasseled ends [24, 25]. This cloth is woven to show a double faced structure (at both sides) or single faced (at one side). It takes approximately seven days to weave depending on the simplicity or complexity [26]. *Akwete* is known for its hand-woven fabrics made by women (Figure 13.5) using a vertical loom [60]. According to [25], the process makes use of equipment and materials such as vertical looms, heddles, beaters, swords, cotton yarns, twines, brass spatula, ropes, shuttles and wax. They further described the *Akwete* loom as a rectangular wooden frame with two horizontal posts. They also reported that, a primary heddle rod and other supplementary heddle rods are used depending on the complexity of patterns to be woven. It is revealed that the *Akwete* cloth has a wider width than most vertically woven counterparts. This is because the tension on the warp yarns are loose and spread apart but as the weaving progresses the tension becomes tighter and the weaver makes the cloth narrower [60]. The weaver picks the appropriate designs on the tensioned warps.

Historical accounts attribute the origin of this traditional weaving from Dada Nwakwata. Through trade activities with the Portuguese in the 14th to 16th centuries, a woven cotton cloth "Acham" was imported into Nigeria and very popular in the Abia State. Dada Nwakwata critically studied the fabric, unravelled the threads and secretly used them to

Figure 13.5 Woman weaving Akwete cloth on the traditional loom [60].

Figure 13.6 *Akwete* cloth produced by the Igbo women in Nigeria.

weave a new design style. Dada Nwakwata had to weave secretly because, women were restricted or prohibited to engage in any economic activities. Her creative activity was only known to his deaf and dump friend, who revealed her craft after Dada Nwakwata died. This therefore engaged the women in the area to take-up the weaving of the craft with less interferences from the men (forbidden to weave the cloth). *Akwete* cloth has uses: dresses, drapes, covers, bedspreads, wall hangings among others depending on the versatility, shape and size [27]. Red and black designs woven in geometric patterns over a white background are widely used by the Igbo men as towel [25]. With several motifs or patterns in existence, these motifs have symbolic meanings that represents the status, circumstances and prestige in the Igbo culture [27]. For example, motifs are reserved for the royals (tortoise motif "ikaki"), as talisman to protect pregnant women and warriors and special occasions.

13.2.1.6 Berber Cloth

The *berber* cloth (Figure 13.7) is woven by the *Amazigh* women in Morocco. These women utilize a vertical indigenous loom made of wooden beams to produce the cloth [28]. The loom is carefully developed for easy dismantle and assemblage. Weavers adopt the flat weaving technique to horizontally insert the weft yarns into the warp. These yarns are then separated with a shed stick and heddle rod. The *berber* cloth features predominantly geometric patterns such as zigzag, triangle, diamonds and chevrons [29] woven widely with red colored threads. In addition, other colors like brown, yellow, white, green and black. As a strictly female craft, patterns or motifs and colors used indigenously emphasize the fertility

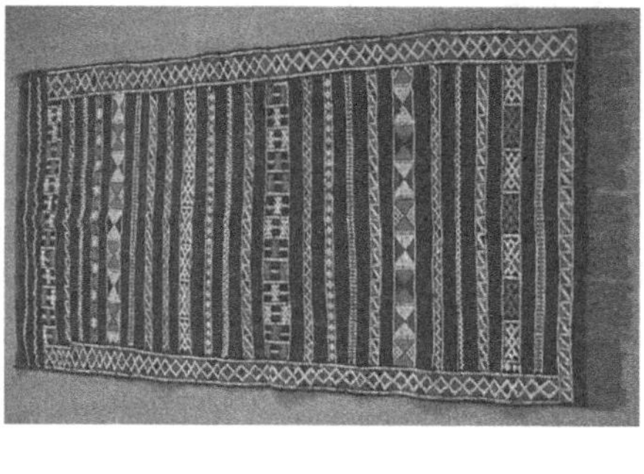

Figure 13.7 *Berber* in Morocco.

of females. Traditionally, the white cloth is used at the mosque for Friday prayers [29]. Symbols woven into the *berber* cloth carry symbolic significance. These include the eye motif (symbol for protection against evil eye), the saw, sickle and scissors motifs (symbol of metalworkers), the seeds motif (symbol for fertility), the finger motif (symbol for protection) and the Lion's paw motif (symbol for strength and protection) among others [5].

13.2.1.7 Shuka Cloth

It is a traditional cotton cloth of the Maasai communities in Kenya and Tanzania. *Shuka* in the Maasai language means "body wrappings," (Figure 13.8) features a plain checkered design (plaid) with predominantly red colors with others such as blue. The reason for the dominant use of red colors in *shuka* is to provide some camouflage and defend to the Maasai people from predators in their reddish savannah environment. The cloth generally known as *African Blanket* as opined by Johnson [30], are colored in red with black stripes, blue, black, checkered and striped fabric. Historically, the *shuka* cloth commonly worn by the Maasai people around their shoulders was indigenously made with animal hides or skin. The pattern characterizing the *shuka* cloth, long existed and was not inspired from the Scottish [30]. As a cultural symbol of East Africa, the wearing of the cloth varies based on the place, age and sex. Essentially, the black *shuka* cloth is popular among the young men after months of circumcision.

13.2.1.8 Kuba Raffia Cloth

Kuba cloth (Figure 13.9) is a traditional textile of the *kuba* people of Democratic Republic of Congo. It is made from intersecting raffia strands on an inclined heddle loom [32] by male weavers in the community. The strands were initially colored with vegetable dyes. After the woven cloth is produced, the *kuba* women finish the cloth using decorative techniques such as applique and embroidery. The cloth features pile surfaces, geometric patterns in shades of tan and black [31]. Traditionally, the *kuba* cloth is joined together to form a larger piece which symbolizes the supremacy and wealth of the rulers [32]. In addition, designs are named after the weaver who created them. *Kuba* ceremonial skirts are truly a community creation from Congo. Naturally, the process begins with stripping and splitting raffia leaves from raffia palm. Some of these fibers are smoothened for embroidery with snail shells.

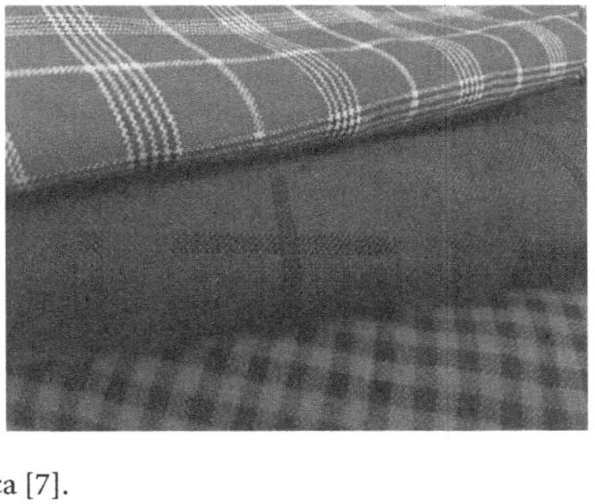

Figure 13.8 *Shuka* in East Africa [7].

Figure 13.9 *Kuba Raffia Cloth* in East Africa [31].

Figure 13.10 (a) Weaver weaving *Kuba* cloth. (b) Woman embroidering Kuba cloth [59].

The fibers are made into skeins for weaving. The raffia cloth is produced on an oblique single-heddle loom at 45-degree angle (Figure 13.10a). Depending on the original size of the palm leaves, a square size of *kuba* is made. Once the cloth square is cut from the loom, it is stiff and course. It is softened by bleaching, wetting with hot water and pounding sometimes in a mortar before embroidery is done [59]. Applique technique is used to cover holes that appear in the fabric due to continuous pounding. Subsequently, the cloth is dyed and embroidered (Figure 13.10b) using techniques such as overstitching and cut-pile, or plush.

13.2.2 African Dye Traditions

Africa has a rich history of indigenous dyeing of fabrics using natural indigo dyes. This ancient method has existed for centuries where the skill has been transferred from generation to generation. The textile produced from this process contains symbols that have cultural and symbolic value that highlights the heritage of the people. Popular textiles from the dye traditions in Africa are *adire, ukara,* and *mud.*

13.2.2.1 Adire

Adire (Figure 13.13) which traditionally means tie and dye, is an indigo-dyed cloth produced with resist patterns by the Yoruba people of Nigeria [33]. Alternatively, [34] established that, àdìre constitute of three words namely; à means to take, *dì* means to tie and *re* means to dye. This produces dark blue hues; very characteristic of *Adire* cloth. The first *adire* was produced in the town of Abeokuta, which according to [33] was dyed on a traditional handwoven cloth *kijipa* with indigo. This coupled with the re-dyeing of faded clothing was practiced by female dyers to produce interesting resist patterns. Patterns or motifs produced on the *adire* cloth reflected the philosophy, religion, nature and life of the Yoruba people. Traditionally, *adire* production involves contributions from decorators (aladire) for creating the resist patterns and two female dye specialists (alaro) for controlling its production and marketing. *Adire eleko* (resist using cassava and starchy maize paste) and *Adire oniko* (resist using raffia) were the oldest form of techniques produced by dyers to produce soft and saturated blue hues against white designs. Tie-dye is a technique that is widely practiced in many parts of the world. This is especially true for Africa esp. Nigeria where *Adire*, an important Yoruba cloth is produced using this technique. This method involves tying or stitching fabric together to prevent the absorption of dye to a particular area and patterns could be made by pleating of cloth [35]. The Yoruba's *Adire* patterns are made by pleating, then knotting or stitching with raffia fiber from banana leaves or thread before immersion for dyeing to happen. Seeds or pebbles are sometimes tied into the cloth to create patterns; some impressions may be sown into the fabric with thread and the stitches drawn up tightly before dying. Yoruba dyers engage the use of mordants to enhance the fastness of dyed fabrics to sunlight and washing and for a brilliant appearance. Forms of Adire include *Adire Oniko*, *Adire Alebere* (stitch resist though considered to be a version of *Adire Oniko*), *Adire Eleko* (the resist agent is starch), *Adire Eleso* (pebbles tied to resist), and Adire Batik (wax resist) [35, 51]. Pits for indigo dyeing are evident across West Africa, including in the Dafina and Samogo regions of Burkina Faso, in the Kolokani region of Mali, in Northern Benin and in Central, Northern Ivory Coast and in Savanah region in Ghana [52]. Indigo dyeing traditions among Maranse of Burkina Faso and Daboya of Ghana are focused on yarn dyeing while Maranse people dye the woven cloth too. To pit-dye textile fabrics, the pit (Figure 13.11) is filled with about 400 liters of water with six large loaves of indigo added plus 20 kilograms of wood ashes (shea tree/Vitellaria paradoxa or marula tree/Sclerocarya birrea) to fix the dye on the fiber [52]. After the dye bath is prepared, the cotton spun yarns are immersed to be colored. This takes a few hours to complete. When dried, the yarns are woven into long narrow strips that are later cut into shorter versions and stitched together to form a fabric. Tie-dye technique is sometimes employed to give an attractive effect when the final piece is woven. The *Maranse* people now combine pot dyeing (Figure 13.12) that is best for synthetic dyes, which are cost effective.

Subsequently, techniques such as *Adire batani* (resist using starch with zinc stencils) and *Adire alabere* (resist using stitching) were employed to produce patterns on the cloth.

Figure 13.11 Dyeing pits at Kofar Mata [53].

Figure 13.12 Dyeing with synthetic indigo in clay pot [52].

Figure 13.13 *Adire* of the Igbo women in Nigeria [8].

Bukola and Olufunmilayo as cited in Gausa and Abubakar [35] are with the position that *adire-eleko, adire-alabere, adire-oniko* and *adire batani* are considered as the different types of *Adire* (Figure 13.13) based on design patterns and depth of dye. They could also be differentiated by the sharp contrast in the techniques of the craft. As elaborated by Olufunmilayo, lighter, tightly-woven commercially made fabric substituted for heavier, coarser homespun; commercially printed cottons, brocades, and other luxury textiles were used, sewing machines were employed to produce *adire alabere* with more detailed patterns.

13.2.2.2 Ukara

The *ukara* cloth (Figure 13.14) is a traditional cloth produced by the Ezillo people for use by the secret societies of the *Ekpe* group in the south East of Nigeria [36]. The cloth is characterized by *nsibidi* symbols, which are patterned on a cotton fabric using grids. These symbols are made of geometric and figurative images that carry symbolic connotations but are also known to the members of the secret societies [37]. Designs produced are first drawn on the fabric, hand-stitch to resist the penetration of dyes and colored in an indigo dye bath. Sultan further revealed that, members of the *Ekpe* society commission the people of *Ezillo* to produce the *ukara* cloth. The *Ukara* cloth is made of plain cotton but transformed into a ritual object when *nsibidi* symbols are inscribed onto it through indigo dyeing. A laborious process is involved in making this *ukara* cloth exclusively used by members of

Figure 13.14 *Ukara* in Morocco [9].

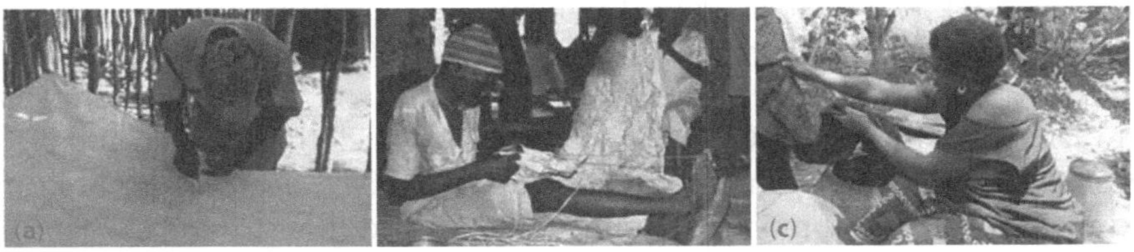

Figure 13.15 (a) Ezillo man drawing a grid pattern to be filled with the required *nsibidi* designs. (b) Omabe Eze sewing an *ukara*. (c) *Ukara* being dyed in a pot; rubbing the cloth to produce deeper color [50].

Ekpe, but *ukara* is produced by a subgroup, Ezillo community in present-day Ebonyi State in the Igbo area in south-eastern Nigeria. After completing a drawn design on the white cloth (Figure 13.15a), the design is sewn with raffia thread (Figure 13.15b) so that parts of the *ukara* (in a pattern of *nsibidi* symbols) will not be exposed during the dyeing process. The cloth is then dipped in indigo dye and hung to dry. The white symbols on the dyed-blue background (Figure 13.15c) are arranged in a grid, giving the cloths an orderly, symmetrical quality that is often reinforced by repeated patterns. When the threads are cut, the *nsibidi* designs stand out, white against a deep blue background [50].

The process of hand-drawing images on a grid and stitching using raffia to resist are done by men, and dyeing is done by women using indigo dyes due to the complexity of the dyeing process. The raffia is then removed by the youth in the community. Designs created are used to communicate information or knowledge, which are only best known to the members of the *Ekpe* group.

13.2.2.3 Mud Cloth

Malian mud-cloth, *bogolanfini* (Figure 13.16) is formed by two words; *bogolan* means something created by using mud, while *fini* means cloth [38]. This cloth is hand woven and hand painted by the Bambara people, who live to the east and north of Bamako in Mali. The cloth is first created by joining or stitching together cotton woven strips and later dyed. Designs are hand painted on its surface using fermented mud, a practice that has influenced its naming. Excess mud is washed off for the painting to repeat for an even darker color. This practice leaves white patterns on brown, black and yellow backgrounds.

Figure 13.16 *Mud cloth* of Mali [10].

For the Malian mud-cloth *bogolanfini*, locally produced cotton spun yarns are employed and woven on a double-heddle loom into a narrow strip (15 cm in width). The long strip (Figure 13.17a) is cut into shorter pieces to the required length. These strips are then joined selvedge to selvedge with a whipstitch (Figure 13.17b). The cloth is washed and dried in the sun to preshrink. *N'Galaman* (Anogeissus leiocarpu) and *N'Tjankara* (Combretum glutin) are two kinds of leaves and branches pounded and soaked in water for a day or boiled in water for a few minutes for a brownish tea, rich in tannic acid. The cloth is soaked in this mordant solution takes on a deep yellow color. The cloth is spread out to dry in the sun and the design outline is painted upon the colored fabric with fermented rich- iron slurry mud (Figure 13.17c) with the help of a piece of bamboo or metal tool. The background surrounding the designs is also filled in with the mud. As the cloth is left to dry, through a complex chemical reaction between the mud and the "leaf dye" the cloth attains terracotta color, which turns gray afterward. Excess mud is removed by washing. The yellow areas are then painted with bleach made from boiled, ground peanuts, water, caustic soda and millet bran. This turns the yellow patterns brown. The cloth is placed in the sun for a week, after which the bleach solution is washed off with water. These leaves give the characteristic white patterns on the dark background [38, 51].

Alternatively, [42] outlined another production process where it starts by first weaving a cotton fabric of narrow width and then sewing the strips together. These fabrics are treated in boiling water to ensure the needed shrinkage to its required size before the coloring or painting process. As a cumbersome process, decorating the sewn cloth with the respective colors takes approximately a week. Subsequently, the cloth is soaked in a solution (a brown liquid made from the leaves of the *n'gallama* tree), giving a very distinctive

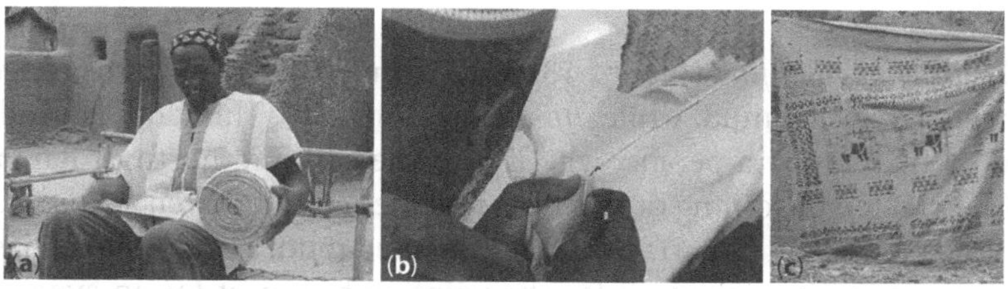

Figure 13.17 (a) Long woven strip. (b) Strip cut to length and stitched at the selvedge. (c) Mud painted on dyed cloth [11].

Figure 13.18 Painting on a woven fabric to produce the *Bogolanfini* cloth.

faint yellow color. This as opined by Kimani and Kanga [42] will act as a fixative to absorb the other colors yet to be painted. Dye paste are extracted from mud collected from ponds and made to ferment. To achieve a black color, tannic acid from tea leaves are mixed with the mud. Using bamboo sticks or feathers, the mud is painted over the patterns created on the cloth (Figure 13.18). Upon drying, the cloth is washed, dried and the mud applied again. This process is repeated until the required color intensity is obtained. Patterns in the cloth symbolizes major events in life and communicates the occupation, character, and social status of an individual.

The use of concentric cycles and twirl as patterns represents the World and life respectively [39]. The mud cloth, characterized by geometric images, have symbolic meanings that seeks to commemorate community heroes and represent historical events [38].

13.2.3 African Print Traditions

Indigenous African textiles are produced using both manual and modern printing methods such as stamping, hand screen and roller printing, and natural and artificial dye paste to create designs that communicate its symbolic significance. Popular print traditions in Africa are *Adinkra* cloth, *kanga* cloth, *Shweshwe and Ankara* or *Kitenge* or African Wax prints.

13.2.3.1 Adinkra Cloth

The *Adinkra* cloth (Figure 13.19) is produced by the Asante craftsmen from Ntonso in Ghana. The cloth is characterized with *adinkra* symbols, which are hand-stamped using calabash or wood carved blocks or printed using manual hand screens [40]. Using dye paste extracted from the *badie and kuntunkuni* tree, the *adinkra* symbols are arranged on a hand woven cotton fabric.

The *Adinkra* cloth in Ghana, which rather employs the stamping technique with a natural print paste. The dye paste (*Adinkra aduro*) is prepared from the bark and roots of a *badie* tree *Adansonia Digitata* (Figure 13.20). They are first soaked for several days to soften, further pounded to soften it and boiled repeatedly to achieve a deep brown color print paste. The fabric is spread and pegged on a flat surface. Stamps prepared from calabash (Figure 13.21) with the *Adinkra* symbols and supported at the back with tiny sticks for easy

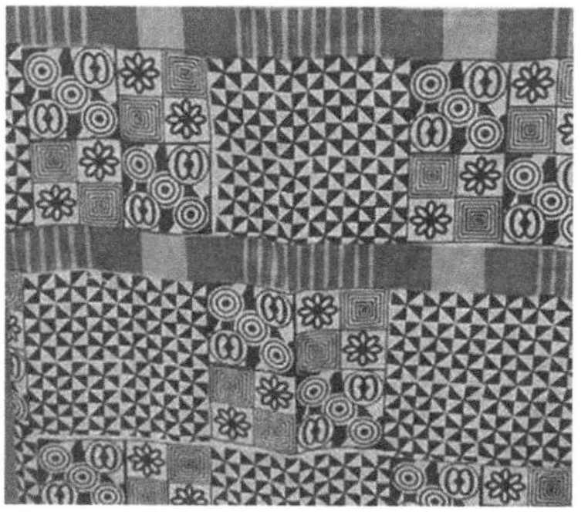

Figure 13.19 Asante *Adinkra cloth* in Ghana [13].

Figure 13.20 *Badie* tree.

Figure 13.21 Stamping process [55].

handling are used for printing. The relief surface of the stamp is coated with dye or print paste (not fast to washing) which are stamped (Figure 13.22) on the cotton cloth. According to [54], designs made from screens are introduced to register the *Adinkra* pattern designs on woven fabrics. This was introduced to make the printing process faster and easier. Other techniques such as embroidery, dyeing and painting are used to enhance the appearance of the *Adinkra* cloth.

Designs created have symbolic meanings that reflects the history, philosophy and beliefs of the Asante people. The use of *adinkra* symbols clearly identifies its naming on the market.

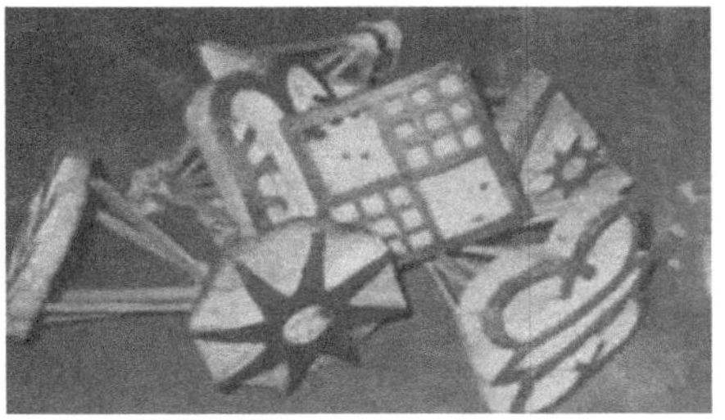

Figure 13.22 Calabash stamps [40].

Due to its cultural value, the *adinkra* cloth is used for specific events including, festivals, funerals, weddings and naming ceremonies.

13.2.3.2 Kanga Cloth

Kanga known as guinea fowl (in Kiswahili), is a printed cloth (Figure 13.23) locally produced in Tanzania and Kenya (known as *leso*). It is so-called *kanga* because the original design patterns resembled guinea fowl's plumage [41] or delivered from the first pattern prints [42]. The *kanga* cloth is characterized with bright colors and design printed in three essential parts of a rectangular shaped cotton fabric. The *kanga* cloth measures approximately 45 to 65 inches in rectangular form [43]. Known as *kanga* in Tanzania and *leso* in Kenya, these textiles display colorfully printed graphic designs worn by women. Cotton cultivation through to fabric production and textile design has merged; factory manufactured cotton yarns are woven into plain-weave cloth. This is followed by bleaching, washing, and mercerizing. Copper plates and rotary screens printed design techniques are applied, and newly printed fabrics are subjected to heat to ensure colorfastness [58]. The three parts of the *kanga* cloth include; designs for the central part (*mji*), border part (*pindo*) and the written text (*jina*) which are either metaphor, slogan, proverbs, riddles among others [42]. The collective use of the symbols and text in the *kanga* cloth symbolically reflects the thoughts, emotions, religion, and health of the wearers.

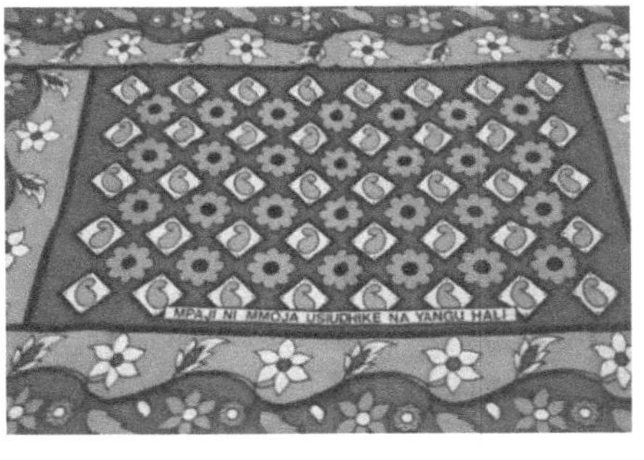

Figure 13.23 *Kanga* cloth in East Africa [14].

13.2.3.3 Shweshwe

Shweshwe (Figure 13.24) is a cotton printed fabric associated with the people of South Africa. Originally, this printed fabric was introduced around the mid-1800s when the French Missionaries came to South Africa [44]. *Shweshwe* was originally created through discharge printing using indigo dyes. Kimani further stated that the presentation of the *Shweshwe* as gifts to King Moshoeshoe around the 1840s was later cherished and preferred by many influencing its popularity in the 19th century [45]. This influenced the naming of the fabric to *shoeshwe* or *isisshweshwe* and later to its present name *Shweshwe.* In South African, the *shweshwe* fabric also known as indigo-dyed discharge printed fabric, is plain woven at a width of 90 cm and dyed with an indigo paste, which is passed through patterns etched on copper rollers. This allows the discharge of a weak acid solution onto the fabric bleaching out the distinctive white designs [57]. With such preference for the fabric, manufacturers, such as Da Gama Textiles, gained the legal rights to produce the *Shweshwe* fabric in South Africa. The fabric features a label of three cats placed on every printed fabric, has distinct textures, geometric patterns, or abstract designs, standard width of 36 inches and tends to be rigid due to its starch content. The starch content is removed by washing the fabric thoroughly with mild soap. *Shweshwe* was originally created in indigo dyes, but varying colors were introduced or incorporated in printing the designs.

13.2.3.4 Ankara or African Wax Prints (West Africa) or Kitenge (East Africa)

Ankara or African Wax prints is so-named in West African (Senegal, Nigeria and Ghana) and *Kitenge* so-named in East Africa. These are colorful designs printed on cotton fabrics using a resist dyeing method. *Ankara* or *Kitenge* or African wax prints (Figure 13.25) are characterized with similar designs at both sides of the fabric, a length of 12 yards for full piece or 6 yards for half piece, a varying width of 46 to 48 inches and bubble or marble effects. African wax prints were coined by European textile firms in Africa, which was based on the Javanese batik print introduced in West Africa in the 19th century [46]. Historically, these prints do not originate from West Africa but from the Europeans, i.e., the Dutch. According to Taylor [47], the Dutch produced imitated wax prints inspired from the Indonesian batik for the Indonesian market. Imperfections from the machine prints were not accepted by the people as alternative to the handmade batik so a decline in sales.

Figure 13.24 *Shweshwe* in South Africa [44].

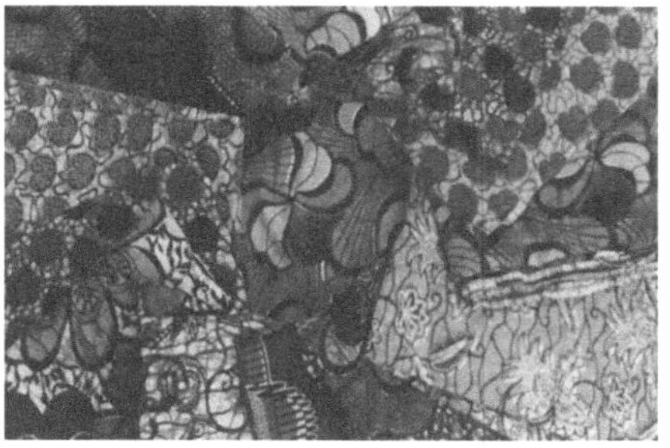

Figure 13.25 *Ankara* or *Kitenge* cloth [15].

The Europeans in their quest to boost their product, turned to the West and East African market. Introduction of these prints were embraced by the people, which was subsequently turned as their own. This influenced the making of the wax prints with symbols and images that have significance meanings within their culture. These prints were scripted with names and symbolic connotation that penetrated the market. [48] opines that designs depicted images of proverbs, symbols, individuals and events.

African wax prints, also known as *Ankara* or *Kitenge*, are omnipresent and common fabric in Africa, especially West Africa. They are a sort of nonverbal communication among African women. They are industrially produced colorful cotton cloths with wax resist printing technique using copper rollers [56]. A distinct feature of this fabric is the fact that both the front and back look alike and cannot be differentiated.

13.2.4 Other African Traditions

Aside the three main traditions mentioned, indigenous textile like the *bark* cloth is produced using traditional and ancient procedures (i.e. beating the bark of a tree for malleability) that are unique to produce the required quality and the *fon appliqué* cloth which is produced by stitching and applique techniques.

13.2.4.1 Bark Cloth

For centuries, Ugandan *bark* cloth (Figure 13.26) is an item of local and regional economic importance in Africa and has been recognized by UNESCO as a masterpiece of the 'Intangible Cultural Heritage of Humanity', to protect the knowledge, traditions and livelihoods associated with its production [49]. Bark cloth is a nonwoven, fibrous textile that has been produced from the mutuba (wild fig) tree by the Baganda people of southern Uganda. The *Bark* cloth making requires careful harvesting of the bark by splitting its outer layer down the length of the tree before gently peeling it off. This is done during the rainy season, when the trees have more sap, and bark is flexible [49]. The naked trunk is protected with banana leaves wrapped around it. The bark is steamed or soaked to soften its texture, before being pounded by a series of well-designed wooden mallets (Figure 13.27a). Through this process, the bark fiber felts, becomes finer, smoother, more flexible and grows in size significantly [61]. It is then spread out in the sun for the natural, rich red-brown color to develop

Figure 13.26 *Bark cloth* in Uganda [16].

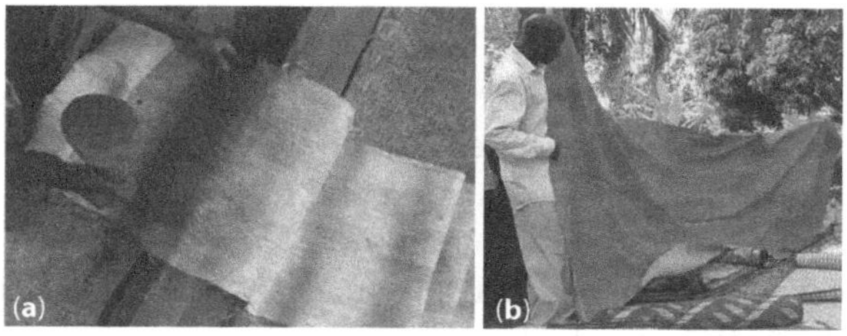

Figure 13.27 (a) Beating the bark. (b) Dried bark cloth [61].

[49] as shown in Figure 13.27b. It could be dyed black by burying in mud rich in iron. A typical *bark* cloth has a rich, terracotta color and is worn by kings and chiefs [49].

13.2.4.2 Fon Appliqué Cloth

Fon Appliqué is a Benin textile made of stitching cut-out shapes (Figure 13.28) onto a contrasting fabric background as banners of battle in the king's palace. Oldest *Fon Appliqué* creators avoided leaving gaps between images by arranging the overlay materials close to each other [51]. Its creation layout, cutting and stitching are involved in making the appliquéd cloth: the shape of the object is cut out and tacked onto a backcloth to hold it in position. This flexibility allows for easy repositioning of the various design elements. With a satisfied composition, the patterns are tacked in position and hemmed at the edges.

Figure 13.28 Stitching and Applique to produce *Fon Appliqué* [51].

13.3 Indigenous African Textiles Techniques

The aboriginals of Africa have a rich heritage of their indigenous craft especially their peculiar fabric fabrication for ritual purposes according to culture. Fabrics assume significant role in the lives and functions of many traditional and indigenous communities. These textiles are made using different techniques that have existed and practiced for centuries. Techniques used to produce indigenous African textiles include; dyeing, weaving and printing among others. This technique is widely used by craftsmen within the society to produce indigenous textiles that are cherished. The dyeing technique (either batik, tie-dye, or tritik) which employs natural dyes is fairly used differentially by the indigenes across Africa in producing their textiles (e.g. *ukara cloth and adire cloth*). Craftsmen employ creative techniques such as stamping or block printing, stenciling and to some extent screen printing to create appealing characters on fabrics e.g. *adinkra cloth, ankara or kitenge or African wax prints.* Weaving remains an oldest method of cloth production in Africa. Weavers employ creative techniques and different loom types as shown in Figure 13.29 (narrow single/double-heddle loom, loom with multiple heddles, loom with supplementary heddles and loom with supplementary warp/weft) to produce their respective fabrics like *smock/fugu, aso-oke, kente, kete, akwete, berber, shuka and kuba raffia.* Indigenes further employ other techniques such as soaking and beating, applique, hand painting, embroidery, stitching in producing their traditional fabrics (e.g. *bark cloth, mud cloth* and *fon appliqué*).

13.4 Museums in African

Museums can be identified by its major collection and target audience, thereby identifying museums as artistry, historical and scientific. Its contributions in respect to continental development of Africa cannot be underrated. Even though amidst its challenges, museums strive to keep the identity and the history of Africans, which attract other continents and people likewise to trace the origin of this wonderful place, called Africa. This is evident in Ghana 2019 Year of Return, which according to [62] welcomed Africans in the diaspora to partake in activities related to the Ghanaian rich cultural heritage, hence contributed about 6.2% of growth to GDP. Museums develop the knowledge of people both in the museum and outside through reading information, lectures and pleasure shown in the numerous

Figure 13.29 Some traditional looms used for weaving.

images and visual displays [63]. Furthermore, the existence of museums in Africa has gone a long way to change the living conditions of communities and regions. In Mali-West Africa, [64] explains how an establishment of a "cultural bank" through the support of World Bank is able to collect and sustain a community museum in Fombori of Mali. Community members are encouraged to bring old and historical artifacts in exchange for money, which is used, in agricultural activities. Additionally, this has served as a local attraction for visitors and a means for artists to meet and sell art works. Museums serve as memory banks, which keep artifacts of a certain period for a generation to identify the kind of art that was executed in past centuries, for example, the art of the cave man [63]. One important task of museums is to prevent the illegal trade of culturally significant items [64], which explains the true meaning and value about the originality of Africans.

13.4.1 Challenges of Museums in Africa

Museums as institutions in Africa faces numerous challenges in its existence and operations. Apparently, one of the major challenges in Africa is poverty, which strongly dwells on finance. [65] opines that lack of funding has been a major setback of Africa museums that depend on international monetary assistance, which has strings, attached thereby affecting their operations. It is also obvious to identify lack of skill personnel in the area of museum education to handle the history and tradition of museums in Africa. Supporting this, [66] stated that, mishandling of information and collection of African relics of antiquity and their origin, meaning and usage in the traditional life of African history, present great difficulty by museum personnel who lacks knowledge in African art. The political instability in Africa has affected museums in the continent to a larger extent. As a tourism center which attracts foreigners and visitors, people are not interested in visiting such places with political instability, before [67] explains that political conflicts in extreme cases decreases the productive and the transactional capacities of the economy. This situation has adverse consequences for investment and thus further weakest its economic growth. Additionally, it is believed that limited research centers in Africa is another obstacle for museum researchers because, continental research leads researchers to the western world to ironically traced their own African history [65]. It further explains practically that, Africa has lost most of its identity to the western world and therefore has to find its image in a reflection.

13.4.2 Contribution of Technology for African Museums

Museums in Africa with the introduction of technology in their operations would attest its contributions. [68] explains that museums can manage collections, give better experiences to visitors and increase physical limits by using online distribution and communication channels via technology. One of the major activities used to showcase antiques in museums is exhibitions. The perpetual existence of natural artifact exhibitions over time is questioned due to the advancement of science and knowledge. With technology, a South African curator and researcher was able to understudy and create extinct plants and animals believed to be in existence around Karoo ecosystem about 270 million years ago [69]. The introduction of technology has improved the strategies adopted by museums to get wider audience to visit and partake in its activities not physically but through technology. Moreover, museum employers utilize information technology to organize activities and to provide guidelines to visitors via the internet. It is believed that the quest to develop African museums imbibe the exploits of technology.

13.5 Conclusion

Indigenous African Textiles are considered to be iconic crafts that are highly cherished by its indigenes which defines their rich cultural heritage. These textiles are produced using different materials, tools and techniques which collectively highlights their cultural pride. The use of materials such as cotton, raffia, silk, wool among others are delivered from natural sources which clearly makes these indigenous textiles fragile. Even though the materials present their own properties which undoubtedly influences the durability, comfort and care, craftsmen employ them in production processes using traditional techniques such as hand weaving, embroidery, dyeing, hand painting and stitching or applique. The use of colors play a central role in the identification of the textiles, its end-use and symbolism. As such, craftsmen are extremely careful creatively to use the appropriate colors in the decoration process to produce indigenous textiles that are used for different events or occasions. Indigenous textile symbolism is a central ingredient that highlights the history, belief, religion and culture of the people that produces the textiles. With such symbolism, historians, indigenes and students have developed great interest to understanding the technicalities and philosophies surrounding the production of indigenous textiles in major museums to help bring such understandings and connections.

References

1. Acquaye, R., Sawyerr, N.O., Kusi, C.A., From Tradition to Innovation: Indigenous West African Textiles in Creative Interior Application. *AJAD*, 3, 2, 18–25, 2018.
2. Strand, E.A., Frei, K.M., Gleba, M., Mannering, U., Nosch, M., Skals, I., Textile Design. *Eur. J. Archaeol.*, 13, 2, 149–173, 2010.
3. Perani, J. and Wolff, N.H., *Cloth, Dress, and Art Patronage in Africa*, Berg, Oxford, New York, 1999.
4. Omatseye, B.O. and Emeriewen, K.O., An appraisal of the aesthetic dimension to the African philosophy of cloth. *J. Language Technol. Entrepreneurship Afr.*, 3, 2, 57–67, 2012.
5. Gema, F., *Berber Carpets: The origin Berber Carpets Weaving*, Bespoke Berber Carpets, Morocco, 2020, https://en.bespokeberbercarpets.com/post/origin-berber-carpets-weaving (accessed Nov. 09, 2020).
6. Seidu, R.K., Howard, E.K., Asinyo, B.K., Assessment of Indigenous and Contemporary Smock Weaves of Bolgatanga in the Upper East Region of Ghana. *ADRRI J. Arts Soc. Sci.* Ghana, 15, 3, 1–14, 2017.
7. Frimpong, C. and Asinyo, B.K., A comparative study of history, equipment, materials, techniques and marketing approach of traditional weaving in Ghana. *Arts Des. Stud.*, 17, 1–18, 2013.
8. Essel, O.Q. and Amissah, E.R.K., Smock Fashion Culture in Ghana's Dress Identity- Making. *J. Hist. Res. Lett.*, 18, 32–38, 2014.

9. Sebold, M., Sourcing & Producing in West Africa with AFIA, 2012. https://chrissie-lam-2p6h.squarespace.com/blog/sourcing-producing-in-west-africa-with-afia (accessed Nov. 09, 2020).
10. Olutayo, A.O., Akanle, O., Fadina, O.A., Aso-Oke (Hand Woven Textiles) of Southwestern Nigeria: A Compact Examination of a Resilient Artifact. *Am. J. Sociol. Res.*, vol, 9–17, 2001.
11. Olutayo, A.O., O., A., Fadina, O.A., The Socio-Economic Developments Of Aso-Oke Of South-Western Nigeria. *Asian Afr. Stud.*, 26, 1, 41–68, 2017.
12. Olajide, M.D., Ajiboye, O.J., Joseph, A.B., Aso-oke production and use among the Yoruba of Southwestern Nigeria. *J. Pan Afr. Stud.*, 3, 55–72, 2009.
13. Aronson, L., Ijebu Y oruba Aso Olona: A Contextual and Historical Overview. *Afr. Arts*, 25, 3, 52–61, 1992.
14. Aremu, P.S.O., Yoruba Traditional Weaving Kíjìpá Motifs and Symbols. *Niger. Mag.*, 140, 3–10, 1982.
15. Obisesan, B.A., *Traditional Fabric Production: A Manifestation of Yoruba Cultural Heritage*, Ile-Ife., Obafemi Awólówò University, 2002.
16. Tyler, M.S., *Meanings of Kente Cloth Among Self-described American and Caribbean Students of African Descent*, University of Georgia, Athens, GA, United States, 2016.
17. Kraamer, M., Origin disputed. The making, use and evaluation of Ghanaian textiles. *Afrique Archéologie Arts*, 4, 53–76, 2006.
18. Boddy-Evans, A., Kente Cloth, ThoughtCo, New York, 2018. https://www.thoughtco.com/what-is-kente-cloth-43303 (accessed Nov. 25, 2019).
19. Lloyd, K., Kentecloth Everything You Need to Know About Kente, KenteCloth.net, 2017. https://www.kentecloth.net/wp-content/uploads/2017/04/kentecloth.pdf (accessed May 07, 2020).
20. Frimpong, C., Opoku, J., Seidu, R.K., The State of Kete Development in the Volta Region. *Textile*, 18, 4, 430–450, 2020.
21. Kwakye-Opong, R., Beyond Ethnic Traditions: Philosophies and Socio-cultural Relevance of the Ashanti and the Ewe Kente cloths. *Res. Humanit. Soc. Sci.*, 4, 26, 150–165, 2014.
22. Ugochukwu, N., Igbo Weaving and Akwete Fabric: The Future of Sustainable Fashion Could be hidden in our past, Eco Warrior Princess, Toronto, Canada, 2019. https://ecowarriorprincess.net/2019/08/igbo-weaving-akwete-fabric-sustainable-fashion-future/ (accessed Nov. 09, 2020).
23. Ikegwuonu, R.N., *Nigerian Textiles*, University of Leeds, Woodhouse, Leeds, United Kingdom, 1979.
24. Oyelola, P. and Onobrakpeya, B., *Nigerian Artistry*, Mosuro Publishers, Ibadan, Oyo, Nigeria, 328 pp., 2010.
25. Nwachukwu, M. and Ibeabuchi, A.O., Akwete Cloth: An Igbo Textile Art, Vanguard Media Limited, Nigeria, 2012. https://www.vanguardngr.com/2012/07/akwete-cloth-an-igbo-textile-art/ (accessed Nov. 13, 2020).
26. Afigbo, A.E. and Okeke, C.S., *Weaving Tradition in Igbo-Land: History and Mechanism of Igbo Textile Industry*, Nigeria Magazine, Dept. of Culture, Federal Ministry of Information, Social Development, Youth, Sports & Culture, Lagos, 1985.
27. Davis, M., Akwete Cloth and its Motifs. *Afr. Arts*, 7, 3, 22–25, 1974.
28. Picton, J. and Mack, J., *African Textiles*, British Museum Press, Great Russell Street, London, 1989.
29. Becker, C.J., *Amazigh Arts in Morocco: Women Shaping Berber Identify*, University of Texas Press, Austin, 2006.
30. Johnson, E.O., No! Scottish missionaries did not inspire the traditional Shuka cloth of the Maasai, Face2faceAfrica.com, New York, 2018. https://face2faceafrica.com/article/no-scottish-missionaries-did-not-inspire-the-traditional-shuka-cloth-of-the-maasai (accessed Jan. 12, 2021).

31. McDonnell, T., A Great African Kingdom Tells Its History In Fabulous Royal Clothes, National Public Radio (NPR), Washington, D.C., United States, 2018. https://www.npr.org/sections/goatsandsoda/2018/10/20/658101128/a-great-african-kingdom-tells-its-history-in-fabulous-royal-clothes (accessed Jan. 12, 2021).
32. Davenport, G., What is Kuba Cloth?, Hearst Magazine Media, Inc., New York, United States, 2020. https://www.housebeautiful.com/design-inspiration/a34304694/what-is-kuba-cloth/ (accessed Jan. 12, 2021).
33. Wolff, N.H., Adire, LoveToKnow Media, Vero Beach, USA, 2020. https://fashion-history.lovetoknow.com/fabrics-fibers/adire. (accessed Nov. 12, 2020).
34. Renne, E.P., Reinterpreting Adire Cloth in Northern Nigeria. *Text. Hist.*, 51, 1, 60–85, 2020.
35. Gausa, S. and Abubakar, E., Tie-dye (Adire) among the Jukun people. *Mgbakoigba J. Afr. Stud.*, 4, 1–13, 2015.
36. Dayo, F., Behind the Scenes: Ukara Cloth, Toghal, UK, 2014. https://toghal.com/2014/07/behind-the-scenes-ukara-cloth/ (accessed Dec. 13, 2020).
37. Sultan, A., At the Hood: The Secret Patterns of Ukara Cloth, altoonsultan.blogspot.com, Groton, VT, United States, 2015. http://altoonsultan.blogspot.com/2015/07/at-hood-secret-patterns-of-ukara-cloth.html (accessed Dec. 13, 2020).
38. Toerien, E.S., Mud cloth from Mali: its making and use. *J. Consum. Sci.*, 31, 52–57, 2003.
39. Azzarito, A., The History of Mud Cloth, Design*Sponge, Ulster County, New York, 2014. https://www.designsponge.com/2014/07/past-present-behind-mud-cloth.html (accessed Nov. 12, 2020).
40. Frimpong, C., Asinyo, B.K., Amankwah, A., Contemporary trends in adinkra cloth production: Design, printing technique, base fabric and printing paste (dye). *Int. J. Fiber Text. Res.*, 3, 1, 43–48, 2013.
41. Nassrulla, T., Khanga–clothing for Africa, Tanzania-Experience.com, Arusha, Tanzania, 2015. https://www.tanzania-experience.com/blog/khanga-clothing-for-africa/ (accessed Nov. 13, 2020).
42. Kimani, N., Kanga: The Talking Cloth of the East African Coast, The Designers Studio blog, Nairobi, Kenya, 2018. https://tdsblog.com/kanga-leso/ (accessed Nov. 13, 2020).
43. Ressler, P., *The Kanga, A Cloth That Reveals-Co-production of Culture in Africa and the Indian Ocean Region*, 2012, [Online]. Available: https://digitalcommons.unl.edu/tsaconf/736.
44. Marshall, R., All hail shweshwe: The history and rebirth of an iconic fabric, News24, Cape Town, South Africa, 2018. https://www.news24.com/w24/Style/Fashion/Style/all-hail-shweshwe-the-history-and-rebirth-of-an-iconic-fabric-20180218 (accessed Nov. 13, 2020).
45. Thompson, A., The Unexpected Origins of South Africa's Most Important Textile. *Culture Trip*, Culture Trip, London, UK, 2018, https://theculturetrip.com/africa/south-africa/articles/shweshwe-the-unexpected-origins-of-south-africas-most-important-textile/ (accessed Nov. 08, 2020).
46. Howard, E.K., Sarpong, G.D., Amankwah, A.M., Symbolic Significance Of African Prints: A Dying Phenomenon In Contemporary Print Designs In Ghana. *Int. J. Innov. Res. Dev.*, 1, 11, 609–624, 2012.
47. Taylor, M.E., The real history of your favourite African prints that will shock you, 2018. https://face2faceafrica.com/article/the-real-history-of-your-favourite-african-prints-that-will-shock-you (accessed Jan. 13, 2021).
48. Clarke, D., *The Art of African Textiles*, PRC Publishing Ltd, China, 2002.
49. Venkatraman, P.D., Scott, K., Liauw, C., Environmentally friendly and sustainable bark cloth for garment applications: Evaluation of fabric properties and apparel development. *Sustain. Mater. Technol.*, 23, 1–26, 2020.
50. Nzewi, C. and Watters, E., *Ukara: Ritual Cloth of the Ekpe Secret Society*, Hood Museum of Art, Dartmouth College, Hanover, New Hampshire, United States, 2018, https://hoodmuseum.

dartmouth.edu/explore/print-archive/ukara-ritual-cloth-ekpe-secret-society (accessed Nov. 16, 2020).

51. Acquaye, R., *Exploring Indigenous West African Fabric Design in the Context of Contemporary Global Commercial Production*, University of Southampton, Hartley Library B12, Southampton, United Kingdom, 2018.
52. Douny, L., From pits to pots: Indigo dyeing traditions of the maranse of Burkina Faso. *Technol. Stories*, 7, 1–17, 2019.
53. Jakobsen, M., An Introduction to the Indigo Dye Styles of Western Africa, Heddels, UK, 2016. https://www.heddels.com/2016/09/an-introduction-to-the-indigo-dye-styles-of-western-africa/ (accessed Nov. 09, 2020).
54. Awayah, I.K., *Exploiting techniques in Asante indigenous hand embroidery for Adinkra cloth production*, Masters Thesis, Kwame Nkrumah University of Science and Technology, Kumasi, Ashanti Region, Ghana, 2012.
55. Adom, D., *The influence of European elements on Asante textiles*, Masters Thesis, Kwame Nkrumah University of Science and Technology, Kumasi, Ashanti Region, Ghana, 2009.
56. Van Kessel, I., Wax Prints in West Africa: Unravelling the Myth of Dutch Colonial Soldiers as Cultural Brokers, in: *Forts, Castles and Society in West Africa*, John Kwadwo Osei-Tutu (ed.), pp. 92–118, Brill, Leiden, Netherlands, 2018. https://doi.org/10.1163/9789004380172_006
57. Bellaafrican, How Shwe Shwe fabric from South Africa came about?, Bellafricana, Lagos, Nigeria, 2015. https://bellafricana.com/shwe-shwe-south-african-textile/ (accessed Nov. 16, 2020).
58. Ryan, M.M., *Kanga Textile Design, Education, and Production in contemporary Dar es Salaam, Tanzania*, 2016, [Online]. Available: http://digitalcommons.unl.edu/tsaconf/1002.
59. Wilson, R., *Wrapped in Tradition: Ceremonial Skirts of Kuba Women in the Western Congo Basin*, 2018, [Online]. Available: https://scholarworks.bgsu.edu/africana_studies_conf/2018/001/3.
60. Chudi-Duru, C.C., *Exploration of Akwete Weaves Motifs for Adaptation on Printed Textiles for Apparel Production*, Ahmadu Bello University, Zaria, Nigeria, 2017.
61. Robertson, L., *Rethinking Material Culture: Ugandan Bark Cloth*, 2014, http://digitalcommons.unl.edu/tsaconf/897.
62. Preko, A.D.K., A Good Museum Experience Pays Off for the Tourism Sector in Ghana. *Conversation*, The Conversation, Melbourne, Australia, 2020. https://theconversation.com/a-good-museum-experience-pays-off-for-the-tourism-sector-in-ghana-146441 (accessed Oct. 18, 2020).
63. Boakyewaa, A., *Educational Value of Manhyia Palace Museum Artifacts*, Masters Thesis, Kwame Nkrumah University of Science and Technology, Kumasi, Ashanti Region, Ghana, 2008.
64. Gänsler, K., How museums in West Africa are saving cultural treasures, Deutsche Welle, Bonn, Germany, 2020. https://www.dw.com/en/how-museums-in-west-africa-are-saving-cultural-treasures/a-53110013 (accessed Oct. 28, 2020).
65. Chikukwa, R., The Dilemma of African Museums, The Herald, Harare, Zimbabwe, 2011. https://www.herald.co.zw/the-dilemma-of-african-museums/ (accessed Nov. 15, 2020).
66. Abungu, L., Museums and Communities in Africa: Facing the New Challenges. *Public Archaeol.*, 4, 2–3, 151–154, 2013.
67. Dalyop, G.T., Political Instability and Economic Growth in Africa. *Int. J. Econ. Policy Studies*, 13, 1, 217–257, 2018.
68. Pop, I.L. and Borza, A., Technological innovations in museums as a source of competitive advantage, in: *Proceedings of the 2nd International Scientific Conference SAMRO 2016: News, Challenges and Trends in Management of Knowledge-Based Organizations*, pp. 398–405, 2016.

69. Browning, C., Ruther, H., Wessels, S., Black, W., Technology and Planning Help Museums manage Outdate Exhibitions. *Conversation*, The Conversation, Melbourne, Australia, 2020. https://theconversation.com/technology-and-planning-help-museums-manage-outdated-exhibitions-143852 (accessed Nov. 06, 2020).

Appendix (Figure sources)

[2] https://riohs.com/the-dress-styles-fabric-from-africa-you-must-know-in-this-season/
[3] https://visitghana.com/attractions/bonwire-kente/
[4] https://www.khanacademy.org/humanities/art-africa/west-africa/ghana/a/kente-cloth
[5] https://www.moroccancaravan.com/item.asp?p=1224
[6] https://craftatlas.co/crafts/akwete-cloth
[7] https://www.junkmail.co.za/household/home-decor/western-cape/cape-town/blouberg/great-range-of-maasai-shuka- blankets/8872f78952924f7699d22167062eca36
[8] https://www.worthpoint.com/worthopedia/old-indigo-dyed-nigerian-adire-cloth-olokun
[9] https://toghal.com/2014/07/behind-the-scenes-ukara-cloth/
[10] https://www.africanfabric.co.uk/fabrics-textiles/large-textiles-cloths/mud-cloth-bogolan/c/mixed-mud-cloth-10044
[11] La Vie à Endé, 2017 at https://www.youtube.com/watch?v=vcI8bWHcypw
[12] readingcloth.blogspot.com/2013/07/bogolanfini-african-mud-cloth-radhika.html
[13] https://african-import.com/product/white-adinkra-cloth-new-design/
[14] https://tdsblog.com/kanga-leso/
[15] https://kitengestore.com/everything-need-know-african-print-fabric/
[16] https://www.africanfabric.co.uk/fabrics-textiles/african-fabrics/bark-cloth/c/bark-cloth

14

Conservation of Textile Immemorial: The Fading Past of Uttarakhand Museums

Pooja Singh[1]* and Alka Goel[2]

[1]Faculty-Fashion Design Department, Footwear Design and Development Institute (FDDI), Chandigarh Campus, An Institution of National Importance, Ministry of Commerce & Industry, Govt. of India, Banur, District SAS Nagar (Mohali), Punjab, India

[2]Department of Clothing & Textiles, College of Home Science, Govind Ballabh Pant University of Agriculture & Technology, Pant Nagar, Uttarakhand, India

Abstract

Antiquities provide an overromantic or personal connection to textiles from the past and ethnicity through objects used or seen in everyday life, as well as indicate cultural and technological encroachment. India is blessed with a hoary tradition of textiles of each particular state having their own costumes, embroideries, and crafts. In India, previously, these antiquities were only hidden under the four walls of house and passed on to the generations to generations; therefore, the general public was unable to know these wonderful antiquities about their existence, with the initiatives of governmental and nongovernmental organizations, museums were started to conserve this art of India. Likewise, there are various museums in Uttarakhand, which are blessed by wonderful artifacts of display, such as costumes, paintings, textiles, jewellery, coins, sculptures and various manuscripts, which speak the golden times of yore representing history of Uttarakhand. The present study explores the status of textile artifacts and existing conservation practices in selected textile museums of Uttarakhand. Six textile museums were selected, and data were collected through interview and qualitative techniques, like observation, interpersonal discussions with respondents, were also used to support the quantitative information collected from the respondents using a questionnaire. The information regarding number of textile artifacts conserved, as well as conservation practices carried out by the respective museum to keep the textile antiquities safe from deterioration were also collected.

Keywords: Antiquities, conservation, deterioration, questionnaire, textiles

14.1 Introduction

A museum is a "non-profit, institution open to the public that collects, helps conserve, research, conveys, and shows the direct and indirect legacy of mankind and its environment for the sake of education, study, and recreation" [1].

**Corresponding author*: singhpooja396@gmail.com

Seiko Jose, Sabu Thomas, Pintu Pandit, Ritu Pandey and Vandana Gupta (eds.) Handbook of Museum Textiles: Volume I Conservation and Cultural Research, (267–290) © 2023 Scrivener Publishing LLC

A museum utilizes a professional staff for better care and exhibits the tangible objects to the public on a regular basis. Such term includes botanical gardens, art museums, children's museums, general museums, historic houses and sites, history museums, natural history, and anthropology museums and specialized museums [2]. Heritage textiles, which are conserved in collections from many types of buildings, serve as links between the past and the present. Many of them are protected and exposed in open-to-the-public heritage buildings that do not have inside microclimate monitoring systems [3].

In India, museums are more traditional in nature, serving as caretakers of collections. The collections are of the country itself, to a degree of near proximity, due to the country's rich indigenous past. The quantity of antiquities found throughout the country is so large that Indian museums are today researched and analysed all over the world [4]. Indian museums house an array of collection, which includes wooden objects, stone sculptures, metal images, manuscripts, textiles, paintings, glass ceramic objects, decorative arts, and articles of other materials but textile antiquities predominate.

Despite being among the world's most fragile items of acquisition, a great amount of textile relics has been collected over the years. Unfortunately, despite massive manufacturing of various textiles over millennia by civilizations famed for superb textiles, only a tiny fraction of fabric samples has survived the ravages of time due to their brittleness.

Relative humidity (RH), fluctuating temperature, light, environmental pollution, non - standard storage and exhibition methods, older inadequate repair, and old types of adhesives such as starch adhesive, animal glue adhesive, and Arabic gum adhesive are among issues faced by archaeological textiles in museums. All of these conditions induce fiber and dye deterioration and decay, resulting in fiber and dye weakening, fiber hardness, separate pieces, lost parts, discoloration, and dust [5].

However, the continued existence of any ancient textile against the adverse effects of climate, light, insects, microorganisms, and inappropriate storage is indeed a miracle. Degradation is major concern for historic textiles causing dimensional changes leading to observable damage which may be cumulative and irretrievable. So, to protect the treasure bequeathed by the forefathers, conservation and restoration of textiles are of paramount importance.

Textile conservation can be defined as the art and science of preserving textiles from natural and man-made deterioration [6]. The responsibility of humans towards historic textiles is to hold these in stewardship for future generations, maintaining them in the best possible conditions and assuring their perpetuity. Hence, these surviving and frail pieces of textile artifacts should be appropriately conserved. Because of the country's uniqueness and cultural value, it has also become a requirement to conserve its richness.

In view of the abovementioned facts, this study was carried out to gather information about various selected museums of Uttarakhand along with their textile antiquities collection. The collected information will benefit the people to make them cognizant about the museums, their collection and conservation practices followed by them. From an education point of view, museum serves as a reference material for students and researchers about the types of textiles, embroideries, costumes of particular place, motifs and various intricate designs. This knowledge will help the students, researchers, and textile/fashion designers to create new articles by getting inspiration from the historic ones. But such treasure may be damaged by the passage of time due to various factors, such as dust, inappropriate temperature and humidity, pollution, insects and microorganisms and various other degrading factors; therefore, conservation of these jewels is the need of hour so that the coming generations would be able to learn and enjoy the beauty of these artifacts.

Considering the abovementioned facts, the present study was undertaken with the following objectives, such as to find out the various categories of textile antiquities well kept-in the museums of Uttarakhand, to determine the display and storage practices used for textile antiquities in the museums, to study the existing conservation practices used to protect the textile artifacts in the selected museums.

14.2 Materials and Methods

14.2.1 Selection of Locale

Six different places of Uttarakhand were purposively selected as locale, which included Ranikhet, Almora, Munsyari, Bhimtal, Nainital, and Ramnagar. Six museums viz; Kumaon Regimental Museum (Ranikhet), Govind Ballabh Pant Government Museum (Almora), Jim Corbett Museum (Ramnagar), Lok Sangrah, Folk Culture Museum (Bhimtal), Gurney House (Nainital) and Tribal Museum (Munsyari) of selected places were covered during the study. These six museums were selected on the basis of information obtained through primary and secondary sources. In the present study, museums displaying textiles were covered.

14.2.2 Tool Preparation and Data Collection

Primary data were collected with the help of developed interview schedule. Books, journals, periodicals, and the Internet were used to gather secondary data. There were two portions to the interview schedule: basic information and particular information.

The general information comprised the name, designation, age, sex, education, employment experience, and trainings obtained by the curators and other people working in the selected institution.

The total number of textile articles in the respective museums, details of the various types of textile antiquities in the museum collection, the basis of categorization for documentation of the museum textiles, pre-treatments given to the textile materials before display, and conservation methods used for textiles, among other things, were all included in the specific information. Close ended and open ended both type of questions was framed to get the detailed information which helped in assessing the existing status and conservation techniques followed in various selected museums. Photographs were also taken to collect the technical information of conservation techniques followed, in the various selected museums.

14.2.3 Data Collection

The information was gathered through personal interaction with museum officials/workers (curators) utilising an interview schedule. Details of displayed and stored items, as well as a list of the amount of objects kept in each museum's stock register, were recorded. Name and quantity of textile artifacts were also documented as per stock register.

14.2.4 Statistical Analysis of the Data

The collected data were tabulated and analyzed statistically. Data were analyzed to check the similarities and differences in conservation techniques followed in the selected museums. A separate table was prepared using frequency and percentage for each question to interpret the data.

14.2.4.1 Weighted Mean Score

In weighted mean score (WMS), different values or weights were assigned to different parameters. If the parameters are in decreasing order of importance like in the questions which are having frequently, occasionally and never options. So, in such questions, museums frequently adopting any conservation technique were more important than those adopting the same technique occasionally and in the same manner; however, later was more important than never adopting the same technique. Therefore, more weight was assigned to frequently, than occasionally and followed by never.

If W1 is true, W2 is true. ———— Wn are the weights that have been assigned to the variables X1, X2, and X3. ———— The weighted mean of Xn was then determined using the formula:

$$\bar{X}w = \frac{W1X1+W2X2----+WnXn}{W1+W2\pm----+Wn}$$

$$= \sum \frac{WiXi}{W}$$

Where W= ΣWi

14.3 Results and Discussion

14.3.1 General Information About the Museums

The museums' general information was gathered, and the results are summarized in the table below. Table 14.1 shows the year of establishment and governing agency of various museums.

It is evident from Table 14.1 that the oldest museum among all selected museums in the present study was Gurney House, Nainital established in 1881 followed by Jim Corbett Museum, Ramnagar established in 1965.

14.3.1.1 The Number of People Who Work at the Museums that Have Been Chosen

Total number of staff members employed are reported in Table 14.2.

Table 14.2 shows that Kumaon Regiment Museum had maximum members, followed by Govind Ballabh Pant Government Museum and Jim Corbett Museums. In Tribal Museum In-charge was also responsible for taking conservation activities of textiles. Local weavers

Table 14.1 General details of selected museums, such as their year of establishment along with the managing agency (N=6).

Museums	Year of establishment	Managing agency
Kumaon Regimental Museum (Ranikhet)	1978	Kumaon and the Naga Regiment of the Indian Army
Tribal Museum (Munsyari)	1985	Private (Run by Dr. Sher Singh Pangtey)
Govind Ballabh Pant Govt. Museum (Almora)	1980	Ministry of Culture
Jim Corbett Museum (Kaladhungi, Ramnagar)	1965	Forest department (Corbett Tiger Reserve) and NGO
Lok Sangrah, Folk Culture Museum (Bhimtal)	1983	Private (Dr Yashodhar Mathpal- international famed archaeologist, artist, scholar and logician Padmshree)
Gurney House (Nainital)	1881	Private (Granddaughter of Jim Corbett, Nilanjana Dalmia)

such as ladies used to weave various traditional textiles such as shawls, mufflers, dann, etc. for displaying in Tribal Museum as well as for sell. The researcher also observed that the staff is according to the hierarchy in all the selected museums. All museums were under staffed consequently single person was attending various tasks.

Horjan suggested that staff is the most crucial resource of regional and local museums. Museum work is extremely complex, and satisfying all of the needs with a small workforce is difficult, despite their best efforts [7].

14.3.1.2 Conservation Laboratories

The information on conservation laboratories in each museum is presented in Table 14.3.

Table 14.3 clearly reveals that none of the museum had conservation laboratory. Kumaon Regiment Museum (Ranikhet) and Govind Ballabh Pant Govt. The National Museum in New Delhi and the National Research Laboratory for Conservation of Cultural Property (NRLC) in Lucknow, respectively, handle the conservation work for the Museum at Almora. It was reported by the respondent that there was a facility for preventive conservation in Kumaon Regiment Museum (Ranikhet) and Govind Ballabh Pant Govt. Museum (Almora).

Govind Ballabh Pant Govt. Museum (Almora) has conservation laboratory but defunct and no conservator was there since last 10 years. Tribal museum, Munsyari is a private museum therefore do not have sufficient funds therefore work of conservator and curator is taken care by single person. Museum conserves their textiles with the help of natural and locally available resources.

Table 14.2 Staff members employed in various museums (N=6).

Post	Museums					
	Govind Ballabh Pant Govt. museum	**Kumaon regiment museum**	**Tribal museum**	**Lok Sangrah, folk culture museum**	**Gurney house**	**Jim Corbett museum**
Director	-	-	-	1	-	1
Curator	1	1	1	-	-	-
Museum incharge	2	1	2	1	1	0
Gallery attendent	1	2	-	1	1	1
Gardener	1	1	1	-	1	1
Store keeper	-	1	-	-	-	-
Technician	1	-	-	-	-	-
Weaver	-	-	3	-	-	-
Helpers	2	2	1	-	2	2
Docent	-	1	-	-	-	-
Artist	-	-	-	1	-	-
Cleaner	1	2	1	1	1	1
Security in galleries	0	0	0	0	0	3
Administration and maintenance	3	2	-	-	1	3
Total	12	**13**	9	**5**	7	12

Jim Corbett Museum (Ramnagar) outsources conservation work. For, Gurney House, conservators from Delhi visit every year to have a view of the collection and to examine the antiquities conserved.

14.3.1.3 Acquisition of Textile Antiquities

The data on textile antiquities acquisition were gathered, and the results are presented in Table 14.4.

Table 14.3 General information of conservation laboratories found in various museums (N=6).

Museum	Conservation laboratory				If no, the place from where the conservation work done
	Yes	**Frequency**	**No**	**Frequency**	
1	Only preventive conservation	n=1 (16.66%)	✓	n=5 (83.33%)	curative conservation was done in National Museum
2	-		✓		-
3	Only preventive conservation		✓		National Research Laboratory for Conservation of Cultural Property, Lucknow
4	-		✓		Tender basis contract
5	Only Preventive conservation		✓		-
6	Only Preventive conservation		✓		Trained officials from Delhi

1. Kumaon Regiment Museum (Ranikhet) 2. Tribal Museum (Munsyari) 3. Govind Ballabh Pant Govt. Museum (Almora) 4. Jim Corbett Museum (Ramnagar) 5. Lok Sangrah, Folk Culture Museum (Bhimtal) 6. Gurney House (Nainital).

Table 14.4 Acquisition of textile articles in different museums (N=6).

Acquisition of textile articles	Museums						F	%
	1	2	3	4	5	6		
Through exploration	-	✓	-	-	✓	-	2	33.33
Donation/gifts	✓	✓	✓	✓	✓	✓	6	100
Purchased	-	-	-	-	✓	-	1	16.66

F=Frequency, %=Percent, 1. Kumaon Regiment Museum (Ranikhet), 2. Tribal Museum (Munsyari), 3. Govind Ballabh Pant Govt. Museum (Almora) 4. Jim Corbett Museum (Ramnagar) 5. Lok Sangrah, Folk Culture Museum (Bhimtal) 6. Gurney House (Nainital).

The textile collections in all the museums were mainly acquired through donation/gifts followed by exploration. Only folk Culture Museum (Bhimtal) had acquired their collection through purchase. Not even a single museum had exchange of articles from other regions.

The museum may acquire objects through purchase, contract, donation, bequest, exchange, field collection, or other appropriate sources, according to the University of Alaska Fairbanks, subject to constraints indicated in particular collection plans [8].

14.3.1.4 Ageing of Textile Articles Placed in Different Museums

Information on effect of aging on textile articles was collected, and results are reported in Table 14.5. Respondents from each museum opined that textile articles do have effect of aging. It can be noted from Table 14.5 that all respondents had observed ageing in textiles due to mechanical stress followed by physical ageing and chemical attack. Ageing due to thermal degradation was observed by only few respondents.

Respondents of Kumaon Regimental Museum (Ranikhet) replied that physically aged textiles become harder, denser, and stiffer, and wet cleaning can minimize physical ageing.

Respondents from Tribal Museum, Munsyari explained that to get rid of chemical ageing, the air from moist environment should be eliminated as soon as possible. It will help in increasing the life of the textiles. To eliminate the effect of mechanical stress on textiles placed in the museum, the unfolding of the artifacts should be done at regular intervals. The respondent of Jim Corbett Museum, Ramnagar reported that chemical attack in textiles showcased in museums due to increase in temperature, it should be minimized by providing the artifacts low temperature so that the chemical reaction will not exaggerate. It is agreed by Singhal and Bhagat that the chief factors which harm the textiles are ageing, light, micro-organisms, air pollution, temperature, humidity, soiling, staining and insects [9].

14.3.1.5 Air Circulation Facilities in Museums

The climate of Uttarakhand is very cold and in upper hills the temperature can drop below freezing point therefore museums at high altitude like Tribal Museum, Munsyari, do not need air conditioners. In Kumaon Regiment Museum (Ranikhet) and Govind Ballabh Pant Govt. Museum (Almora) exhaust fans were used for air circulation. All the museums have ceiling fans for air circulation. In Tribal Museum and Lok Sangrah, Folk Culture Museum (Bhimtal) only first gallery has ceiling fan otherwise all the other two galleries were without any air circulation facility because according to the respondent of Tribal Museum there is no need for any air circulation facility because cooler temperatures are suitable for textile

Table 14.5 Museums are distributed based on what is known about the effects of ageing on museum textiles (N=6).

	Museums							
Types of ageing effects	**1**	2	3	4	5	6	F	%
Physical ageing	✓	✓	-	✓	-	✓	4	66.6
Photochemical degradation	-	-	-	-	-	-	0	0
Thermal degradation	✓	-	-	✓	-	-	2	33.3
Chemical attack	-	✓	-	✓	-	✓	3	50
Mechanical stress	✓	✓	✓	✓	✓	✓	6	100

F=Frequency, %=Percent, 1.Kumaon Regiment Museum (Ranikhet), 2. Tribal Museum (Munsyari), 3. Govind Ballabh Pant Govt. Museum (Almora) 4. Jim Corbett Museum (Ramnagar) 5. Lok Sangrah, Folk Culture Museum (Bhimtal) 6. Gurney House (Nainital).

Table 14.6 Museums are distributed according to the air circulation facilities in various museums (N=6).

	Museums							
Air circulation facilities	**1**	**2**	**3**	**4**	**5**	**6**	**F**	**%**
Exhaust fan	✓	✓	-	✓	✓	✓	5	83.33
Ceiling Fans	✓	✓	✓	-	✓	✓	5	83.33

F=Frequency, %=Percent, 1.Kumaon Regiment Museum (Ranikhet), 2. Tribal Museum (Munsyari), 3. Govind Ballabh Pant Govt. Museum (Almora) 4. Jim Corbett Museum (Ramnagar) 5. Lok Sangrah, Folk Culture Museum (Bhimtal) 6. Gurney House (Nainital).

collection conservation. Jim Corbett Museum and Gurney House are equipped with ceiling fans and exhaust fans.

Table 14.6 briefs about the air circulation facilities in the museums. Ceiling fans were installed in all the museums except Jim Corbett Museum, Ramnagar and five museums had installed exhaust fans and not even a single museum had installed air coolers and air conditioners. According to National Property Board, Sweden use of exhaust air systems with heat exchange is the standard requirement for heat recovery which should be installed in the museums [10].

14.3.1.6 Protective Measures Used to Protect the Windows/Ventilators From Sunlight and Dust

Various protective measures used for protection of windows/ventilators from sunlight and dust in different museums were also enquired, and results are reported in Table 14.7. It was found that in Uttarakhand, only in the *Kumaon Regimental Centre Museum* (Ranikhet) blinds were installed. It was found that Govind Ballabh Pant Government Museum (Almora) and Tribal Museum (Munsyari) covered the windows of museum with cloth curtains.

It was noticed that museums were using cloth curtains because draperies were the most traditional method to cover the windows and ventilators and also easily affordable to every level of museums.

Table 14.7 Museums are distributed in accordance to the methods used to protect windows/ventilators from sunlight and dust (N=6).

	Museums							
Protective furnishing	**1**	**2**	**3**	**4**	**5**	**6**	**F**	**%**
Cloth curtains	✓	-	✓	Ventilators covered with glass			2	33.3
Blinds	-	✓	-				1	16.6

F=Frequency, %=Percent, 1.Kumaon Regiment Museum (Ranikhet), 2. Tribal Museum (Munsyari), 3. Govind Ballabh Pant Govt. Museum (Almora) 4. Jim Corbett Museum (Ramnagar) 5. Lok Sangrah, Folk Culture Museum (Bhimtal) 6. Gurney House (Nainital).

14.3.1.7 Methods Used for Identification of Fibers

The information on methods used for identification of fibers before starting conservation work of textile materials was composed and results were documented. Every textile which adds to the collection of artifacts in the museum will firstly identify for the type of fiber. So that the documentation and other conservative measures could be taken according to the type of fiber the textiles is composed off. Identification of textile artifact was done by taking the fibers from the fabric seam allowance. Method used for identification of textile fibers before adopting conservation treatment in all the museums was visual inspection. No museum used microscopic examination, X-ray diffraction, burning and chemical tests.

According to Goodway, fibers from anthropological or archaeological sources are often fragmentary or degraded, fossilised, or burned, and are occasionally fragmentary or deteriorated. The simplified methods of preparation for microscopic inspection were shown to be more successful than the traditional biological processes of soaking, cleaning, and staining when fibers were in such poor condition [11].

14.3.1.8 The Details of Temperature and Relative Humidity Ranges in a Variety of Museums

The curator of Gurney house and other museums revealed that they had installed relative humidity measuring instruments in their display areas to measure the macro climate of the museum building though none monitored the micro climate, i.e., climate inside the showcases or closed display space nor practiced any method to control the same which is a threat to the physical integrity of the organic artifacts. All museums make arrangements to control the macro climatic conditions (temperature and relative humidity). According to the curators, all the stated methods assisted as preventive measures to control climatic fluctuations, which were the greatest threat to textile materials in the tropical countries like India. A respondent of Tribal Museum, Munsyari opined that much effort to control humidity was not required in their case as the museum had favourable geological location whereas, most of the museums of Uttarakhand did not adopt any method to control temperature and humidity. They usually display textiles in general conditions as per weather.

In both summers and winters, the temperature and relative humidity at Gurney House Museum in Nainital were kept at 18°C to 20°C and 45% to 5%, respectively. Govind Ballabh Pant Government Museum, Almora; *Kumaon Regimental Centre Museum*, Ranikhet; Tribal Museum, Munsyari; *Lok Sangrah, Folk Cultural Museum, Bhimtal*; National Corbett Museum, Ramnagar were displaying their textiles in the general conditions as per weather.

Museums' room temperatures should be kept between 18°C and 20°C, and relative humidity should be between 45% and 50% for proper conservation of displayed artifacts, according to Nalini. Lower humidity levels are consistent with conservation. At 10°C temperature 30% relative humidity is acceptable for textiles displayed in museums [12].

14.3.2 Types of Display Techniques Used for Textile Antiquities

The table showcasing and upright showcasing display technique is mainly practiced by all the museums studied because according to the respondents in the museums showcasing is the safest method, which protects textile articles from dust, light, touch of the visitors and other environmental factors.

Wooden hangers were also used by many of the museums for display of antique textiles. In Tribal museum there were no showcases for bigger items such as lehenga, coat, choli and carpet. These articles were hanged with the help of wooden hangers (Figure 14.1) on the polythene type material which is mounted on the wall.

Traditional costumes in Tribal Museum were displayed on mannequins, which depict the real image of culture of Uttarakhand. In Govind Ballabh Pant Museum, Almora special showcase was made of wood covered with glass used to display the daily wear costumes and costumes of bride and groom of Hilly areas of Uttarakhand.

In Folk Art Museum, Bhimtal, saris and traditional costumes of Uttarakhand were folded and hung with the help of rope inside a showcase in which glass was only in the front and the back of the showcase was open (Figure 14.2). Tibetan wall hangings were hung directly on wall with the help of nail. The last room of the museum was so beautifully decorated with the help of nisaad patakaaen (flags) of Tibetan people on which mantras were written in their own language. By entering this room visitors will definitely have the feeling that they had entered in the region of Tibetan people.

Woollen coat was hung on a self-prepared stand of wooden sticks in a cross to balance the three dimensional structure of coat. Various clothes of children were displayed in a large basket hung from the wall. The articles were very roughly displayed in a basket due to the lack of infrastructure. Some articles were kept folded and displayed in a table and upright showcases in Folk Art Culture Museum, Bhimtal. When asked by the curator of this museum, then he mentioned that due to lack of finances the antiquities could not be able to manage and display properly in a correct manner.

In Govind Ballabh Pant Museum, Almora (Figure 14.3), the daily wear costume of farm men and women and bride and groom costumes of people belonging to hilly areas of Uttarakhand were displayed with the help of miniature dolls like structure enclosed in a wooden showcase covered with glass. Jewellery and other small articles were displayed on table showcases.

Gurney house museum, the entire house of Jim Corbett was conserved as a museum. There were no special kind of showcases to display the artifacts but the artifacts were displayed in their original form as they were at the time of Jim Corbett.

Lehanga and Woollen coat hung by wooden hanger	Carpet hung on wall with plastic sheet at the back	Dhoti, Aangudi and Patka displayed on mannequins	Lehenga woollen jacket and muffler worn by farm women on mannequin

Figure 14.1 Types of display techniques used for textile antiquities in Tribal Museum, Munsyari.

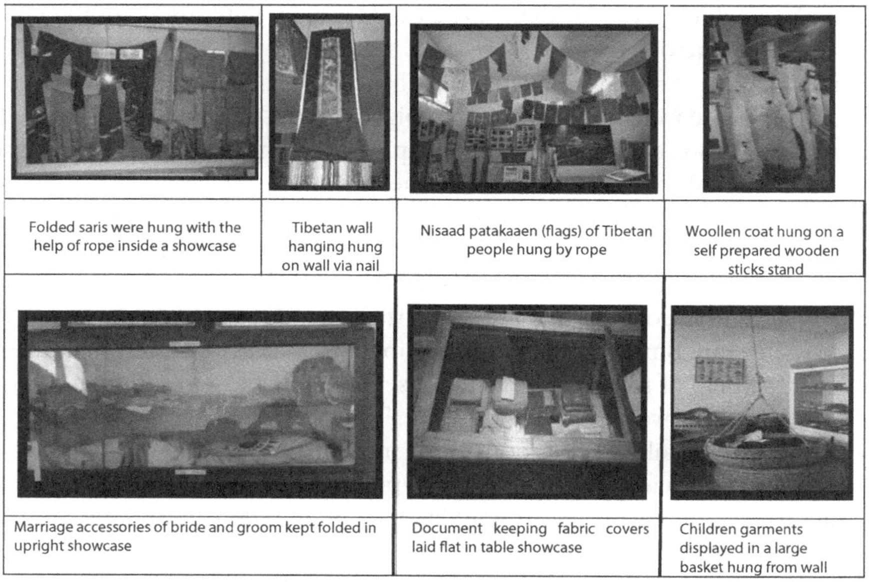

Figure 14.2 Types of display techniques used for textile antiquities in Lok Sangrah, Folk Culture Museum, Bhimtal.

Figure 14.3 Types of display techniques used for textile antiquities in selected museums.

In Jim Corbett Museum, Ramnagar various personal belongings of Jim Corbett such as hat were showcased in an upright showcase. Some articles such as bedding folder, net used to catch fishes were kept in an open in a room without the showcase to show the visitors.

In Kumaon Regimental Museum, Ranikhet most of the articles were showcased in a upright showcases, table showcases and pedestal showcases. Some articles, such as protective wear dress of sergeant was displayed on dummy to make the viewers more understandable about the features of this dress and also for better display and also one more reason behind dummy display was that such kind of protective wears are very big and large in size therefore difficult to accommodate in table showcases so correct way of displaying such king of three-dimensional structures are through dummy display. Various articles such as flags were laminated and displayed on the wall.

Templeton suggested that It's critical that the materials used to display textiles are of conservation quality. Because textiles naturally become acidic as a result of the environment, only acid-free fabrics should be used for exhibition. The most important consideration when selecting an appropriate material for displaying textiles is that it must be compatible with and non-harmful to the object [13].

14.3.2.1 Labeling Methods Carried Out for the Displayed Artifacts

It was observed that laminated frame on wall method is practiced by Kumaon Regiment Museum and Govind Ballabh Pant Museum frequently. Metal pins were not used by any of the museum. Respondents of Kumaon Regiment Museum replied that the method of labeling they used to practice occasionally in their museums was pasting on showcases. For displaying articles in Jim Corbett Museum and Gurney House, pasting on showcases and labels kept near antiquities were practiced occasionally.

The investigator also observed that in Tribal Museum information regarding any textile antiquity were written on paper and then pasted with the help of tape on the showcases or packaging material of particular textile.

Table 14.8 shows that the most popular way of labelling for the presented articles was, kept near antiquity (Figure 14.4) which got the first rank followed by laminated frame on wall (IInd rank). A label pasted on showcases was the next preferred method (IIIrd rank). The use of metal pins was not preferred by any of the museum because the respondents of the respective museums replied that metal pins cause rusting to the textile antiquity, create stains and ultimately deteriorate the strength of the fabric by breaking the chemical bonds at that specific area of contact with the metal pins.

Table 14.8 Museums are distributed according to the labeling procedures used for the displayed articles (N=6).

Labeling procedures	Frequently	Occasionally	Never	Weighted mean score	Rank
Stitching on the textiles	0	1	5	1.16	IV
Using metal pins	0	0	6	1	V
Laminated frame on wall	2	4	0	2.33	II
Pasting on showcases	2	3	1	2.16	III
Kept near antiquity	3	3	0	2.5	I

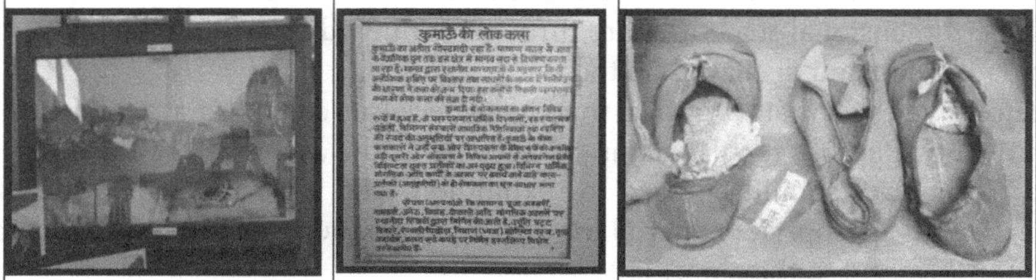

Labels pated on showcases, Lok Sangrah Folk Culture Museum, Bhimtal	Laminated frame on wall, Govind Ballabh Pant Museum, Almora	Labels kept near antiquity, Tribal Museum,Munsyari

Figure 14.4 Labeling methods used in selected museums for textile artifacts.

The recommended material for labels is a densely woven, white cotton fabric (e.g., sheeting material) according to Canadian Conservation Institute (CCI) Notes 13/8 [14].

14.3.3 Storage Equipments Used in Selected Museums

The keepsakes were stored in wooden cupboards by Tribal Museum, Munsyari, and Folk Culture Museum, Bhimtal (Figure 14.5).

A representative from the Kumaon Regiment Museum responded that museum objects react similarly to the materials. Hazardous gases are most likely to be released by modern organic materials. When objects are maintained in a confined space (such as a display case, polythene bag, or plastic box), where a concentration of the toxic vapour can build up, the process of deterioration accelerates. As a result, modern organic materials should be placed in enclosures that allow for air circulation. They should be wrapped in acid-free tissue paper that absorbs the released gases and stored in a separate section of the storage facility with appropriate air circulation.

After interrogating the respondent of Tribal Museum about storage of museum textiles in polythene bags, replied that they used to open polythene bags once or twice in a year to unfold the textiles and to determine condition of particular textile. Due to lack of space they used to fold all the textiles except carpets (Rolled) and place them in storage area. According

Wooden cupboards in Tribal museum, Munsyari (A) and Folk Culture Museum, Bhimtal (B)

Figure 14.5 Storage equipments used in selected museums for textile artifacts.

to respondent of Govind Ballabh Pant Museum, Jim Corbett Museum and Gurney House all textiles they had in their museum were displayed. Not even a single article was stored.

The results are supported by Alig *et al.*, that Wood's hygroscopic properties act as a buffer against sudden changes in relative humidity. However, once wood products are sealed as directed, their buffering effect is diminished [15].

14.3.4 The Collection of Textile Artifacts Collections in Various Museums of Uttarakhand

14.3.4.1 Details of Stored Textile Materials

On observation, results pertaining to the storage methods practiced by all museum curators revealed that small fragile artifacts were laid flat in the show cases whereas, sturdy artifacts (garments like jackets and choli) were hanged on padded hangers. Carpets were rolled and stored depending on the condition.

Woven, knitted, dyed/printed textiles were kept folded and interleaved with neem leaves. Textiles with surface decoration were laid flat during storage and interleaved with tissue paper and neem leaves. Carpets were mainly rolled or folded in most of the museums and interleaved with Acid free paper by all the museums. Costumes were kept folded or hung in most of the museums and interleaved in tissue paper and packed in brown paper.

14.3.4.2 Govind Ballabh Pant Museum, Almora

The museum collection displays daily wear costumes of men and women of Uttarakhand (Figure 14.6) and costumes of bride and groom that speak about our culture of Uttarakhand. Museum also displays paintings of Aipan (Aipan is a type of traditional folk art created by Uttarakhand women. With white paste created from rice flour, this art is done on the floor over a brick red background. On all major events and home ceremonies and rituals, such as Ganesh Chaturthi, Makara sankranthi, Kark sankranthi, Initiation ceremonies, Weddings, Shivratri, and Laxmi puja, the traditional art is performed. These symbols are said to inspire divine power, bringing good fortune and warding off evil) and wide collection of jewellery of Uttarakhand men and women.

Goddess of Uttarakhand wearing sari	Daily wear costumes of men and women of hilly areas of Uttarakhand	Costumes of bride and groom of hilly areas of Uttarakhand

Figure 14.6 Artifacts displayed in Govind Ballabh Pant Museum, Almora.

14.3.4.3 Tribal Museum, Munsyari

This museum housed various articles of textiles of Uttarakhand region depicting our glorious past. This includes traditional costumes such as lehnga, large trade bags, blouse, veils men's dhoti, kurta, headgear, woolen jackets, carpets, shawls, coat, skin of various wild animals, chutku (A type of floor covering), paul (shoes), muffler, blankets and chamtaul (fabric used to hold hot utensils at the time of working on traditional chulhas) (Figure 14.7).

14.3.4.4 Kumaon Regiment Museum, Ranikhet

Flag of the 8th regiment Nizam Infantry (now 5 Kumaon) which traces its origin back to the 2nd regiment Salabat Khan's Ellichpur Brigade in 1788; Belt of a soldier of the Gwalior Infantry; India's independence medal awarded to Boy Karan Singh; Fur cap of a Chinese soldier, captured by 'c' company od 13 Kumaon during the battle of Rezang La, Sino-India war 1962; Pouch of a soldier of 13 Kumaon who fought against the Chinese in the famous battle of Rezang La, Sino-India war 1962; Coat Parka of L. N. K Raghunath Singh of 13 Kumaon worn during the battle of Rezang La, Sino-India war 1962; Indian national flag which is used to drape the body of Gen. Raina, MVC, Padma Bhushan when he passed away on 19 may 1980 during his tenure as the Indian High Commissioner to Canada; Jacket of a Kumaon rifles officer; Dark blue sleeveless jacket ornamented with black and green mohair braid; Tunic of Lt. Col ATB Gaig, 19th Hyderabad Regiment comprising of red superfine wool short jacket, having blue and green kummerband; During the advance of the 1/19th Hyderabad regiment (now 3 PARA) along the river Irrawady, a Japanese party

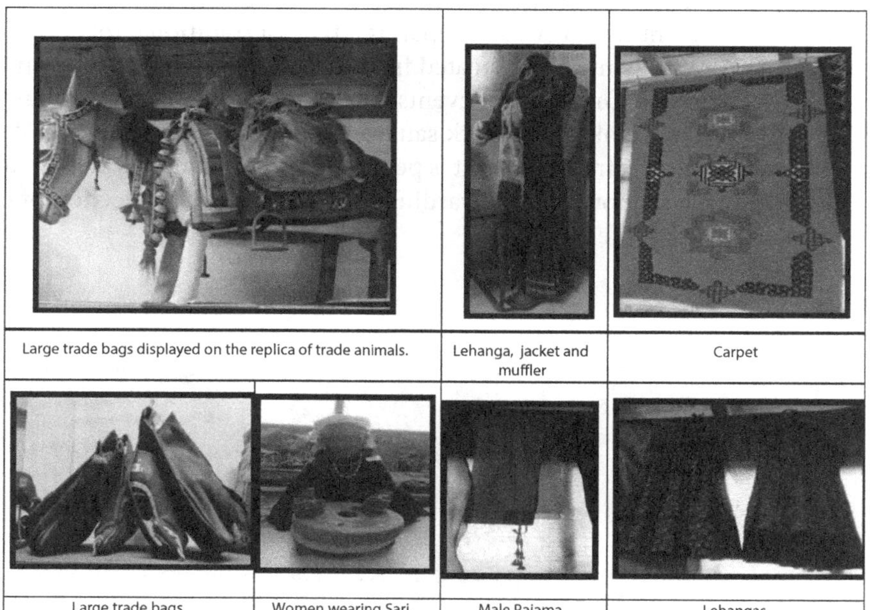

Figure 14.7 Artifacts displayed in Tribal Museum, Munsyari.

war ambushed by a company of the battalion commanded by Captain Later General T. N. Raina, in which "thousand stitch belt" stained with a blood of a Japanese Sergeant was captured. Later Gen. T. N. Raina, MVC Padma Bhushan presented it to museum.

When a Japanese soldier was called to active duty, the women in his family usually made the "1000 stitch belt" as good luck momento. She would implore passers-by to add a stitch, and once a thousand stitchers had been summoned, the belt was said to have extraordinary powers to protect the bearer from war dangers. All these articles were so beautifully displayed that the visitor will be easily understood the pain and struggle and more importantly courage of Indian freedom fighters. In this museum, photographs of artifacts were not allowed due to some governmental rules and security reasons.

14.3.4.5 Lok Sangrah, Folk Culture Museum, Bhimtal

Folk Culture Museum houses excellent compilation of costumes and arts of hilly regions of Uttarakhand. The museum comprises of one-man collection "Dr. Yashodhar Mathpal." This includes traditional costumes of Uttarakhand such as "pichhora," "Sari," "farm dress of male and females," "nisaad patakaaen," "mukut," "choli," "lehanga," "woollen coat," "fiber made articles, such as robes, caps, wall hangings and shoes," various kinds of baskets made by "Tharu" and "Buxa" tribe of Uttarakhand, "paintings of Folk art of Kumaon," "200 years old documents keeping bag" and "carpets" (Figure 14.8).

14.3.4.6 Jim Corbett Museum

Jim Corbett Museum, Ramnagar various Corbett belongings such as fishing rod with net, upholstered furniture, cap and paintings were displayed (Figure 14.9).

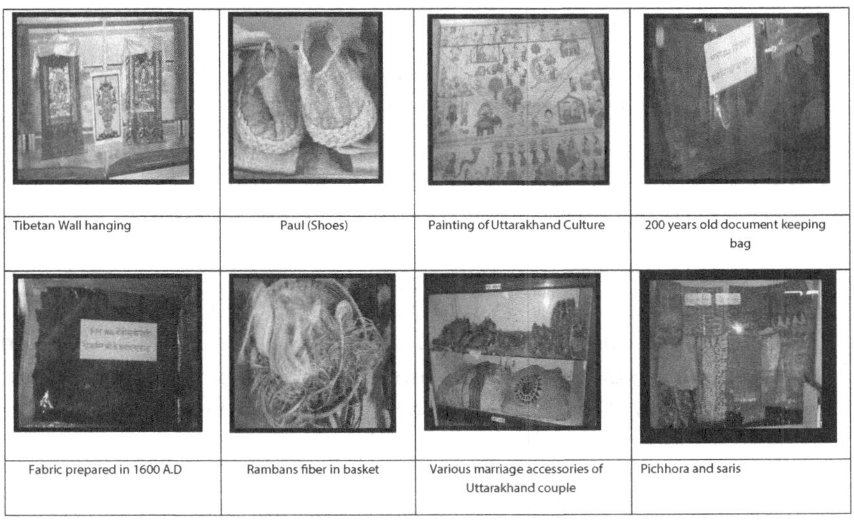

Figure 14.8 Artifacts displayed in Lok Sangrah, Folk Culture Museum, Bhimtal.

Figure 14.9 Upholstered furniture, bedding and fish net.

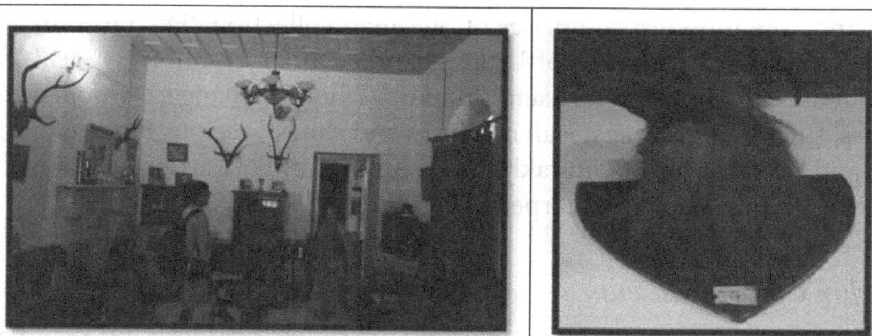

The complete house of Jim Corbett is conserved	Head of Deer hunted by Mr. Jim Corbett

Figure 14.10 Artifacts displayed in Gurney House Museum, Nainital.

14.3.4.7 Gurney House Museum, Nainital

The complete house of Jim Corbett was conserved (Figure 14.10) including upholstered furniture, draperies, wall hangings, mattresses, floor coverings, and many more priceless possessions of house.

14.3.5 Various Methods of Prevention Used in Various Selected Museums

Preventive methods such as proper cleaning, lighting, control of humidity & temperature and periodical fumigation are necessary for textiles so that the textiles displayed in museums will get protected from dust and other contaminants in museums which are emanating from the footsteps of visitors or many other sources. Therefore, it is important for museums to keep an eye on museum textiles to get rid from the damaging factors such as dust, insects, pollution and more importantly temperature & relative humidity.

Proper cleaning to reduce the effect of polluting agents and proper lighting arrangement was done in all the museums frequently. Fumigation was not done frequently in any of the museum because textiles do not require fumigation very often but on a regular basis at certain time interval to kill the microbes living on the surface of antiquities.

Table 14.9 Museums are distributed in accordance to the preventive methods used in various selected museums (N=6).

Preventive methods	Frequently	Occasionally	Never	Weighted mean score	Rank
Proper cleaning to reduce the effect of polluting agents	6	0	0	3	I
Proper lighting arrangement	6	0	0	3	I
Control of humidity and temperature	2	0	4	1.66	II
Periodical fumigation for prevention of fungal growth and insect attack	0	2	4	1.33	III

Table 14.9 evidently tells us that in the museums the most frequently favored preventive method was proper cleaning to reduce the effect of polluting agents and proper lighting arrangement which secured first rank (W.M.S =3), the next maximum chosen method was control of humidity and temperature, secured second rank (W.M.S =1.66) followed by periodical fumigation for prevention of fungal growth and insect attack which got third rank (W.M.S =1.33).

The results are in accordance with Action Pest Control fumigation targets a big number of bugs in a large area. Second, fumigation is completely safe for artwork, literature, wooden objects, furniture, and other such items. Finally, even the most fragile exhibitions and collections can be safeguarded without risk of damage [16].

14.3.5.1 Covering Materials Used for Various Artifacts Displayed in Selected Museums

Covering displayed articles with glass cases (Figure 14.11) is the most frequently used method for textiles in all museums because this is the safest method for textiles which protect them from dust and other impurities from outer environment.

Artifacts displayed in Glass cases in Tribal Museum, Munsyari | Lamination of artifacts in Tribal Museum, Munsyari | Painting with protective transparent sheet in Folk Culture Museum, Bhimtal

Figure 14.11 Various types of covering materials for displaying articles in selected museums.

Table 14.10 Museums are distributed according to the usage of covering materials for exhibited artifacts (N=6).

Covering materials	Frequently	Occasionally	Never	Weighted mean score	Rank
Glass cases	6	0	0	3	I
Protective transparent sheets	1	3	2	1.83	III
Lamination	1	4	1	2	II

Protective transparent sheets are used by some museums. According to respondent of Kumaon Regimental Museum flags were very big in that museum so they use to mount that on hardboard covered with protective transparent sheets. Lamination method by acrylic is very limited used in museums only for letters like in Kumaon Regimental Museum.

In most cases, acrylic is preferred to glass. Acrylic is a tough material to break. If glass fractures and shreds through a textile, it can cause irreversible damage. Furthermore, acrylic is considerably lighter than glass, making it easier to handle when hanging bigger framed items [17].

Table 14.10 shows that glass cases were the most chosen covering material for displayed textile artifacts (W.M.S=3), followed by protective transparent sheets (W.M.S=2) and lamination (W.M.S=1.83), which was the least preferred option (W.M.S=1.83).

14.3.5.2 *Special Kind of Lighting System in the Museum to Protect the Textiles/ Garments From Fading/Ageing*

All the museums were practicing diffused lighting for the textile's antiquities. While passing through the galleries of the Tribal Museum, the investigator observed that they had installed tube lights and LED bulbs. Upon enquiry curator of the museum, they utilise LEDs because they are constructed of benign materials, unlike fluorescent lighting, which contains mercury and may be harmful to the environment. LEDs are also recyclable and classified as "green" or "environmentally friendly." Respondents also replied that LED's do not wash out colors of the artifacts like other light sources such as fluorescents, therefore LEDs are perfect for display of textile artifacts.

In Tribal Museum,apart from using artificial lights, they used to open the windows of museum when any visitors visits and that daylight is sufficient for the visibility of antiquities to the viewers. In most of the galleries of Govind Ballabh Pant Government Museum and Gurney House the textiles were kept in complete darkness. When viewers visit to the galleries then only, they used to switch on the lights. In Kumaon Regimental Museum and Folk Culture Museum according to the respondent's florescent tube lights with UV filters were used. They had also replied that textiles should be stored or displayed in as little light as possible, and preferably in total darkness.

Michalski and Druzik also suggested that light from halogens, classic incandescent lamps, or UV-filtered sunshine will cause some extra light damage to items or artworks in comparison to LED bulb if chosen based on good or exceptional light (that is, CRI over 90) [18].

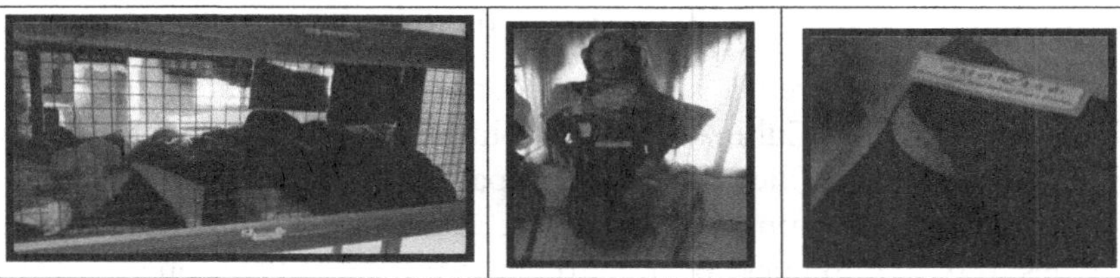

Caption written on showcase "See, but don't touch" in Tribal museum	Mannequin encased with wire case in Tribal museum, Munsyari	Interesting caption on mannequin to avoid from touching

Figure 14.12 Different captions on showcases to protect artifacts from visitor's touch.

14.3.5.3 Touching on Museum Antiquities

After a long interview of respondents and observation of investigator, it was concluded that touching was not permitted on museum textiles because the human body emits acids and oils through its pores, which can harm museum textiles. Respondent of Kumaon Regiment Museum replied that abrasion, grease and sweat will also deteriorate textiles. All museums under study place tags on various showcases that "Touching is not permitted." In Tribal Museum, to protect their collection from visitors, they encased the mannequins with wires case so that visitors could not be able to touch the fragile textiles and they also put attractive captions on various dummies about "Do not touch" (Figure 14.12).

14.3.6 Conservation Techniques Used in the Museum

Museum collection conservation is a critical step in the recovery, preservation, and advocacy of the past. Part of the museum's job as a keeper of the past is to protect things that have been removed from their original setting, as well as the knowledge linked with that context. The proper administration of museum collections includes recording and preserving the past for future generations, as well as displaying it for the enjoyment and education of current communities [19].

14.3.6.1 Pretreatments Given to Textile Antiquities and Display Boards

Kumaon Regiment Museum (Ranikhet) and Govind Ballabh Pant Museum (Almora) used to place small fabric bags contain paradichlorobenzene in their showcases. Vacuum cleaning of materials on open exhibit, textiles being returned to storage, and newly acquired textiles before they are placed in storage or on show is a frequent practise at the Kumaon Regiment Museum.

In Tribal Museum, Munsyari crushed leaves of various locally available species of high hill areas were diluted with water and then through spraying technique they used to protect their museum textiles.

Jim Corbett Museum (Ramnagar), Gurney House (Nainital) and Folk Culture Museum (Bhimtal) regularly practice dusting and unfolding their textiles to protect from deterioration.

The presence of airborne dust and grazing insect pests can be reduced by regularly removing debris from around the museum [20].

14.3.6.2 Methods of Reinforcing the Deteriorated Textile Antiquities

As mentioned earlier, not even a single museum had conservation laboratory. Tribal museum, Munsyari and Folk Culture Museum, Bhimtal, if any article requires strengthening, they get it done by local ladies expert in this particular work.

Govind Ballabh Pant Museum get their conservation work done by NRLC, Lucknow. Museum textiles were sent to Lucknow for strengthening and other conservation activities. Apart from all these methods of strengthening only netting and pasting were not practiced in Tribal Museum. While interviewing the conservator of Kumaon Regimental Museum it was noticed that netting was done on the weaker areas of the fragile textile to stabilize it temporarily. Netting increases the study life of the textile and also improves the strength to withstand the tension during display.

While interviewing officials of Jim Corbett Museum (Ramnagar), it was revealed that the Government Forest Department (Corbett Tiger Reserve) organizes public tender meeting. The conservation contract was given on the basis of experience of agency related to heritage conservation and lowest price for a given period of time. During this period the particular conservators of selected agency get the conservation work done.

In Gurney House, conservators from Delhi visit every year to have a view of the collection and to examine the antiquities conserved. The conservators were sent by the owner of Gurney House.

The results are in accordance with Liu *et al.*, that the initial stage in these conservation measures was to choose reinforcing materials. To offer suitable support for textiles, reinforcement materials should have the right structure, weight, and fiber. As a result, the type of fibers and level of damage were investigated using various methods [21].

14.3.6.3 Backing Material Used in Conservation of Museum Textiles

Backing material used in conservation of museum textiles for cotton, linen, wool, silk and metallic was same as that of the antiquity in all three museums. The respondent of the Tribal Museum informed that they used to wash the backing fabric before use to remove finishes due to which the fragile textile had severe risk of damaging. They also prefer like with like that means match the backing fabric with the textile antiquity.

While interviewing the respondent of the Tribal Museum, he also replied that they do not follow any fixed rule. They select the fabrics required for backing which best suits the needs of the antiquity under study. Most importantly, the respondent had mentioned that the weight of the backing fabric should be half of the weight of the object under study or less than the weight of the fragile textile or depending on the capacity of the particular textile on how much weight it can bear.

In order to provide support, backing materials must be sturdy and able to endure tension, but not so heavy that they put pressure on the primary textile. When used in conjunction with local repairs and visual compensation techniques, the fabric should extend beyond the damaged areas, with adequate stitching in weak places [22].

14.4 Conclusion

Selected museums house exclusive textile antiquities including traditional costumes of bride and groom and farm working men and women of Uttarakhand, marriage accessories like pichora, mukut, etc., carpets, mufflers, skin of various animals, shoes, large trade bags, potli bags, Tibetan wall hangings, paintings on Uttarakhand culture, children's garments and personal belongings of Jim Corbett were conserved. The use of showcase was the most commonly used storage and display technique practiced by all the museums. Articles, such as sari, pichhora, muffler, and floor covering, were folded and kept in plastic sheet but the caretakers of museums unfold these textiles at a regular period of time to increase the life of conserved artifact. Various preventive measures, including periodical cleaning and dusting, keeping naphthalene balls, spraying powdered solution of local plant species, lighting arrangement, were used by the museum's caretakers.

References

1. Sandahl, J., The Museum Definition as the Backbone of ICOM. *Mus. Int.*, 71, 1-2, vi-9, 2019.
2. Oklahoma Museums Association, 2018. https://www.okmuseums.org/resources/starting-a-museum.
3. Camelia, I.D., Vasile, H.G., Tudor, C., Alexandru, I., Liliana, I., Jan, W., Anamaria, A., Mamadou, D., Cristina, L., Zharas, B., Adina, A., Considerations Regarding the Research for the Conservation of Heritage Textiles in Romania, in: *Waste in Textile and Leather Sectors*, A. Körlü, (Ed.), IntechOpen Limited, London, United Kingdom, March 14th 2020, Available from: https://www.intechopen.com/chapters/71457.
4. Gaur, S. and Bhagat, S., Conservation of Printed Textiles: to preserve what is established. *Int. J. Innov. Res. Dev.*, 5, 14, 228–233, 2016. Retrieved from http://www.internationaljournalcorner.com/index.php/ijird_ojs/article/view/136835/95958 on 18 Aug 2021.
5. Ahmed, H.E. and Ziddan, Y.E., A new approach for conservation treatment of a silk textile in Islamic Art Museum, Cairo. *J. Cult. Herit.*, 12, 4, 412–419, 2011. https://doi.org/10.1016/j.culher.2011.02.004.
6. Annette, What is Textile Conservation? What is Textile Restoration. Chetwynd Textile Conservation Services, 2018. Retrieved from https://www.chetwyndconservation.com/blog/what-is-textile-conservation-textile-restoration on 19 Aug 2021.
7. Horjan, G., Towards the education we really need in Regional Museums, in: *Staff and Training in Regional Museums*, 2011, Retrieved from http://network.icom.museum/fileadmin/user_upload/minisites/icr/pdf/StaffAndTraining_WEB.pdf.
8. University of Alaska Fairbanks, 2009. Acquisitions and Accessioning. https://www.uaf.edu/museum/collections/ethno/policies/acquisitions/. Retrieved on 15 June 2018.
9. Singhal, D. and Bhagat, S., Cleaning of textile artifacts in museums. *Int. J. Appl. Home Sci.*, 2, 7&8, 247–252, 2015.
10. *Ventilation in Historic Buildings*, National Property Board, Sweden, 2018, https://www.sfv.se/globalassets/bygg-pa-kunskap/byggnadsvard/ventilation_in_historic_buildings_b-2.pdf. Retrieved on 19 June 2018.
11. Goodway, M., Fiber Identification in Practice. *J. Am. Inst. Conserv.*, 26, 1, 27–44, 2013.

12. Nalini, V., *Humidity control in museums and libraries*, CAREL Industries, Via dell'Industria, Brugine PD, Italy, 2017, Retrieved on 13 July 2018. http://www.carel.com/blog/-/blogs/humidity-control-in-museums-and-libraries.
13. Templeton, R.H., Display of Textiles in Museum Collection. *Conference: Heritage Conservation Workshop At: Chandigarh, India. Retrieved from-*, 2012, https://www.researchgate.net/publication/318014663_Display_of_Textiles_in_Museum_Collection on 19 Aug 2021.
14. *Canadian Conservation Institute (CCI) Notes 13/8*, CCI Textile Lab, Ottawa, ON, Canada, 2008, Retrieved from- https://www.canada.ca/en/conservation-institute/services/conservation-preservation-publications/canadian-conservation-institute-notes/applying-accession-numbers-textiles.html on 19 Aug 2021.
15. Alig, D., Heard, T., Spicer, G., Storage of Textiles: Issues and Methods Textile Conservators Face when Planning for Textile Storage-Section B. Storage Furniture, Textile Specialty Group, 2021, Retrieved from- https://www.conservation-wiki.com/wiki/TSG_Chapter_VIII._Storage_of_Textiles:_Issues_and_Methods_Textile_Conservators_Face_when_Planning_for_Textile_Storage_-_Section_B._Storage_Furniture on 17 Aug 2021.
16. Action Pest Control, Why Museums turn to Fumigation Services, 2017. Retrieved from-https://www.actionpest.com/blog/post/why-museums-turn-to-fumigation-services on 19 Aug 2021.
17. The George Washington University Museum-The Textile Museum, Displaying Textiles, 2021. https://museum.gwu.edu/displaying-textiles.
18. Michalski, S. and Druzik, J., LED Lighting in museums and Art Galleries, in: *Technical Bulletin 36*, CCI, Ottawa, Canada, 2020, Retrieved from-https://www.canada.ca/en/conservation-institute/services/conservation-preservation-publications/technical-bulletins/led-lighting-museums.html on 19 Aug 2021.
19. Depcinski, M.C., Conservation in Museums, in: *Encyclopedia of Global Archaeology*, C. Smith, (Ed.), Springer, New York, NY, 2014, https://doi.org/10.1007/978-1-4419-0465-2_796.
20. North Carolina Museum of History Foundation, How to care for your textiles, 2021. https://www.ncmuseumofhistory.org/collections/artifact-care/textiles.
21. Liu, J., Li, Y., Hu, D. *et al.*, A new method for the conservation of ancient colored paintings on ramie textiles. *Herit. Sci.*, 9, 13, 2021. https://doi.org/10.1186/s40494-021-00486-4. https://heritagesciencejournal.springeropen.com/articles/10.1186/s40494-021-00486-4.
22. Montegut, D., Sanford, M., Phipps, E., Knutson, T., Karsten, I., Mayhew, F., Murray, A., Stevens, S., Commoner, L., TSG Chapter V. Analysis and Testing Methods for Textiles-Section C. Fiber Identification-Wiki, Textile Specialty Group: Textile Conservation Wiki, 2021, www.conservation-wiki.com.

15

The Conservation and Display of Indian Textiles at the Victoria and Albert Museum

Elizabeth-Anne Haldane[1]*, Lynda Hillyer[1] and Deepshikha Kalsi[2]

[1]*Textile Conservation Studio, Victoria and Albert Museum, Cromwell Road, London, United Kingdom*
[2]*Textile Conservation Studio, B10, Green Park Main, New Delhi, India*

Abstract
This chapter focuses on the care and conservation of the Indian textile collections at the Victoria and Albert Museum (V&A) in London. There are approximately 10,000 Indian textiles in the collection, encompassing a comprehensive range of textile and costume manufacturing techniques and historical periods. The principles of conservation, which underpin the treatment of all objects, are discussed, and the importance of understanding the causes of deterioration to assess condition. Preventive conservation is stressed, including maintaining stable environmental conditions, good housekeeping and integrated pest management, and rotation of light sensitive objects. Exhibiting textiles at the V&A is a collaborative effort in which the conservator works with curators, designers, technicians, scientists, registrars, and exhibition organizers to realize the display. Drawing from the preparation of Indian textiles and costume for permanent and temporary display the conservator's role in object research, including scientific investigation are discussed and the importance of understanding context and object interpretation particularly regarding display. Common conservation issues and treatment solutions are examined through case studies. Topics include surface cleaning, humidification, wet and solvent cleaning, stabilization and support methods. Finally, strategies for the handling and display of Indian textiles and costume, which can be very large, embellished, and heavy, are discussed.

Keywords: Indian, textiles, costume, V&A, conservation, exhibition, display

15.1 Introduction to the V&A and the Indian Textile Collections

The Victoria and Albert Museum (V&A) is the UK's National Museum of Art, Design and Performance and holds a collection of over 2.3 million objects. The museum evolved out of the groundbreaking and hugely popular Great Exhibition, the first international display of design and manufacturing, held in a temporary glass building in London's Hyde Park in 1851. The aim of the exhibition was to educate the public and inspire designers and manufacturers by displaying exemplary products from around the world. At the end of

**Corresponding author*: e.haldane@vam.ac.uk

Seiko Jose, Sabu Thomas, Pintu Pandit, Ritu Pandey and Vandana Gupta (eds.) Handbook of Museum Textiles: Volume I Conservation and Cultural Research, (291–314) © 2023 Scrivener Publishing LLC

the exhibition, a selection of objects was purchased to form the core of a new museum of manufacturers that would later be named the Victoria and Albert Museum, acknowledging the role of Prince Albert in the creation of the Great Exhibition and the founding of the museum. The mission of the V&A today is still to enrich people's lives by promoting research, knowledge, and enjoyment of the designed world to the widest possible audience, whether that is in person visiting one of the V&A sites, an international touring exhibition or remotely through publications or digital content available on the V&A website [1]. The V&A's founding mission has a direct bearing on the conservation and display of the collections. As a museum of design, the aim is, if possible, to display the objects as the designer or artist originally intended, which is different for example, to the ethos of a social history or ethnographic museum.

The collections displayed in the permanent galleries of the main V&A site in South Kensington, London are generally organized and exhibited by material type, for example, metalwork, sculpture, paintings, etc., or by geographic area or period, for example, the British Galleries, and Medieval and Renaissance Galleries, where a range of different types of objects are displayed together. Highlights of the V&A's collections of Indian art and design are primarily displayed in the permanent South Asia Gallery. The collections are also often requested for loan to other museums, and feature in temporary exhibitions organized by the V&A, including the major international touring exhibitions *Maharaja: The Splendour of India's Royal Courts* (2009–2013) and *The Fabric of India* (2015–2018) (Figures 15.1, 15.2).

There are over 10,000 Indian textiles in the collection including 155 textiles, which were displayed at the Great Exhibition of 1851 [2]. Throughout the second half of the 19th century, the museum continued to acquire textiles that had been selected from all over India for display at further international exhibitions. The collection was significantly enhanced in 1879 when the V&A acquired a large portion of the textile and decorative art collections

Figure 15.1 Maharaja: The Splendour of India's Royal Courts, 2009 © Victoria and Albert Museum, London.

Figure 15.2 *The Fabric of India*, 2015 © Victoria and Albert Museum, London.

of The India Museum, established in 1791 by the East India Company, including approximately 6000 textiles. Between 1881 and 1882, almost 1000 textiles were purchased in India to add to the collection. Although the focus in the 19th century was on collecting contemporary textiles, the collection now encompasses a broad range of historical periods and a comprehensive range of textile and costume manufacturing techniques (Figure 15.3). Highlights include indigo resist and block printed textiles from the 14th century excavated at Fostat in Egypt, Mughal velvets, Dacca muslins, embroideries, bandhani, and an extensive and important collection of kalamkari chintzes made for both Western and Eastern

Figure 15.3 Cased display of 19th and 20th century objects to illustrate stitch techniques in The Fabric of India © Victoria and Albert Museum, London.

Figure 15.4 Contemporary saris acquired for *The Fabric of India*, 2015 © Victoria and Albert Museum, London.

export markets [3]. The collection continues to be added to today, with an emphasis on acquiring contemporary Indian fashion and textile art (Figure 15.4).

The V&A has a long connection with Indian institutions. The Nehru Trust for the Indian Collections at the V&A was founded in India in 1990 to make the collections at the V&A more accessible and to facilitate the study, preservation and display of India's art and cultural heritage. Curators and conservators from the V&A have often visited India to collaborate on research and to offer advice, and individual scholars and museum professionals have regularly come to study at the V&A, including in the conservation department, on scholarships supported by the Nehru Trust and other philanthropic organizations [4].

15.2 Care of Collections

The care of collections at the V&A is the responsibility of many different departments in the Museum each contributing from their own area of expertise. The conservation department is comprised of studios specializing in the treatment of different materials including metals, ceramics and glass, sculpture, furniture, paper, books, paintings, and textiles, supported by conservation scientists with expertise in environmental conditions and material analysis. The textile conservation studio is the largest, there are currently fifteen permanent members of staff, nine textile conservators and six costume specialists whose role is to focus on interpretation and display methods. The department has a long tradition of teaching and supports conservation students and postgraduate fellows by offering placements in the conservation studios, working alongside V&A conservators, so that they can gain hands on practical experience of all aspects of conservation in a museum environment. During the last 10 years, the V&A has hosted several interns from India in the Textile Conservation studio [5]. Conservators have also been involved in delivering short training courses to raise awareness of Textile Conservation in India in conjunction with SUTRA, a Kolkata-based nonprofit organization with a mission to increase awareness of India's textile heritage and to encourage research and conservation of textiles in India [6].

Drawing on this experience, the focus of this chapter is on the basic principles of conservation, looking at the routine work of the textile conservation studio, which is vital to good collections care. The work to prepare Indian textiles and costume for both permanent and temporary display will be discussed, looking at common conservation issues and treatment solutions through case studies, complemented with suggestions for further reading of more complex treatments already published. The chapter will emphasize the role of preventive conservation measures including maintaining stable environmental conditions in both galleries and stores, good housekeeping to prevent pest infestation and the importance of object rotation. Only a small percentage of the collection can be displayed at one time so good storage is vital, underpinning the long-term preservation of the objects.

15.3 Conservation

15.3.1 Principles of Conservation

Conservation is a complex and multifaceted discipline, a combination of art and science, that at its core is about the management of change. The conservator needs to know how to slow down the inevitable process of change and deterioration to prevent further damage. To achieve this, conservators need to understand the physical world in which the object exists, the science of materials and how they deteriorate [7]. They also need to understand the interaction between the materials that they use to stabilize the object and the materials from which the object is constructed. Technical skills are only part of the conservator's work. A conservator must understand the multiplicity of meanings and values that an object can present and the role of the object in a collection. Treatment should not affect the significance of the object, either in its structure or in its more intangible meaning, i.e., the integrity and authenticity of the object. Conservators adhere to strict codes of ethical guidelines developed by professional conservation institutes [8]. Conservation seeks to stabilize what is extant, the minimum amount of work should be undertaken to achieve a satisfactory result and where possible any new materials used should be reversible. The mission of the V&A conservation department, working in a design museum context, is to ensure that the collections are stable, in optimum visual condition and technically understood and to achieve and demonstrate the highest standards of conservation practice, training and research.

15.3.2 Assessing Condition, Causes of Deterioration

To assess and control risk, it is necessary to have a thorough understanding of the condition of the object and the causes of its deterioration. Textiles are some of the most sensitive objects in a collection. Understanding the deterioration of textiles is the foundation of the conservator's skill. It underlines every decision made when the object is in the conservation studio. The agents of deterioration are well known, inherent vice, photochemical degradation, humidity, chemical degradation, biodeterioration, atmospheric pollution, physical or mechanical damage. Inherent vice refers to the pretreatment of a textile that can influence the fiber's ability to resist the agents of deterioration. Processes used in the manufacture of a textile object, for example, bleaches, dyes, mordants, finishes and pigments can initiate and accelerate chemical degradation. Light (visible and UV) causes damage that is cumulative and irreversible, it causes chemical and mechanical changes in a textile object that makes it more vulnerable to the presence of oxygen, humidity, and heat. Degradation begins at the fiber surface and fiber ends causing yellowing and embrittlement of fibers resulting in loss of tensile strength and elasticity. Textiles are generally very hygroscopic, and their large surface area means that they respond very rapidly to changes in humidity. High humidity can swell fibers and leave them more open to attack by microorganisms, such as mould and mildew that grow in these conditions. Conversely, prolonged low humidity can result in desiccation of the fibers and the loss of natural elasticity. Because of their fibrous nature and relatively open structure, textiles are very efficient collectors of airborne pollutants such as dust. Coupled with this, deterioration involves changes in the molecular structure of the textile, a deteriorated textile has voids and cracks in its structure that make it very permeable to all types of pollutants. Textiles, particularly wool, silk, and soiled textiles, can be an attractive food source for insects such as clothes moth or carpet beetle that can cause severe damage to the object.

All deteriorated textiles are more vulnerable to mechanical or physical damage and may no longer be strong enough to be used in the way they were originally intended. Textiles are sometimes damaged through use or poor storage before they became museum objects. Repeated laundering can cause a loss of strength and color, and inappropriate cleaning methods can cause dyes to run and leave stains. Textiles can become distorted and worn through wear, sewing threads securing seams or decorative elements to a textile can fail resulting in damage or loss of parts. Textiles and costume are often repaired by their owners during use, sometimes these repairs are exquisitely executed but often they are clumsy repairs that cause more damage.

The V&A's Indian textile collection is an interesting combination of objects that were collected new and never used except for exhibition purposes, and textiles and costumes that have clearly been used and worn. Many of the former collected in the 19th century, were initially displayed in less-than-ideal environmental conditions, have suffered from light damage, and often have a layer of sooty soiling from open display, the mark of an

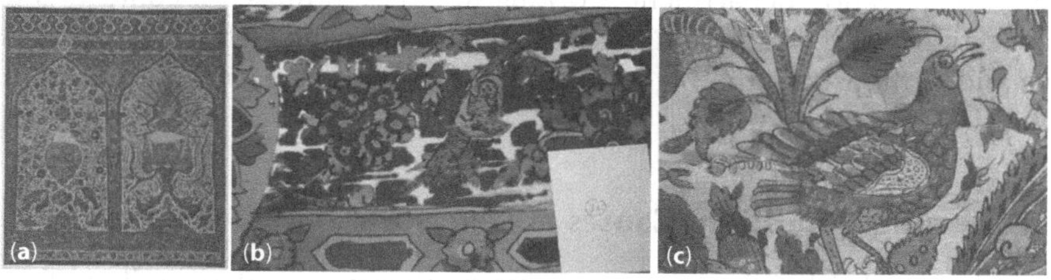

Figure 15.5 (a) Kalamkari tent hanging IS.19-1989 after treatment, stitched to a fabric covered display board © Victoria and Albert Museum, London. (b) Iron mordant damage from dye process, before treatment IS.19-1989© Victoria and Albert Museum, London. (c) Historic repairs to patch bird before treatment IS.19-1989 © Victoria and Albert Museum, London.

industrial city. In contrast the main problem with an early 17th century kalamkari tent hanging (IS.19-1989) was degradation of the cotton exacerbated by iron mordants used in the dye process (Figures 15.5a, b). The hanging also had multiple repairs from use including patches that were of intrinsic interest and added to the history of the object (Figure 15.5c). The conservation treatment was extensive, posing not only technical challenges but ethical decisions regarding removal of the historic patches. The treatment, which has been published [9], illustrates the importance of detailed documentation to record current condition, original construction, and modifications through use. This is particularly important if historic repairs must be removed for the long-term stability of the object.

15.3.3 Preventive Conservation

Most damage to the object comes from the environment, which is why preventive conservation is the most important element in the care of our heritage. Good storage and display conditions are vital to prevent further deterioration [10, 11]. It is essential to control light exposure and to maintain a stable environment. Standardized guidelines for light levels, temperature, and relative humidity have been developed by the worldwide museum community and are a valuable benchmark; however, controlling the environment can be very difficult and expensive to achieve and of increasing concern from a sustainability viewpoint [12]. It is widely recognized that maintaining a stable environment without dramatic fluctuations is key, at the V&A temperature, relative humidity, and light are continually monitored using a networked electronic system with an alert function that facilitates a rapid response to problematic changes. It is usually easier to maintain a stable environment within a well-sealed showcase, which also offers protection from dust, pollution and handling and can help in the control of pests. It is vital that materials used in the storage and display of objects are not contributing to their deterioration, where possible it is best practice to use inert materials or if this is not feasible to use barrier layers or protective coatings to prevent direct contact [13]. Following the phasing out of chemical deterrents due to their impact on human heath, the control of pests is achieved through a combination of good housekeeping practice and regular monitoring. There is a wealth of up-to-date information and advice available online through groups including Pest Odyssey UK [14] and The MuseumPests Working Group based in the USA [15]. Good housekeeping is also critical to keeping objects free from dust. Floors, plinths, cases, etc. are cleaned regularly to reduce the amount of dust in the atmosphere that can transfer to objects. Although it is preferable to case textiles, some are simply too large, such as the textiles shown in Figures 15.1 and 15.2, another way to control risk is to limit the time on open display to temporary exhibitions. Any textiles placed on open display must be stable enough to withstand vacuuming carried out by a conservator, it is also useful to remember that vertically displayed textiles attract less dust than those displayed horizontally. They must also be placed out of reach of visitors interested to touch and feel the texture of textile exhibits but adding to accumulated soiling.

15.3.4 Understanding the Object—Context and Scientific Investigation

Collections at the V&A are used in a variety of ways, for study, publications, exhibition at the V&A or loan to another institution. The conservator works in close collaboration with the curator to understand the context and significance of the object prior to any work.

When the condition of a textile is being assessed the conservator will discuss with the curator what the expectations are for the object's display, ask how the object should be interpreted, and discuss how the mounting and display can help to convey this information. For example, a sari could be displayed on a mannequin if it was important to understand how the textile looked when worn (Figure 15.4) or it could be displayed rolled with just a small section visible if the focus was purely on the textile techniques (Figure 15.6). It is important at this stage to ascertain if the object will be on temporary display or out for a longer period and whether the object will be cased or if it will go on open display. V&A temporary exhibitions are often lent to other museums, which adds another level of risk as the objects must be able to withstand the repeated handling involved in transporting them around the world [16]. With this knowledge, the conservator assesses the condition of the object and its suitability for the proposed display and advises the curator if the display is appropriate or if it poses a risk to the stability of the object and whether treatment or alternative display methods can mitigate this risk.

The conservator is in a unique position to record technical details of an object's construction and to add to the historical perspective of the object. Firstly, it is necessary to understand what the object is made from to place it in its correct curatorial context and to aid treatment. Sometimes, this is not straightforward, and it may require scientific analysis to help identify the materials. The V&A is fortunate in having several instrumental analytical techniques available in house, some are classed as destructive requiring the removal of small samples such as a fiber for microscopic identification, others are nondestructive such as Energy Dispersive X-Ray Fluorescence (EDXRF) a technique used for elemental analysis. When the specialist equipment required is not available in-house, such as for the identification of dyestuffs, samples are sent to external laboratories, although this comes at an additional cost.

The Fabric of India was a major V&A exhibition that explored the world of handmade textiles from India from the 3rd to the 21st century and featured over 200 objects [17]. The first section of the exhibition, Nature and Making, focused on textile production in India with displays of raw materials such as cotton bolls and silk cocoons next to corresponding

Figure 15.6 Display of weaving techniques, including several partially displayed saris in *The Fabric of India* © Victoria and Albert Museum, London.

textiles, and alongside all sorts of nontextile embellishments including a variety of metal threads in both sample form and woven into textiles that were all identified using EDXRF analysis (Figure 15.7) [18]. The section on dyes was curated to provide an overview of the wide variety of natural dyes native to India. Each of the textiles shown was from the V&A's own collection of 19th century Indian everyday fabrics paired with raw samples of dye materials borrowed from the Royal Botanic Gardens, Kew (Figure 15.8). If there was any doubt,

Figure 15.7 *The Fabric of India* raw materials display 19th century textiles and metal thread samples © Victoria and Albert Museum, London.

Figure 15.8 *The Fabric of India* dyes display 19th century textiles and raw dyestuffs © Victoria and Albert Museum, London.

dye analysis was commissioned to ensure the textiles were accurately identified prior to inclusion. Correct identification is necessary not only for conservation and educational purposes, but it is also critical to know if an object contains materials from endangered species that are regulated by the Convention on International Trade in Endangered Species of Wild Fauna and Flora, which is an international agreement between governments [19]. This is particularly important for objects going on loan to an international venue as CITES permits are required for transportation.

The V&A is increasingly using its website to share information about the collections and aims to digitize the whole collection and make it available online through the "Explore our collections" search function. There is also a wealth of articles by curators and conservators on the website alongside blog posts that are written in a more informal style. A series of weekly blog posts was written to document the development of *The Fabric of India* exhibition, including curatorial research, object analysis, conservation treatments, exhibition design, planning, and installation of some of the largest textiles displayed at the V&A. These posts are archived on the website [20].

15.4 Object Treatment

At the V&A object documentation and digital photographic records are stored on a centralized computer database and supporting material such as diagrams and material samples are also held with the individual hard copy object files. A treatment proposal and an estimate of the time required to do the work is written for each object. The treatment is agreed in consultation with the curator, sometimes with advice from other material specialist conservators as textile artifacts often incorporate a large variety of nontextile materials. The treatment record, including details of analysis, and any materials and chemicals used is added to the object database. Interventive conservation treatment can be broadly divided into two categories, cleaning, and support, a third related category, often regarded as noninterventive, is mounting for display. In the case of textiles, it is not always possible to be truly noninterventive as it is often necessary to stitch through an object to attach it to a support for display.

15.4.1 Object Treatment—Cleaning

When considering cleaning it is first necessary to determine what type of soiling is present on an object. Soiling can be categorized into loose dust and particulate matter, ingrained dirt, accretions of solid matter such as adhesives or paints, stains, and discoloration caused by degradation of the fibers that is generally yellow in appearance. Cleaning is an irreversible process. The next step after soiling identification is to consider if it is appropriate to remove it, and if it is feasible to do so without causing damage to the object. Soiling on an object can provide valuable evidence of an object's history and provenance, particularly in the case of archaeological textiles [21]. Advancements in analytical techniques are opening new areas of research, for example soil analysis is routinely carried out on American Civil war flags as pollen can be used to locate where the flag was flown. Stains are often the most important part of an object, the blood-stained clothing worn by British naval hero Admiral Lord Nelson when he died in battle is preserved at the National Maritime Museum in London. Soiling, creases, and stitch marks can provide evidence of original construction or alterations and could easily be lost during a wet cleaning treatment. Loosely bound dyes can be washed out and colors changed during wet cleaning processes, and it can alter textural effects that have been applied to fabric using water and pressure. Cleaning can also remove surface finishes such as sizes, glazes and water-soluble decoration such as gold leaf attached with gum Arabic,

15.4.2 Surface Cleaning and Humidification

Cleaning can be broken down into different categories, surface cleaning, wet cleaning, and solvent cleaning. Surface cleaning is used to remove loose soiling prior to any other cleaning method. A variable suction vacuum cleaner allows the process to be controlled, low suction is used for delicate surfaces and soft brushes are useful to dislodge dust from three-dimensional surfaces into the vacuum cleaner. In the case of this Indian jama 05643:2(IS) vacuuming made little visible difference to the dark gray soiling typically found on V&A textiles collected in the 19th century, so mechanical cleaning with a latex sponge or "smoke sponge" was tested (Figures 15.9a–c). The sponge was gently wiped across the surface of the textile successfully picking up the soil particles. Sometimes, soiling is in the form of heavy deposits, which can be removed mechanically with scalpel or tweezers.

When a textile is creased or folded the intramolecular bonds that hold the polymers together are stretched and can break reforming in areas of less strain to hold the creases in place. When water is introduced, it acts as a "plasticizer," and the polymers can rearrange themselves into their original position; however, sometimes, the textile has been creased for so long that the fiber structure has started to break down resulting in permanent deformation [22]. This jama is an interesting case study as ideally it would have been wet cleaned to remove waterborne stains and cellulose degradation; however, as discussed, water is very good at removing creases, so it was decided not to risk cleaning the jama as the cotton fabric was deliberately creased when it was made to produce a delicately textured surface that must be retained (Figure 15.9d). The added complexity of a colored silk lining and gold thread decoration cemented the decision (Figure 15.9e). Controlled humidification is a valuable tool that not only improves the visual appearance of an object but also aids its long-term preservation. Water is introduced to the textile in a vapor state to relax the fibers, the textile becomes more pliable but does not become wet. It is a much less aggressive treatment than wet cleaning, but it is not without risk. Particular care should be taken when humidifying any textiles with fugitive dyes, painted surfaces and/or binding media, metal or plastic decoration. It is not advisable to humidify very soiled objects because humidification opens up the fiber structure and soiling can become driven further into the fiber or irreversibly bound to the fiber.

Following surface cleaning, the jama was treated to reduce creasing from storage without affecting the texture of the cotton using a combination of ultrasonic humidification, where a fine mist of cold water was directed at the textile (Figure 15.9f), and for more stubborn creases, contact humidification (Figure 15.9g). In contact humidification, the vapor is locally applied to the textile through a high tech "breathable" waterproof fabric (Gore-tex or Sympatex) developed for outdoor clothing use. A damp blotter provides the source of the water and the whole is covered in polythene to seal the moisture in. This must be checked frequently to ensure that the textile never becomes wet.

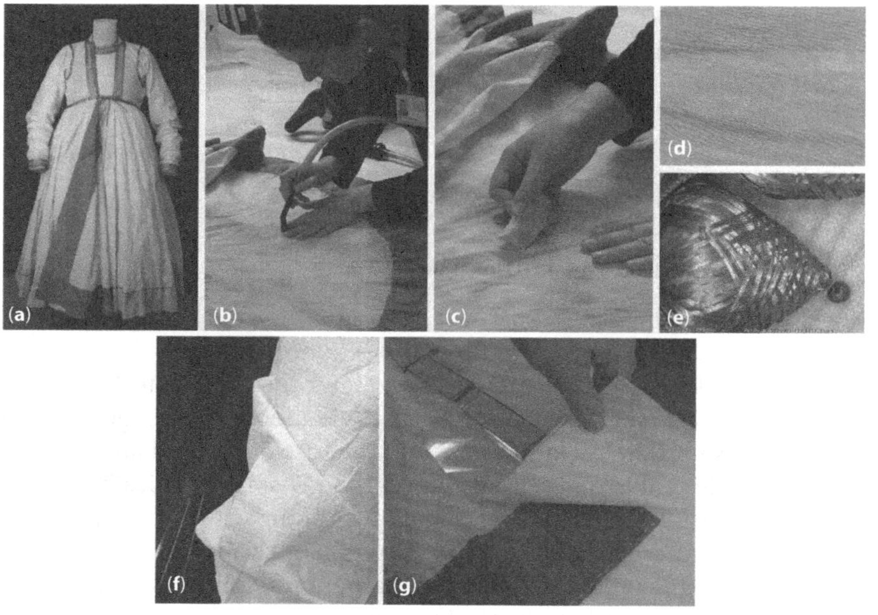

Figure 15.9 (a) Jama 05643:2(IS), ca.1855, after treatment. (b) Vacuuming the jama 05643:2(IS). (c) Removing soiling with a smoke sponge, 05643:2(IS). (d) Above, deliberately creased cotton, 05643:2(IS). (e) Detail of button wrapped in metal strip, 05643:2(IS). (f) Ultrasonic humidification of jama 05643:2(IS). (g) Contact humidification of jama 05643:2(IS), using Sympatex © Victoria and Albert Museum, London.

Another method that is very useful when the whole textile requires treatment is a cold poultice (Figure 15.10). In this method, a soft board wrapped in polythene is covered with one or two layers of slightly damp muslin, topped with a dry muslin layer. The textile is placed directly on top and covered with polythene to create a vapor chamber. Once the textile is pliable the creases can be eased out and if appropriate the textile can be pinned in place to the board and left to dry naturally.

Figure 15.10 Cold poultice humidification method used for cotton and wool shawl 4807(IS) © Victoria and Albert Museum, London.

15.4.3 Wet Cleaning

Water is highly polar; it is very good at dissolving substances and is very effective at removing soiling that can be detrimental to the long-term stability of textiles and aesthetically unpleasing. Water also dissolves most of the acidic degradation products of natural fibers that are noticeable as yellowing of the textile, by doing so the condition of the textile and its visual appearance are both improved, As previously discussed it is very good at relaxing creases and distortions and improving overall flexibility, it is also readily available and does not pose a health and safety risk. Water that is used for conservation treatments must be purified before use, deionized water is treated to remove metal ions, reionized is a form of softened water that has the harmful ions removed then different ions introduced to buffer the water and slow down its reactivity. Very degraded fibers are much less resistant to swelling on wetting and contraction on drying so wet cleaning can be very damaging. De-ionized and distilled water can be a very aggressive solvent as it naturally wants to replace the water-soluble compounds removed in processing. It readily absorbs carbon dioxide, which lowers its pH. The pH of the wash bath should be monitored before and during use as soiling as released from the object. In conservation treatments washing solutions should be prepared specifically for each object depending on the fiber being cleaned, and the type of soiling that must be removed. At the V&A surfactants which have been tested for conservation use are used, nonionic is suitable for most textile fibers but anionic detergents are also used [23, 24]. Additives to the washing solution usually include a soil suspension agent that prevents soil from being redeposited during the wet cleaning process. Other additives include compounds that buffer the wash bath to prevent it becoming too acidic as the soil or degradation products are released into the water, some also enhance the cleaning process, such as chelating agents which react with metal ions present in soiling and aid in their removal [25]. The wet fastness of all the colors in the textile must be tested before any wet cleaning processes. Testing can either be *in-situ* or preferably by removing tiny samples of threads. The test must replicate the detergents plus any additives used and the length of time that the textile will be wet. The V&A has a specialist wash table which can be tipped in two directions to aid in the removal of water. It is sectioned off to the size of the object with a plastic roller and a shallow bath is created from plastic sheeting clipped to the sides of the table. The plastic is opened at the corner to let the water out. Preparation is key to successful wet cleaning. It is important to think about how the object will be handled when it is wet and more vulnerable to damage, for example objects can be sandwiched between two pieces of clear polyester film (Melinex) to turn them over in the bath and a plastic roller is useful for turning larger objects. Sometimes layers will need to be separated with some soft net or sheets of Melinex to prevent them sticking together. Drying must be planned, ensuring that there is a large enough space to dry the object. The surface should be smooth and waterproof. If the object needs to be pinned out to dry, then a soft board covered with a polythene barrier layer should be preprepared. Three dimensional objects need special consideration.

A Kashmir shawl, 0804(IS), was wet cleaned by Indian conservators Madhura Wairkar and Deepshikha Kalsi as part of their internship training at the V&A in 2011. The shawl, white goat-hair (pashmina) with multicolored patterned ends, was made for display at the Paris Universal Exposition in 1867. It was acquired from there by the India Museum and transferred to the V&A in 1879. It was requested for display in the permanent gallery as part of a regular rotation programme of light sensitive objects. The shawl was assessed and found to be slightly yellowed, indicating some degradation of the fibers. There were bands of dark grey soiling, some staining and small areas of insect damage. It was concluded that the shawl would benefit from wet cleaning, small samples of all colors had to be tested first in the proposed washing solution. The colors were fast in the nonionic washing solution chosen, but some dyes bled when tested with a chelating agent that would have been helpful in removing the dark soiling, so this additive was not included. The wet cleaning process takes most of the working day. The textile is first wetted out with deionized water then the detergent solution is sponged onto the textile, which helps to loosen the soiling, the detergent is then rinsed off, the object turned over and the process repeated on the reverse (Figures 15.11a–c). Depending on the degree of soiling the process is repeated and interspersed with periods of soaking in a shallow bath. Thorough rinsing is crucial to remove detergent residues, this can take several hours of repeated running rinses and soaking in a bath. After the final rinsing the shawl is blotted dry with towels (Figure 15.11d) rolled up and transferred to the preprepared drying space.

The damp shawl was then manipulated to the correct shape and pinned to a softboard, with particular attention paid to the patterned ends. Glass weights were used to tension large plain areas, these are periodically moved to allow the textile below to dry. After pinning, cool air dryers were used to speed the process as there is always a risk of dye bleed just as the textile dries (Figures 15.11e, f). The shawl was much less yellowed after wet cleaning and although some small stains remained the dark lines of soiling and hard creases had disappeared. The holes were supported with polyester Stabiltex (a sheer plain weave fabric) patches stitched in place using fine polyester threads (Figure 15.11g). The shawl was

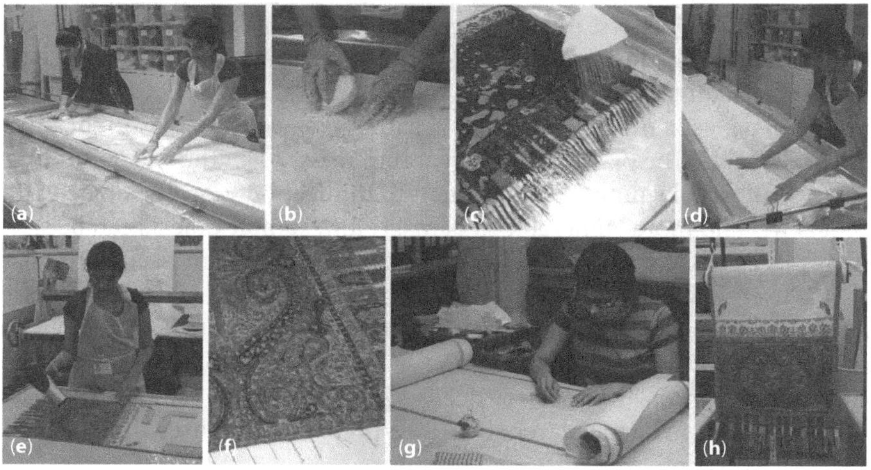

Figure 15.11 (a) Kashmir shawl 0804 (IS) during wet cleaning, sponging detergent. (b) Detail of sponging process. (c) Rinsing detergent away (0804(IS). (d) After wet cleaning, removing excess water with towels. (a-d) © Victoria and Albert Museum, London. (e) Drying process, Kashmir shawl 0804(IS). (f) Detail of textile pinned to soft board during drying (0804(IS). (g) Supporting holes with patches stitched using a curved needle (0804(IS)). (h) Test hang for shawl (0804(IS)) display method (e-h) © Victoria and Albert Museum, London.

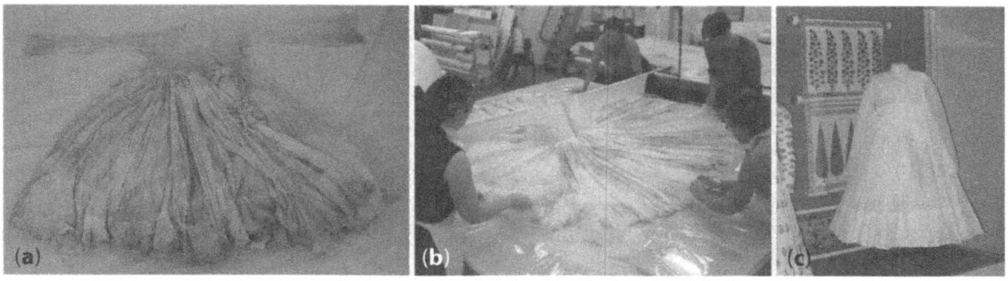

Figure 15.12 (a) Jama IS 05567 before treatment © Victoria and Albert Museum, London. (b) Drying jama IS 05567 after wet cleaning © Victoria and Albert Museum, London. (c) Jama IS 05567 after treatment on display.

prepared for display by rolling it onto a padded and fabric-covered roller that is suspended from brackets in the display case (Figure 15.11h).

The wet cleaning of an 19th jama IS 05567 was very long and labour intensive due to the large volume and weight of the skirt, the process began at 7am and ended at 9.30pm and involved a team of Seven conservators. Before treatment the jama was extremely creased with ingrained soiling all over, stains and tears in the bodice and skirts (Figure 15.12a). It was wet cleaned with an anionic detergent that has been found to be successful for cellulose-based textiles. Following the wet cleaning process, excess water was removed from the jama with towels and acid-free blotting paper It was moved to a large table and the skirts spread out to dry. The bodice, much weaker in its wet state, was supported by passing a tube through its long sleeves and suspending this at either end of the table. The volume of the skirt meant was essential that it was not left damp overnight. Cellulose degradation products can be highly water-soluble; if any remain in the object, they can migrate to form concentrated lines of discoloration as the textile dries and can be difficult to remove later. The jama was dried using cool air dryers, at the same time the layers of muslin were carefully separated from each other and manipulated to straighten the weave (Figure 15.12b). After drying, the muslin was much cleaner and less creased. Following further stitched treatment and mounting it was suitable for display (Figure 15.12c).

15.4.4 Solvent Cleaning

When it is not possible to clean an object with water, organic solvents are an alternative, but they must also be spot tested on the object before use. The choice of solvent will depend on the soiling that needs to be removed, for example white spirit is a nonpolar solvent that is good at removing greasy soiling [26]. Solvents can be applied as a spot treatment or the textile can be immersed in a bath; however, they carry an added health and safety risk requiring suitable extraction systems to remove harmful vapors and personal protective equipment. This 19th Century Indian umbrella IS.17-1991 (Figure 15.13a), is a complex composite object consisting of multiple textile layers secured together with decorative brass nails with velvet appliqué and silver foils. On the advice of a metal conservator, it was partially cleaned with solvents. The tarnished silver foils and heavily soiled fringe were gently swabbed with Industrial Methylated Spirits (IMS) an ethanol/methanol mix that will dissolve similar substances to water but does not swell the fibers as much as water. It evaporates very quickly, making it very suitable for cleaning metals. It was not possible to humidify the

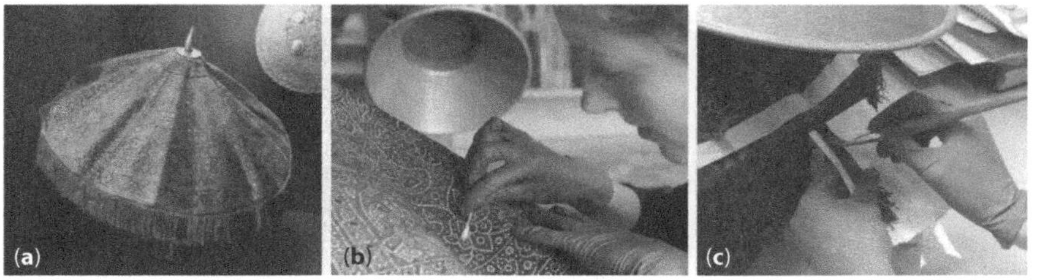

Figure 15.13 (a) 'Thousand nail' umbrella IS.17-1991 on display © Victoria and Albert Museum, London. (b) Metal areas of IS.17-1991 being solvent cleaned under extraction. (c) Swabbing fringe of IS.17-1991 with solvent onto blotting paper to absorb soiling.

textile elements as there was a risk of corrosion to the parts of the nails submerged inside the layers of textile. If the specialist equipment is available, it is also possible to carry out spot cleaning with water or solvents.

At the V&A there are two sizes of vacuum table that allow textiles to be cleaned in a very controlled way because the liquid applied is very quickly drawn through the perforated tabletop and away from the object [27]. The smaller of these tables is similar to that used by a commercial dry cleaner, the larger is a multipurpose low-pressure table that has been specifically made for conservation use and can also be used to heat set adhesive films.

15.4.5 Stabilization and Support

Following cleaning, the next step is to consider if there are any loose elements or seams requiring stabilization, or areas of structural damage that need some form of support. It is important to consider the condition of the whole object and plan where support is required. Patches can be used to support small areas of damage but if there are many damaged areas it may be more appropriate to consider a full support where a new material is introduced to stabilize an entire object [28]. The most common support material for textiles is another textile, and the usual method of applying the support is to stitch it to the object. If possible, the new material will be made from the same fiber as the original, so that they respond to environmental conditions in a similar way; however sometimes, it is more appropriate to use a sheer open weave fabric such as silk crepeline or polyester Stabiltex or a fine nylon "conservation" net that is less obtrusive and is not heavy in weight. Sometimes treatment may involve the apparent restoration of the original appearance, usually because the method of support takes the focus away from areas of loss so that the object appears more whole. Dyeing support fabrics to achieve a good color match for the damaged areas of the object is common in textile conservation. A full restoration would mean restoring something to how we think it looked at an earlier time in its history. This is rare in textile conservation but there are occasions when it happens. However, there is never any intention to deceive. Any restoration of the original appearance is carried out in the context of conservation, any additions are clear on close inspection and any additional material is different from what is extant—this is also recorded meticulously in the documentation of the object.

A spectacular 19th century red cotton jama (05563(IS)) from Bikaner, Rajasthan was conserved for display in 2019 by Samuel H. Kress Foundation Conservation Fellow Lauren Osmond. The jama decorated with metal thread ribbons, tinsel and jewel beetle wing case sequins was also a case study in a wider research project investigating changes to the structural color of the iridescent wing cases of jewel beetles [29]. A short film was made, documenting Lauren's research and the conservation treatment, for the V&A website [30]. The jama needed a significant amount of work before display, it was very soiled, creased, had multiple small holes and loose seams, and some of the decorative trimmings were loose or damaged (Figures 15.14b, d). Following extensive surface cleaning and humidification (Figure 15.14c) a variety of different support methods were used to treat the jama including a supportive display mount. The loose trimmings were secured first using a fine polyester sewing thread. Loose sequins were stitched down following the original sewing pattern. Areas of gold work ribbon that were unravelling were attached to the support fabric below with regular lines of laid thread couching, a stitching technique commonly used in textile conservation to secure and strengthen damaged areas. As the gold work was broken in some places, springy and vulnerable to snagging, patches of fine dyed nylon net were stitched over the ribbons to secure them (Figures 14d, e).

Fine polyester Stabiltex was chosen for the support patches because it blended in well with the semi-transparent cotton muslin. The patches were heat cut to size to prevent the edges unravelling (Figure 15.14f). The patch is placed behind the area of damage but always extends to an area of strength (Figure 15.14g). The warp and weft of the patch should be carefully aligned with those of the object, the patch is pinned in place then the edges secured to the object with zig-zag stitches. Laid thread couching was applied to the front side to secure the damaged area to the patch using sewing threads pulled from Stabiltex fabric to stitch with (Figure 15.14h).

Where there were areas with multiple holes next to each other, larger patches were used. When scaling up the size of patches or applying a full support, it is important to allow sufficient excess in the support fabric so that the support does not exert tension on the object itself. The amount of fullness required in the support fabric depends on the nature of both the object and the support fabric, and the density of the stitching required is determined by the degree of damaged areas on the textile. Large flat textiles requiring a full support are usually worked on a frame with the textile partially rolled up. The advantage of this method is that the conservator can see the underside of the textile if necessary and when the area of the textile showing has been conserved, the object can be "rolled on" to the next area safely and with minimal handling.

The final part of the jama treatment was to construct a structural support lining for the bodice to help carry the weight the heavy skirt. This was made from a double layer of polyester Stabiltex as it needed to be strong and sheer (Figures 15.14i, j). An exact pattern was taken from the bodice as it was crucial that the support fitted perfectly to take the weight without causing distortions. The support lining was stitched in place along the seamline between the skirts and the bodice. A cotton tape with soft loop Velcro attached was also stitched to the top of the skirt and fastened to a corresponding hard hook Velcro attached to the waist of the mannequin torso (Figures 15.14k, l). The mannequin is a simple torso and stand that has been padded up to the correct profile with layers of polyester wadding covered with cotton jersey [31]. Soft silk-covered arms and layers of net and silk petticoats provide support. Rare earth magnets covered with fabric were stitched to the torso at strategic points, once the jama was dressed corresponding magnets were put in place to hold the covered section of the left cross-over front, and the heavy decorative ties in place (Figures 14m, n).

Figure 15.14 (a) Reattaching loose elements on jama 05563(IS). (b) Ties before treatment, on left. (c) Ties during humidification, on right. (d) Bodice trim before treatment, on left. (e) Bodice trim after treatment, on right. (f) Heat cut Stabiltex patch. (g) Slight shadow where patch is positioned. (h) Detail of laid thread couching. (i) Stabiltex support. (j) Support *in situ*. (k) Velcro stitched to waistband. (l) Detail of stitching. (m) Mannequin for jama. (n) Jama 05563(IS) on display © Victoria and Albert Museum, London.

15.5 Display

The display of the jama was a collaborative project between the conservator and a costume mounting specialist. Most costume at the V&A is mounted on commercially available mannequins that have been adapted to support the costume with underpinnings to present it in its correct profile. The sari in Figure 15.15 is displayed on a fiberglass mannequin with discreet padding to cushion the object and to make the surface less slippery, and an elasticated petticoat to hold the sari folds in position [32]. Textile conservators also work in collaboration with technicians who fabricate bespoke display mounts. The wall mounted costumes in Figure 15.3 all have an inner textile mount made from a stiff polyester felt (Reemay) covered with silk that helps to hold the object's shape, an acrylic former is slipped inside the mount and attached to the wall with metal brackets at the wrist edge of the garments. The mount for the pink embroidered dress on the far left of Figure 15.3 is shown in Figure 15.16a, b [33]. The internal metal armature of the thousand nail umbrella in Figure 15.13a was broken so the cover was supported on a new metal stand covered with perforated aluminium panels that could be stitched through allowing padding to be added to support the folds in the object (Figure 15.17). The bespoke angular fiberglass elephant and horse in the

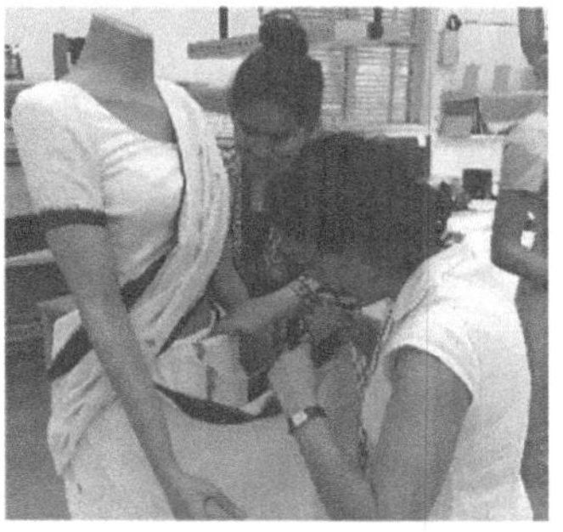

Figure 15.15 Sari IS.14:1,2-2015 being dressed on a fibre-glass mannequin © Victoria and Albert Museum, London.

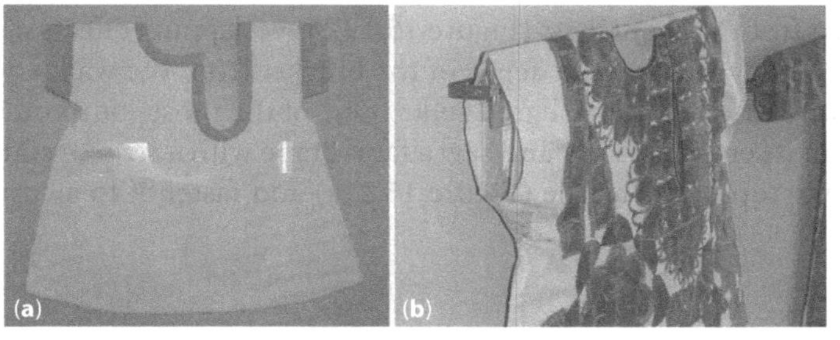

Figure 15.16 (a) Textile mount shown with shaped acrylic support on top. (b) Side view of dress IM. 40-1908 showing metal bracket attached to acrylic suppor © Victoria and Albert Museum, London.

Figure 15.17 Metal umbrella mount with mesh panels © Victoria and Albert Museum, London.

Maharaja exhibition shown in Figure 15.1, were cushioned with polyester wadding pads covered with cotton before the textiles were draped on top; a separate platform supported the howdah so that it did not sit directly on the textile (Figure 15.18).

Due to the large size of many of the Indian textiles in the collection, it has been necessary to devise various methods of safely reducing the size, usually the length, of the textiles to display them. The fabric-covered rollers, padded display boards and formers that can be seen in many of the images in this chapter, all originated with the methods developed to create a modular system for rotating Indian textiles in the permanent gallery. The flat textiles

Figure 15.18 Fiberglass elephant mount with soft padding © Victoria and Albert Museum, London.

pictured in Figure 15.6 are partially rolled and displayed on padded and fabric-covered rollers suspended from wall brackets (Figure 15.19). Overlapping effects can be achieved as seen in Figure 15.20 by varying the depth of the bracket from the wall, and in Figure 15.8, multiple textiles are draped over a shared roller. One of the most common methods for displaying a textile in its entirety is to hand sew a fabric tape with a strip of soft Velcro machine stitched to it to the top of the textile (Figure 15.21a) and fasten it to a corresponding hard

Figure 15.19 Detail of bracket to suspend roller © Victoria and Albert Museum, London.

Figure 15.20 Overlapping textiles on rollers with brackets at different depths © Victoria and Albert Museum, London.

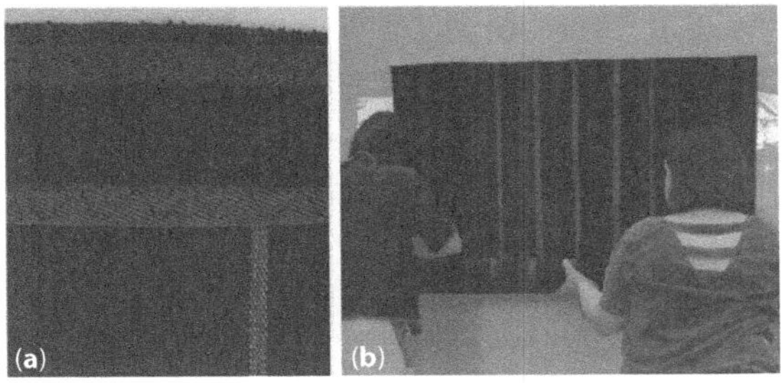

Figure 15.21 (a) Detail of cotton tape with Velcro stitched to object with zig zag stiches. (b) Installation of textile IS.115-1989 being attached to Velcro batten using a tube for support © Victoria and Albert Museum, London.

Velcro batten attached to the wall. To facilitate safe handling the textile is usually rolled on a tube for installation (Figure 15.21b).

Temporary exhibitions, such as The Fabric of India, bring their own challenges as this is often an opportunity to display objects that are too large for permanent display. The large size and delicate nature of the 17th century cotton floor spread IM.77-1938 (Figure 15.2) required a different installation method. It was unrolled on the floor and attached with Velcro to a split batten, this was slowly hoisted up the wall, using a temporary pulley system operated from behind the wall, and locked into the corresponding batten already fixed in position (Figures 15.22a–c).

In the case of an early 20th century appliqué room hanging IS.24-1994 (Figure 15.23a), which measures 16.58 m wide × 2.45 m high, it was necessary to install the textile using an upright roller as it was displayed on all four walls of a bespoke room [34]. The textile was fully conserved and considered robust enough to withstand the handling involved [35]. Velcro was sewn to the top of the textile (Figure 15.23b), and it was rolled up for installation. The roller slotted over a plug with a slippy base that could be easily turned on the floor as the installation and deinstallation progressed (Figure 15.23c). Multiple connecting tabs of hard Velcro were used to hold the textile to the Velcro sewn to the roller top (Figure 15.23d). This method was successful and was employed when the exhibition toured to two further venues, full instructions were produced, and experienced textile conservators travelled to the host venues to work with local staff to ensure safe handling of all the objects.

Figure 15.22 (a) Textile IM.77-1938 attached to Velcro batten on floor. (b) Slowly hoisting the batten and textile. (c) Locking the Velcro batten to the corresponding batten on the wall © Victoria and Albert Museum, London.

Figure 15.23 (a) IS.24-1994 appliqué room hanging on display. (b) Attaching Velcro to the hanging in the Textile Conservation studio. (c) Using an upright roller to safely handle the textile. (d) Detail of padded roller with Velcro at top © Victoria and Albert Museum, London

15.6 Conclusion

This chapter has focused on the care and conservation of the Indian textile collection at the V&A, just one part of a wider collection of textiles and costume held by the museum. The principles of conservation hold the same for all objects, the conservator is responsible for ensuring that the treatment of any object is carried out to the highest possible standard, working within ethical guidelines, and using methods and materials that have been stringently tested. Conservation decision making is fundamentally about minimizing risk to the object, so it is vital that the conservator considers and proactively engages with the wider aspects of preventive conservation and collections care. The maintenance of stable environmental conditions in both galleries and in storage, good housekeeping, integrated pest management and rotation of light sensitive objects are all vitally important to the long-term preservation of collections.

The conservator will often spend more time with an object than any other museum professional, and they have an important role in object research, recording technical details of an object's construction, and sometimes commissioning scientific investigation. It is also particularly important to work with curatorial colleagues to understand the cultural context of an object, and how it should be conserved and interpreted for display. Although conservation issues and treatment solutions are often very similar whether a textile is from India or Europe, it is important to assess each object individually and carry out necessary tests prior to commencing any treatment. Exhibiting textiles at the V&A is a collaborative effort in which the conservator works with curators, designers, technicians, scientists, registrars, and exhibition organizers to realize the display whether it is for a temporary exhibition or longer term in the Museum's galleries. Indian textiles and costume, which can be very large, embellished, and heavy have necessitated the development of strategies for their safe handling and innovative display methods. The success of these projects, which have resulted in some of the most spectacular exhibitions at the V&A [36], relies very much on the professionalism of dedicated V&A staff, attention to detail, and excellent teamwork.

Acknowledgments

We thank our colleagues in the V&A Textile Conservation Studio both past and present, including all our students and interns, Madhura Wairkar who conserved the shawl, Jenny

Barsby who conserved the umbrella, Lauren Osmond, and Gill MacGregor for their work to conserve and mount the red jama. Thanks to Joanne Hackett, Head of Textile & Fashion Conservation for comments on this paper. All images used are courtesy of the Trustees of the Victoria and Albert Museum, London.

References

1. https://www.vam.ac.uk/collections/the-va-story accessed 24/08/2021.
2. Fotheringham, A., *The Indian Textile Sourcebook*, p. 14, Thames & Hudson, London in association with the Victoria and Albert Museum, London, 2019.
3. Crill, R., *Chintz: Indian Textiles for the West*, V&A Publishing, London, 2008.
4. http://www.nehrutrustvam.org accessed 24/08/2021.
5. Kalsi, D., Creative pathways: developing a career in textile conservation in India, lessons learned textile conservation-then and now. *12th North American Textile Conservation Conference*, Ottawa-Gatineau, Canada, pp. 341–355, September 23-27, 2019, http://www.natcconference.com accessed 24/08/2021.
6. Haldane, E.A., Glenn, S., Hunter, S., Hillyer, L., RAKSHA: Raising awareness of textile conservation in India. *The Textile Speciality Group Post-Prints of the American Institute for Conservation (AIC), 40th Annual Meeting*, vol. 22, pp. 19–31, 2012.
7. Timar-Balazsy, A. and Eastop, D., *Chemical Principles of Textile Conservation*, pp. 3–156, Butterworth-Heinemann, Oxford, UK, 1998.
8. https://www.icon.org.uk/resources/standards-and-ethics/icon-ethical-guidance.html, accessed 15/08/2021.
9. Gentle, N., The examination and conservation of two Indian textiles. *Conservator*, 17, 19–25, United Kingdom Institute for Conservation (now ICON icon.org.uk), 1993.
10. Lennard, F. and Ewer, P. (Eds.), *Textile Conservation: Advances in Practice*, pp. 197–226, Elsevier, UK, 2010.
11. Robinson, J. and Pardoe, T., *An Illustrated Guide to the Care of Costume and Textile Collections*, Museums & Galleries Commission, UK, 2000, https://collectionstrust.org.uk/resource/an-illustrated-guide-to-the-care-of-costume-and-textile-collections accessed 27/08/2021.
12. http://www.icom-cc.org/332/-icom-cc-documents/declaration-on-environmental-guide lines/#.YRkxQohKiUk accessed 24/08/2021.
13. Pretzel, B., Materials and their interaction with museum objects. V&A *Conserv. J.*, 44, 9–13, 2003. Also available online http://www.vam.ac.uk/content/journals/conservation-journal/issue-44/materials-and-their-interaction-with-museum-objects/ accessed 27/08/2021.
14. https://www.pestodyssey.org accessed 15/08/2021.
15. https://museumpests.net accessed 15/08/2021.
16. Haldane, E.A., Moving Maharaja: the display and transportation of Indian costume and textiles. *Material in Motion, Pre-Prints of the 10th North American Textile Conservation Conference*, New York, pp. 145–158, November 16-20, 2015.
17. Crill, R. (Ed.), *The Fabric of India*, V&A Publishing, London, 2015.
18. https://www.vam.ac.uk/blog/caring-for-our-collections/all-that-glitters-part-2-is-most-defi nitely-not-all-gold, accessed 24/08/2021.
19. https://cites.org/eng/disc/what.php, accessed 15/08/2021.
20. https://www.vam.ac.uk/blog/fabric-of-india, accessed 22/08/2021.
21. Eastop, D.D. and Brooks, M.M., To clean or not to clean: the value of soils and creases (1996), in: *Changing Views of Textile Conservation*, M.M. Brooks, and D.D. Eastop, (Eds.), pp. 228–235, Getty Conservation Institute, Los Angeles, 2011.

22. Cooke, B., Creasing in ancient textiles (1988), in: *Changing Views of Textile Conservation*, M.M. Brooks, and D.D. Eastop, (Eds.), pp. 221–227, Getty Conservation Institute, Los Angeles, 2011.
23. Fields, J., Wingham, A., Hartog, F., Daniels, V., Finding substitute surfactants for synperonic N. *J. Am. Inst. Conserv.*, 43, 1, 55–73, Spring 2004.
24. Sato, M. and Quye, A., Detergency evaluation of non-ionic Dehypon® LS54 for textile conservation wet cleaning. *J. Inst. Conserv. (ICON)*, 42, 1, 3–17, 2019.
25. Timar-Balazsy, A. and Eastop, D., *Chemical Principles of Textile Conservation*, pp. 194–225, Butterworth-Heinemann, 1998.
26. Timar-Balazsy, A. and Eastop, D., *Chemical Principles of Textile Conservation*, pp. 163–184, Butterworth-Heinemann, 1998.
27. Osmond, L., Effects of wet cleaning beetle elytra embellished textiles. V&A *Conserv. J.*, 65, 12–13, 2018. https://www.vam.ac.uk/blog/caring-for-our-collections/effects-of-wet-cleaning-beetle-elytra-embellished-textiles accessed 27/08/2021.
28. Brooks, M.M., Eastop, D.D., Hillyer, L., Lister, A., Supporting fragile textiles: the evolution of choice. *Lining and Backing: The Support of Paintings, Paper and Textiles: UKIC Conference*, United Kingdom Institute for Conservation (now ICON icon.org.uk, pp. 5–13, November 7-8 1995.
29. Osmond, L., Distorted iridescence: investigating the microstructural changes in jewel beetle elytra on textiles. *The Nature of Textiles, Postprints of the Forum of the Icon Textile Group*, London, pp. 15–27, 2018, https://www.icon.org.uk/, accessed 24/08/2021.
30. https://www.vam.ac.uk/articles/jama accessed 24/08/2021.
31. Flecker, L., *A Practical Guide to Costume Mounting*, Routledge, UK, 2007.
32. https://www.vam.ac.uk/blog/caring-for-our-collections/slippery-sari-sagas-and-delightful-draping accessed 27/08/2021.
33. https://www.vam.ac.uk/blog/caring-for-our-collections/guest-post-conservation-close-up accessed 27/08/2021.
34. https://www.vam.ac.uk/blog/fabric-of-india/a-room-with-a-view accessed 24/08/2021.
35. https://www.vam.ac.uk/blog/fabric-of-india/guest-post-a-room-with-a-view-part-2-the-textile-conservators-story, accessed 27/08/2021.
36. https://www.vam.ac.uk/blog/fabric-of-india/guest-post-raising-the-roof-mounting-tipus-tent accessed 24/08/2021.

16

Between Science and Art: Activities of the Natural Dyeing Laboratory

Katarzyna Schmidt-Przewoźna

Institute of Natural Fibers and Medicinal Plants, Poznan, Poland

Abstract

Since 1993, the "Natural Art" Studio has been operating at the Institute of Natural Fibres and Medicinal Plants (INF&MP) in Poland to promote various aspects of natural fiber and dye applications. The present paper lists the various activities to promote and identify ancient natural dyeing practices, including projects, workshops, conferences, exhibitions, and developing color catalogs. Successful dye recreation attempts were made through color trials to match the produced dye with ancient fabric dye. Ancient dye reproducibility was tested through scientific tests for objective evaluation. The reconstruction and adaptation of old dyeing techniques made it possible to perform dyeing compatible with the colors and raw materials used in museum artifacts. A case study to analyze antique dyes of ancient Polish Kontush sash is presented in detail. The research developed ecological methods of applying natural dyes to historical fabrics used in conservation procedures.

***Keywords*:** Natural dye, ancient dye, INF&MP, color catalog, Polish Kontush sash

16.1 Introduction

A natural consequence of working with natural fibers will be to increase their value with colors obtained from natural sources and recreation of old technologies. Researchers at Natural art studio at INF&MP pointed out that recreating ancient color pellets was not easy. The first attempt to dye flax with plant dyes in 1997 showed how difficult it was and failed. The obtained colors were faint and not fast, far from expected. Further research was devoted to collecting information from scientific and archival sources and searching for dyeing plants. After a thorough study, the systematized coloring trials were resumed in 1999. The research aimed to define the so-called color triad, which allows one to obtain a wide range of colors when combining colors. As a result, dyeing plants and other natural resources were divided into three color groups: yellow, red and blue [1]. Hundreds of trials with the use of natural dyes on fabrics and fibers, such as flax, wool, jute, hemp, and silk, finally allowed to create a wide range of colors [2, 3]. An original method was also developed to obtain several colors from a single dye using color modifiers/mordants, such as

Email: k_schmidt@interia.pl

Seiko Jose, Sabu Thomas, Pintu Pandit, Ritu Pandey and Vandana Gupta (eds.) Handbook of Museum Textiles: Volume I Conservation and Cultural Research, (315–326) © 2023 Scrivener Publishing LLC

soda, citric acid, copper sulphate and iron sulphate, and natural mordants: galls, sumac, rhubarb leaves, and tree bark (oak, willow, alder).

When carrying out dyeing work, the crucial issue was to obtain raw materials, *i.e.* plant dyes. It was not an easy task. Many plants are not available in offers and catalogues of herbal companies, as they are no longer commonly used. This was the case, among others, with the madder *Rubia tinctorum* L.—the source of red color and the woad *Isatis tinctoria* L.—for blue color. These difficulties resulted in the idea of creating our own garden of dyeing plants. In 2003, this idea was implemented at the Experimental Research Station of the INF&MP in Petkowo near Sroda Wielkopolska.

An increasing range of works and experiments and the experience gained led to establishing the Natural Dyeing Laboratory at the Department of Innovative Textile Technologies at INFMP in 2009. It is the only research center of this kind in Poland. Intensive research is being carried out here linked with recreating old dyeing technologies and their application in unique collections of fabrics. Experimental dyeing trials are conducted on linen, hemp and silk fabrics [4]. They have shown the possibility of obtaining a broad color palette and technological standards of color fixation. Several scientific projects involve studies of the health-promoting properties of natural dyes, including their anti-allergic and aseptic properties and UV protection effect [5–8]. The scientific output of the Natural Dyeing Laboratory comprises more than 100 publications on issues related to Polish and world fabrics, natural fibers, processes of restoration and application of natural dyes [1].

In 2016, dyeing plants were also planted in Plewiska near Poznań as part of the INF&MP Garden of Medicinal Plants (Figure 16.1). The garden was founded in 1946 on the initiative of Prof. Wacław Strażewicz, a pioneer of Polish herbalism. In 1980, the Garden of Medicinal Plants was recognized as a cultural monument and entered monuments' register. About 1500 taxons were collected here, more than 400 of which were the subject of detailed research on acclimatization, breeding, cultivation and phytochemistry, and which were also the basis for the development of ready-made forms of herbal medicines [9]. The work carried out here aims to collect medicinal and spice plants and provide materials for research on plant medicines (https://iwnirz.pl/strona,64.html). A part of the Garden is devoted to cultivating dyeing plants for the Natural Dyeing Laboratory. Plants grown in the garden result from cooperation with numerous herbal companies, botanical gardens, scientific institutes in Europe and worldwide. Thanks to wide contacts, the seeds of dyeing plants were exchanged. In this beautiful place, the cultivation of marigold, plains coreopsis, safflower and reseda, for the implementation of the Multiannual Programme

Figure 16.1 (a) The garden of dyeing plants, INF&MP, Experimental Research Station in Petkowo (b) woad plantation at Petkowo.

"Reconstruction and Sustainable Development of Production and Processing of Natural Fibrous Materials for Agriculture and Economy," financed by the Ministry of Agriculture and Rural Development.

Organic, plant and animal dyes have played a large role in dyeing natural fiber fabrics. There are several divisions of dyes, taking into account their chemical structures, methods of application on fabrics and the colors obtained. Tables 16.1 to 16.4 below present the division of dyes by color [12].

Table 16.1 Plants yielding yellow color.

Common name	Latin, Botanical name	Part used	Color obtained
Alder	*Alnus* L.	Bark	Yellow
Apple tree	*Malus* Mill.	Bark and leaves	Yellow, reddish, green-yellow
Annato	*Bixa orellana* L.	Seeds	Yellow, orange
Barberry	*Berberis vulgaris* L.	Bark, leaves	Yellow
Birch	*Betula alba* L.	Bark, leaves	Yellow
Burdock	*Arctium lappa* L.	Herb	Yellow
Golden chamomile	*Anthemis tinctoria* L.	Flower	Yellow, lemon, green, olive
Chebulic myrobalan	*Terminalia chebula* Retz.	Nut-like fruits	Yellow
Common yarrow	*Achillea millefolium* L.	Herb	Yellow, green
Common juniper	*Juniperus communis* L.	Fruits	Yellow, olive,
Common fumitory	*Fumaria officinalis* L.	Herb	Yellow
Crocus	*Crocus sativus* L.	Stigmas of the pistils	Yellow, red
Dahlia	*Dahlia variabilis* L	Flowers	yellow, reddish, brown
Dandelion	*Taraxacum officinale* L.	Herb and flowers	Yellow, beige, olive lemon
Dyer's broom	*Genista tinctoria* L.	Young shoots	Lemon, yellow, yellow olive
False hemp	*Datisca cannabina* L.	Root, stem, leaves	Yellow, yellow-green
Field horsetail	*Equisetum arvense* L.	Herb	Pale yellow-green
Garden rhubarb	*Rheum rhabarbarum* L.	Root	Yellow

(Continued)

Table 16.1 Plants yielding yellow color. (*Continued*)

Common name	Latin, Botanical name	Part used	Color obtained
Herb-Robert	*Geranium robertianum* L.	Herb	Yellow
Heather	*Calluna vulgaris* Hull.	Herb	Yellow and light red
Jackfruit	*Artocarpus heterophyllus* Lam.	Fruit	Yellow
Kamala	*Mallotus philippensis* Muell. Arg.	Resin and bark	Yellow, orange
Knotweed	*Polygonum persicaria* L.	Herb	Yellow, brown
St. John's wort	*Hypericum perforatum* L.	Flowers	Yellow, brown, red
Oak	*Quercus robur* L.	Bark	brown
Onion	*Allium cepa* L.	Peels	Yellow, olive, reddish brown
Plaints coreopsis	*Coreopsis tinctoria* L.	Flower	Yellow, beige, brown,red
Early Sunrise Coreopsis	*Coreopsis grandiflora* L.	Flower	Yellow, dark - yellow, olive
Pomegranate	*Punica granatum* L.	Peels	Yellow
Safflower	*Carthamus tinctorius* L.	Petals of flowers	Yellow, red
Saw-wort	*Serratula tinctoria* L.	Leaves	Yellow
Solidago	*Solidago serotina* Ait.	Flower	Yellow, olive
St. John's wort	*Hypericum perforatum* L.	Flower	Yellow, brown, red
White willow	*Salix alba* L.	Bark	Yellow, olive
Weld	*Reseda luteola* L.	Herb	Yellow, olive
Tansy	*Tanacetum vulgare* L.	Flower	Yellow, olive, green
Three-Lobe Beggarticks	*Bidens tripartitus* L.	Herb	Yellow
Tumeric	*Curcuma longa* L.	Rhizome	Light yellow, yellow, lemon
Yellow chamomile	*Anthemis tinctoria* L.	Flower	Yellow, lemon, green, olive
Yellow Flag Iris	*Iris pseudacorus* L.	Flower	Yellow

Table 16.2 Plants yielding red color.

Common name	Latin, Botanical name	Part used	Color obtained
Alkannet	*Alkanna tinctoria* L.	Root	Red, pink, blue
Brazil wood	*Caesalpinia brasiliensis* L.	Core	Red, rust
Bilberry	*Vaccinium myrtillus* L	Fruit	Pink, blue, gray, and brown
Blackberry	*Rubus* L.	Fruit	Purple, pink
Crocus	*Crocus sativus* L.	Stigmas of the pistils	Red, orange, yellow, carmine
Cutch tree	*Acacia catechu* L.	Bark, fruit	Brown
Elder	*Sambucus elbucus* L.	Fruit	Pink, blue
Grapevine	*Vitis vinifera L.*	Fruit	Dark red, purple, bluish
Hedge bedstraw	*Galium mollugo* L.	Root	Red, pink
Henna	*Lawsonia alba* Lam.	Leaves / roots	Red, brown
Indian mulberry (noni)	*Morinda citrifolia* L.	Root, bark	Red, brown, pink
India madder	*Rubia cordifolia* L.	Bark	Red, ginger. cherry, burgundy
Lady's bedstraw	*Galium verum* L.	Root	Pale red, red
Madder	*Rubia tinctorum* L.	Root	Red, salmon, brown
Mallow pink	*Althaea rosea* L.	flower	Yellow, light pink
Marjoram	*Origanim vulgare L.*	Tops of shoots	Red
Red-root sage	*Salvia miliorrhiza* Bunge	Root	Salmon, peach
Sacred bo tree	*Ficus religiosa* L.	Bark	Brown
Safflower	*Carthamus tinctorius* L.	Petals of flowers	Carmine red, coral, yellow. orange
Sandalwood	*Pterocarpus santalinus* L.	Wood	Red, rust
Wild madder	*Rubia peregrina L.*	Root	Red, brown, pink

Table 16.3 Red dyes obtained from insects.

Common name	Latin, Botanical name	Part used	Color obtained
Polish Cochineal	*Porphyrophora polonica* L.	insect larvae	Red, crimson, scarlet
Kermes	*Kermes vermilio* Planchon	insect larvae	Red
Cochineal	*Dactylopius coccus* Costa	insect larvae	Red, crimson, scarlet, violet
Laka	*Kerria lacca* Kerr.	l insect larvae	Red, burgundy, pink

Table 16.4 Plants yielding blue color.

Common name	Latin, Botanical name	Part used	Color obtained
Bilberry	*Vaccinium myrtillus* L.	fruits	Pink, blue, gray and brown
Chinese-indigo	*Persicaria tinctoria* (Aiton) Spach	leaves	Blue, navy blue, indigo
Common privet	*Ligustrum vulgare* L.	berries	Blue, sky blue, gray-blue
Sambucus	*Sambucus nigra* L.	fruit	Pink, blue,
Indigo	*Indigofera tinctoria* L.	leaves	Blue, navy blue, indigo
Java indigo Tarum	*Marsdenia tinctoria* R. Br.	leaves	Blue, navy blue, darc blue
Logwood	*Haematoxylum campechianum* L.	Extract from the heart of the tree	Blue, navy blue, violet, black, lilac
Vine	*Vitis vinifera* L.	Black fruil	Dark red, purple, bluish
Woad	*Isatis tinctoria* L.	leaves	Blue, navy blue

16.2 Promotion of Antique Dyes, Pigments, and Prints

16.2.1 Projects

The head of the laboratory constantly participates in numerous research projects of the specialist groups affiliated with the dyes in history and archaeology, Textile Institute in Manchester, and International Symposium and Exhibition for Natural Dyes (ISEND). These conferences are attended by outstanding scientists, ethnobotanists, anthropologists, ecologists, fashion designers, clothing manufacturers, authors, filmmakers, and many artists. Enthusiasts of natural dyes and ecological technologies meet to exchange experiences,

Figure 16.2 (a–b) The festival *Lets Kids Design*, Szczecin 2016 (c) workshop in Podlasie (d) the shades of Campeche tree.—an installation, shibori, naturally dyed silk, National Silk Museum, Hangzhou, China (e) the exhibition "Thread, Space, Color, the Museum Górków Castle in Szamotuły."

talk about dyes produced from plants, insects, soil, and shells used in textiles, cosmetics, food, and art. It is also an opportunity to meet artists and craftsmen who present old technologies in the workshops.

Apart from technological problems, the history of weaving, color, design, aesthetics, fashion, style, and various contexts of natural coloring is widely discussed here. The Laboratory's scientific and promotional achievements were recognized internationally by assigning Katarzyna Schmidt-Przewozna to organize the Dyes in History and Archaeology (DHA) World Conference at the Institute of Natural Fibres and Medicinal Plants in Poznań in 2009. European, non-European, developed ties in the course of the scientific projects (Figure 16.2).

16.2.2 Workshop and Exhibitions

The exhibition titled "In the circle of colors," which was the result of a research project financed by the Scientific Research Committee, was held at the National Museum in Poznań (2002), the Regional Museum in Kościan (2003), the Regional Museum in Śrem (2004). These exhibitions presented the results of research concerning various dyeing methods, presented dyes, herbs used in dyeing trials and color catalogs, samplers showing effects obtained in dyeing experiments (Figures 16.3, 16.4, and 16.5).

The exhibitions also showed artistic fabrics made of naturally colored linen yarns, realized in 1999-2002. These exhibitions were supplemented with fabrics from various cultures from the collections of the Ethnographic Museum in Poznań [10]. In 2005, as part of the 8th Poznań Festival of Science and Art, the INF&MP Natural Dyeing Lab prepared a multimedia show and an exhibition of natural dyes and fabrics dyed with them at the Faculty of Biological Sciences of Adam Mickiewicz University in Poznań.

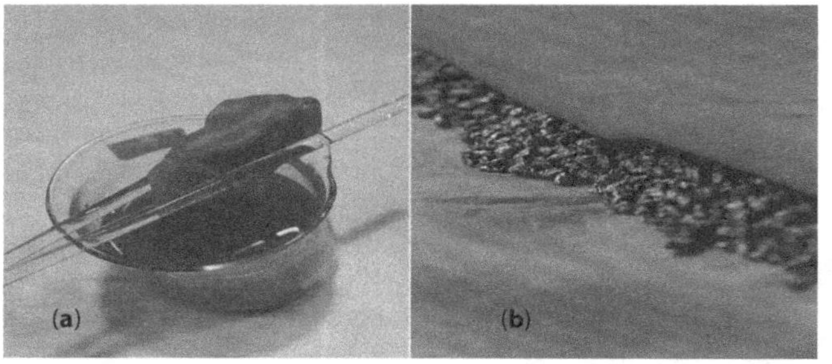

Figure 16.3 (a) Kermes dye on silk (b) cochineal dye on silk.

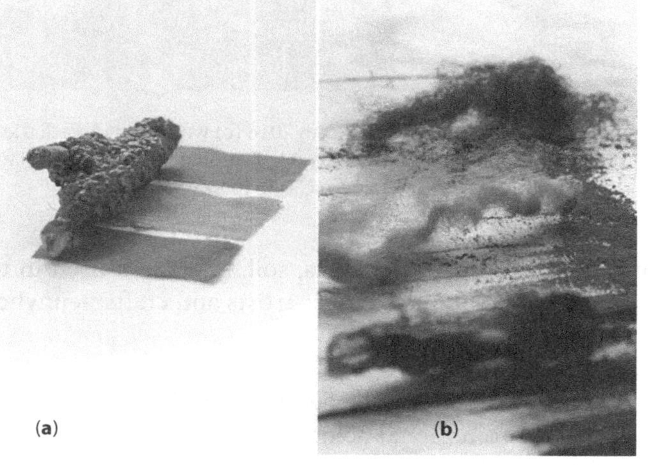

Figure 16.4 (a) Lac dye on silk. (b) logwood on wool.

Figure 16.5 Exhibition in ethnographic museum.

In 2018, the studio manager took part in the project *The art of dyeing silk: Colour reproductions of historical recipes and dialogue about practical skills* organized in Amsterdam by the Netherlands Institute for Cultural Heritage. For 2 days, eight participants recreated the old dyeing recipes. They dyed textiles with saffron, safflower, brazil wood, logwood, lac-dye, annatto, cochineal and kermes. The next milestone was the participation in the first Biennial of Natural Dyes in China "Bond," Hangzhou 2019. The symposium was accompanied by the exhibition Natural Dyeing: Contemporary Art and Design Works. These events took place in the world's largest National Silk Museum (China National Silk Museum). The exhibited works were included in the Museum collection.

16.2.3 Color Catalog of Ancient Dye and Its Reproduction

As a result of research related to applying natural dyes onto fabrics, the Poznań center has gained knowledge and experience, which are not the result of random effects, but the result of specific long-term experiments. The hundreds of tests performed allow us to achieve the expected results. This work is determined by diligence and precision but it also requires creativity and imagination. For several years now, coloring experiments have been recorded in the form of color catalogs. They are documentation of a complex process of color searching. Each vegetable dye is described separately according to a fixed code.

More than 50 different plant substances have been tested at the Natural Dyeing Laboratory, INF&MP to obtain a rich color palette.

At the same time, extensive research is conducted into the ecological properties of naturally dyed fabrics. Studies have shown that many natural dyes have excellent properties affecting human health. Among other things, some dyes have anti-allergenic properties and provide natural protection against harmful UV rays [7]. These studies have been recognized as essential and relevant in the contemporary world and awarded numerous international prizes. This is the result of the cooperation with Prof. Małgorzata Zimniewska, current director of the Institute.

16.2.4 Reconstruction of Ancient Dyeing Techniques

In 2021 as part of the module Dyes and Finishing Processes at Bern University of Applied Sciences, various ancient dyeing technologies have been reconstructed with the following natural dyes: Logwood (*Haematoxylum campechianum* L.), American Cochineal (*Dactylopius coccus,* COSTA 1835), Madder (*Rubia tinctorum* L.), Annatto seeds (*Bixa orellana* L.), Walnut shells (*Juglans*—species), Weld (*Reseda luteola* L.), Dyer's broom, Dyer's greenweed (*Genista tinctoria* L.), Safflower (*Carthamus tinctorius* L.) using cold and hot methods, dyeing black with gallnuts, indigo (*Indigofera tinctoria* L.). In the project the students over dyed with weld of blue (with indigo) dyed wool and silk (mordanting with alum) and over-dyeing with madder of blue and others. In the processes of dyeing linen, wool and silk, the dyeing time, water temperature and pH were controlled. Such applied sciences recreate the ancient dyes into modern textiles and help in awareness regarding ancient eco-friendly dyeing practices [11].

16.3 Analysis of Antique Polish Kontush Sash Dyeing Material: A Case Study

Kontush sash was an ornate fashionable accessory of Polish garments worn around the waist. It was a decorative sash (350 mm wide and 4 m long) with colorful weaving and fringes at the edges. Kontush sashes, once an indispensable fashion statement for noblemen, are now an artifact and part of museum display (Figure 16.6). Sashes were assessed for their dye content.

In 2002 to 2004, the Natural Dyeing Studio identified natural dyes used in antique Polish fabrics from the 17th and 18th centuries in cooperation with the Institute of Dyes and Organic Products from Łódź. The Polish *kontush* sashes were subjected to visible and near-infrared light spectrophotometric analysis. To determine which dyes were used in these belts, a spectral database of textile raw materials and coloring agents used in them was created and then assessed. Spectrophotometric analysis of the kontush sash dye shows the presence of several natural dyes listed in Table 16.5.

The identification was based on the dyeing made on silk, linen, wool and cotton with natural dyes and their spectral graphs with the original materials. The museum artifacts were subjected to reflective spectrophotometric tests in the spectral range covering the visible and near-infrared range using measuring equipment: spectrophotometer DK-2A by Beckman, spectrophotometer M 2020 PL by Macbeth, spectrophotometer CM-508c by MINOLTA. Trials to apply plant dyes on fabrics were used to create color catalogs and a database of natural textile raw materials and dyes applied on these raw materials. The research made it possible to perform dyeing compatible with the colors and raw materials used in museum artifacts and to develop ecological methods of applying natural dyes to historical fabrics used in conservation procedures [12, 13]. The reconstruction and adaptation of old dyeing techniques using more than 50 plant and animal substances has become one of the forms of the Institute's activities in a broad approach to the whole context of the "fiber culture." The results of the research work perfectly harmonize with the scientific search within the framework of eco-technology and artists connected with ECO Design [14]. This theme, which seemed to be quite niche, has found interest among outstanding scientists,

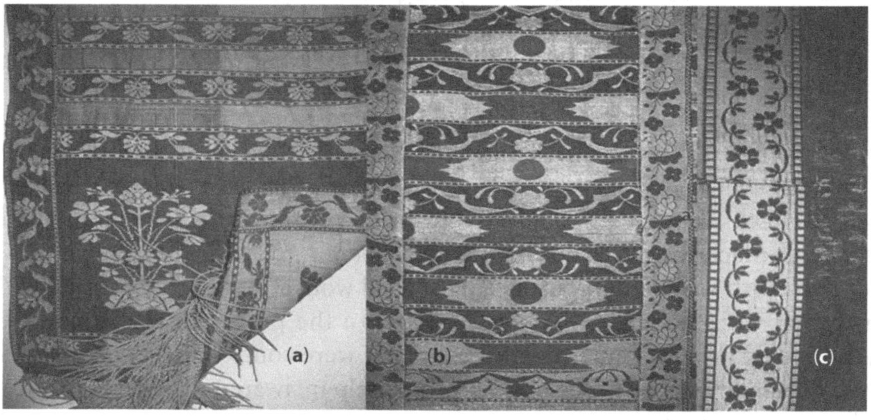

Figure 16.6 Polish antique textiles dyed in natural dye (a) Lipkow sash, (b) Sluck sash, (c) Paschalis.

Table 16.5 Analysis of colors obtained on silk fabrics dyed with natural sources—comparison with the colors of the Polish *kontush* sashes.

	Spectrophotometric analysis						
Natural dyes	**L***	**a***	**b***	**c***	**h***	**Color**	**PANTONE®**
indigo *Indigofera tinctoria*	44,8	-5,6	-9,5	11,1	239,4	Classic blue	19-4052 TCX
indigo *Indigofera tinctoria*	19,6	-1,2	-9,3	9,4	262,8	Dark blue	19,3984 TCX
Woad *Isatis tinctoria*	33,1	-6,1	-11,6	13,1	242,4	Light blue	17-4033 TCX
Cochineal alum	37,0	44,4	0,5	44,4	0,7	Pink	17-2227 TCX
Cochineal citric acid	53,1	35,7	5,1	36,1	8,1	Light pink	17-2230 TCX
Madder *Rubia tinctorum*, alum	38,4	37,2	21,4	42,9	29,9	Madder red	17-1553 TCX
Weld Reseda luteola, alum	79,4	0,9	58,5	58,5	89,1	Yellow	13-0648 TCX
Weld Reseda luteola, iron	79,7	-2,5	46,9	47,0	93,1	Olive green	17-0613 TCX
Dyers'broom *Genista tinctoria* copper	60,4	4,2	48,0	48,2	85,0	Dark yellow	15-0743 TCX
Safflower Carthamus tinctorius	63,0	25,0	13,1	28,2	27,7	Coral	17-1544 TCX

as evidenced by invitation to participate in international projects such as COST Action 628 and Corona financed by the European Union.

16.4 Conclusion

Parallel to scientific works, the Natural Art Studio and Natural Dyeing Laboratory have been promoting its experiences at the interface of art and science in many forms already for many years. They participate in fairs, workshops, presentations, shows, and exhibitions, for example, in "The International Cooperation of the Polish Academy of Sciences." It would seem that the world of natural dyes is gone forever, replaced by chemistry and modern technology. However, fortunately, return to the old knowledge and skills can be observed more and more often in contemporary culture. This makes the world of natural dyes not only a phenomenon from past centuries but also awaits further exploration of its secrets. So maybe it is worthwhile to look at a nearby meadow, pick herbs, and try it out for yourself.

Acknowledgment

Thank you to the Institute of Natural Fibers and Medicinal Plants—PiB, Poznan, Poland for the possibility of creating a Natural Dyeing Laboratory and conducting research there related to the issues of natural dyeing. Special thanks to the Ministry of Science and the Ministry of Agriculture and Rural Development for sponsoring several research grants.

References

1. Schmidt-Przewozna, K., The nature as inspiration for art. *J. Nat. Fibers*, 5, 1, 88–93, 2008.
2. Pastuszak, W., Color in computer graphics, PWN, Warszawa, 2000.
3. Schmidt-Przewoźna, K., Villavicencio, A.M., Kicińska-Jakubowska, A., Zajączek, K., Brandys, A., Coreopsis tinctoria nutt. as a source of many colours. *Int. J.*, 6556, 4, -61, 2019.
4. Schmidt-Przewozna, K., Kozlowski, R., Zimniewska, M., Influence of corona treatment on linen fabric dyed with natural dyestuffs, in: *Textiles for Sustainable Development*, p. 263, 2007.
5. Jose, S., Pandit, P., Pandey, R., Chickpea husk–a potential agro waste for coloration and functional finishing of textiles. *Ind. Crops Prod.*, 142, 111833, 2019.
6. Schmidt-Przewozna, K. and Kowalinski, J., Yellow natural dyestuffs and their light fastness properties and UV protection on natural fabrics, in: *Renewable Resources and Biotechnology for Material Applications*, pp. 49–59, Nova Science Publishers, Hauppage, 2011.
7. Pandey, R., Patel, S., Pandit, P., Nachimuthu, S., Jose, S., Colouration of textiles using roasted peanut skin-an agro processing residue. *J. Clean. Prod.*, 172, 1319–1326, 2018.
8. Pandit, P., Jose, S., Pandey, R., Groundnut testa: an industrial agro-processing residue for the coloring and protective finishing of cotton fabric. *Waste Biomass Valorization*, 12, 6, 3383–3394, 2021.
9. Schmidt-Przewoźna, K. and Brandys, A., Utilization of contaminated lands for cultivation of dye producing plants, in: *Bioremediation and Bioeconomy*, pp. 329–359, Elsevier, Woodhead Publishing, U.K., 2016.
10. Schmidt-Przewozna, K. and Przewozny, W., In the circle of colour-exhibition in the ethnographic museum. *J. Nat. Fibers*, 1, 2, 93–98, 2004.
11. Rzepińska, M., *History of Colour in European Painting*, Wydawnictwo Literackie, Kraków, 1983.
12. Vankar, P.S., Isolation and characterization of the colorant molecules from each dye plant, in: *Natural Dyes for Textiles: Sources, Chemistry and Applications*, p. 45, 2017.
13. Vankar, P.S. and Shukla, D., Isolation and characterization of the colorant molecules from each dye plant, in: *New trends in Natural Dyes for Textiles*, Elsevier, Woodhead Publishing, U.K., 2019.
14. Pargai, D., Jahan, S., Gahlot, M., Functional properties of natural dyed textiles. *Chemistry and Technology of Natural and Synthetic Dyes and Pigments*, pp. 1–19, London, U.K., 2020.

17

Visitor Interactions and Museum Textiles

Kanika Sachdeva[1]* **and Divya Singhal Gupta**[2]

[1]*Conservation Scientist, London, UK*
[2]*Amity School of Fashion Technology, Amity University, Uttar Pradesh, India*

Abstract
The display of textile artifacts is often tricky as it frequently challenges the biggest objective of an exhibition, i.e. education. Apparel and clothing are an integral part of our physical existence and understanding one without touch and feel can be compared to understanding a cuisine without tasting it. Considering the organic nature of textiles, it is a category at the highest risk of damage with slightest physical interaction. Balancing the challenges on both sides has often been an uphill task for curators and exhibitors who have tried to fill the gap created by lack of touch and feel, supplementing adequate descriptions and theoretical information about the objects. However, there are novel technologies available today that can considerably bridge this gap by producing replicas that are strong enough to interact and close enough to provide the required education.

Keywords: Light, textile artifacts, display, tactile interaction, museum textiles

17.1 Introduction

Museums have been important tools of public education since centuries. They essentially communicate through objects in their possession while shaping public knowledge and opinions [1]. Museums worldwide spread this education by telling stories beyond the realms of time, location and size. This process has changed considerably over the years with visitor interactions being enhanced by intelligent additions and technological advancement. The museums no longer limit their role to safeguarding and displaying the tangible heritage, rather enhance visitor experience by focusing simultaneously on the intangible value of the objects. Museums are actively restructuring exhibitions to attract new audiences while encouraging self-paced learning experiences. Exhibitions now are a complicated process with involvement of designers, academicians, historians, technologists, interpreters along with curators and conservators. An exhibition is set up as a curated experience manipulating color, light, and mixed media resources to create an entertaining learning experience.

The amount of information available on fingertips nowadays has changed the context in which people perceive things. The function of museums is not limited to exhibiting artifacts at present times. Visitors are now curious to know about the details and story of an

**Corresponding author*: dr.kanikabhatiasachdeva@gmail.com

Seiko Jose, Sabu Thomas, Pintu Pandit, Ritu Pandey and Vandana Gupta (eds.) Handbook of Museum Textiles: Volume I Conservation and Cultural Research, (327–340) © 2023 Scrivener Publishing LLC

object. This has resulted in development of museum exhibition planning and design not only to meet these needs, but to anticipate them. Audience research reflects that museum need interactive arrangements that help the audience to get experience and learn actively in exhibitions, instead of just being an observer from a distance. Arrangements that allow active participation of the audience to be a part of museum experience with opportunities provided by technological development gives the audience the opportunity to participate in the museum environment [2, 3]. Active audience participation can go a long way in providing an enriching learning experience, motivating visitors to make repeat visits to the museums.

17.2 Textile Exhibitions—Challenges in Display

An effective & comprehensive textile exhibit should be able to appeal to four of five human senses, i.e., sight, smell, sound, and touch. The traditional exhibition formats have done everything to appeal to a customer's sight, the other three dimensions of experience have been largely left unaddressed. This could be largely due to the fact that museum artifacts have always been considered too sacred to be allowed for direct visitor interaction. This was largely justified considering the fragile nature of museum textiles. Also, the fact that most of the museum pieces either have a history that makes them exclusive or are some of the last pieces of a dying art. This makes them irreplaceable in every sense and tends to increase the veil of inaccessibility around them. In recent times, the dialogue about education has shifted towards a more holistic approach. Museums being an important medium of public education cannot be sidelined from the debate. Where formal education moved from theory to a more practical based approach that encourages learning by doing and understanding with experience, museum education for a long time emphasized on sharpening the visual aspects of exhibition to make up for the vacuum in other sensory experiences. The discussion in this chapter would largely focus around addressing these questions:

1. Is it right for textile museums to limit public education to visual aspects?
2. Does every museum artifact need to be deemed untouchable?
3. Is lack of access impacting customer experience negatively in textile museums?
4. Can textile museums design exhibitions to curate more experiential journeys for their audience as compared to just visual ones?

Answer to the above-mentioned questions would largely revolve around three major aspects of textile exhibitions: (a) display methods, (b) display lighting, and (c) exhibition protocols.

17.2.1 Display Method

This discussion would largely revolve around the presence or absence of display cases; modes of display in terms of flat, hanging; how close can a visitor go to the artifact, etc. There are three major stakeholders in any exhibition; the artisans who created the object, the exhibitors of the object and the audience of the object. The perspective of each of these is equally important and results in a dynamic and complex relationship. The artisan makes an object with his viewpoint, museum exhibitors and curators display the object in the context of its user and the visitor will have their own perception of the object. Thus, the viewpoint of each element is differently directed and might be contradictory at times. It is the role of museum professionals to build the context of the object in a clear and objective fashion, so that the audience would see it from an unbiased perspective. An example of the conflicting perceptions of these elements is Her Majesty Queen Elizabeth II's comments on the wedding dress of *Catheriene Middelton* (Designed by Sarah Burton of Alexander McQueen), which was exhibited at Buckingham Palace in 2011. She called the exhibit "horrible" and "disturbing." The tiara and floating veil displayed on a headless mannequin failed to impress Her Majesty [4]. This example highlights the fact that the perspective of the audience often gets ignored in the balancing act between the exhibit's vulnerability & curatorial limitations. It is important here to note that textile exhibition curators are bound with certain guidelines that arise from extreme vulnerability of fragile textile objects. This implies that while curators and conservators are in principle agreeable to the wider experiential aspects of museum education, the substrate does not provide them enough room for interactive innovations in exhibitions. To understand this further, we need to first take a look at the set of guidelines that are laid down for museum professionals working on textile artifacts.

Guidelines for display of textiles and costumes: As per the guidelines laid down by ICOM, the display cases should be designed in a way that the visitors can appreciate the artifact in its totality. In order to enhance the visitor experience, the display of the artifact should be designed to bring out its functionality and purpose, keeping in mind its condition. Only the objects that are stable should be displayed after consultation with the conservators. As per the ICOM guidelines, textiles artifacts should not be put up for permanent or long-term displays. Also, the guidelines do mention that a safe distance should be maintained between the audience and the object to avoid contact (especially textiles) [5].

The display of textile artifacts in museum environment can be discussed under 3 factors:

i. Open display or Display cases
ii. Display design (Flat, hanging, slanting, draped etc.)
iii. Level of interaction between audience and the display

17.2.1.1 Open Display or Display Cases

Traditionally, it was a said norm to exhibit textile artifacts in glass display cases. While the material suitability of display cases is an important conservation concern, the discussion in this chapter focuses upon the role of these glass cases in visitor interactions. Display cases have been hailed for a very long time for their advantages of protecting the artifacts from dust and pollutants [6]. Although early display cases were plagued with the problem of harmful microenvironments inside the chambers, accelerating the process of degradation; the problem was corrected by development of environmental control display cases that further cushioned the artifact from fluctuations in temperature, humidity etc. While environmental control technology takes care of the conservation aspects of the textile, it cannot take care of the psychological barrier created between the visitor and the object due to the glass wall in between. It has been repeatedly noted by exhibition experts that audience interaction can be enhanced if there are fewer physical and intellectual barriers between them and objects of interest. Apart from the benefit of closer interaction, the gallery also looks more informal and relaxed with open displays. Open display also provides a means of displaying large objects that would otherwise never be seen by the public as the costs and means needed to construct and transport custom sized showcases are extremely high. Mapping this discussion back with the questions posed at the beginning of the chapter, it is important here to note that environmental control display cases have clear advantages and are imperative for historic, aged textiles. However, the museums at some point of time might be displaying textiles, which are neither old nor fragile. The focus of these exhibitions could be to educate the audience about the art and technique and the artifact could be strong enough to sustain a short-term display. Therefore, it is not mandatory to display every textile exhibit in a suitable transparent enclosure, immaterial of its individual details. It is interesting here to note that there have been successful experiments with textile exhibitions where open displays have gone a long way in enhancing the audience interaction, furthering the cause of public education. However, it is important to follow certain guidelines for open display of objects in order to minimize chances of damage [7]:

- Temperature and relative humidity levels in the gallery space need to be stabilized to meet the requirements of the objects on display.
- Direct daylight should be stopped from entering the gallery and there should be appropriate and flexible lighting system that caters for the light and UV sensitivities of the objects on open display.
- Conservation grade materials should be used to construct internal walls or furnishings.
- Paint for the building and internal structures should have minimal volatile organic components. Minimum 4 weeks should be allowed after fresh paint for the paint to off gas before installing objects.
- The air filtration system should be efficient in saving objects from possible pollutant sources, including visitors who will be in close proximity to the objects on open display.
- Save open display objects from damage through inappropriate handling by the general public. Objects should be kept out of touching distance by the use of either psychological barriers such as low level or sloping plinths that prevent the public from getting too close or physical barriers, such as roped off areas, acrylic, or glass barrier screens.
- Objects should be stable enough to withstand the vibrations that may occur around it with people walking or running near it.
- There should be ample security around the object to save it from theft. This could include visible security personnel and strong attachments to display bases so that the object cannot be easily lifted from the surface.
- Objects on open display require more maintenance and cleaning than objects in showcases as there is higher exposure to the dust, pollution & pests.

17.2.1.2 Display Design

Display design is one of the most significant factors that creates lasting impact on exhibition design and audience interaction. Curators and exhibition designers have used innovative ways of displaying textile artifacts where each has its own merits and limitations. There cannot be one standard suitable design for all textile artifacts as each exhibit comes with its own complex combination of shape, weight, fiber type, embellishment, and vulnerability. Some of the basic display techniques have been discussed here in reference to their suitability to the object and its vulnerability [8].

- **Framing of Textiles**: Textile artifacts are sometimes framed using archival materials. Although, this technique considerably limits the interaction of audience with the artifacts; it can still be deemed suitable where the artifact is to be presented in combination of accessories to be able to make sense or in situations where the artifact is divided in certain parts, which can be only displayed in unity by framing. This is one of the seldom used display methods considering that the frame might become the permanent home of that object and the micro-environment of the frame will affect its longevity. Decision about framing a textile is taken after careful consideration of different aspects like fragility, construction method, type of ornamentation present, etc. It is imperative to assess the special needs of any object before framing it, otherwise it can undergo rapid deterioration if framed in improper condition. Expert opinion needs to be considered while framing.
- **Free-Hanging Textiles**: Since times immemorial, textiles have been used to cover doors and windows offering protection, adornment and controlled visibility. This principle has been used for the display of textiles in exhibitions as free hanging artifacts, subject to certain conditions. Flat textiles like tapestries, rugs, saris have often been displayed in this manner. It needs to be noted here that "free-hanging textiles display" comes with its own set of challenges. The most important requirement for a free-hanging display is the condition of the exhibition space. Light, temperature and relative humidity, footfall, and insects, can cause damage to free-hanging textiles. Moreover, Indian textiles are heavily ornamented and embroidered. These textiles tend to get torn, cracked or damaged under their own weight, if left hanging for longer durations. If there is no barrier between the free hanging textile and the visitor, then the chances of touching the artifact becomes quite high, which might lead to further staining or deterioration of the artifact. However, the technique has proven to be very helpful in case of short duration exhibitions of tapestries, rugs & other drapes, provided the exhibit material is strong enough to sustain the weight.
- **Draped Textiles**: There are numerous examples of flat textiles, which are ornamented according to a specific draping pattern. For instance, the *pallu* of a sari is much heavier than its body, the phulkari *bagh* is designed to cover one's face, the *pagdi* has to be worn in a certain manner. If these textile artifacts are displayed just hanging freely, the visitors would never be able to relate their story in context and not appreciate the craftsmanship. Thus, draping the textile over a shaped mount, which could be suggestive of the wearer is an efficient way to display the textile in its usable form.
- **Rolled Display**: In most cases, the textile is quite long and due to space constraints, the whole object cannot be displayed at once. Such textiles are

rolled and only a small part is exposed for exhibition. If exposed for a long time, such textiles will get damaged from the visible area. Rolled display has clear benefits in terms of managing space constraints as well as making sure that the hanging hem does not bear the weight of the whole of the textile. However, if displayed in this manner for longer durations; the exposed part of the textile might suffer deterioration at a much faster pace as compared to the rolled part. Therefore, a rolled display should not be used for long duration exhibitions as far as possible.

- **Slanting Method**: In slanting method, costumes, made-ups or open-width textiles are laid out flat on a slanting board or surface. Thus, the exhibition design is not as space consuming as a flat horizontal display and also helps in dividing the gravitational force all along the textile and not stressing a particular area. This method is most suited for display as well as storage of fragile textile artifacts. The materials used for making the slanting board should not be harmful for the object. If wood is being used, then it should be varnished to avoid off-gassing. The angle of the slant ideally should be between 20 and 60 degrees. The fabric used for covering the slant mount can either be unbleached, undyed and de-sized cotton or muslin. In some cases, polyester fleece might also be used. In order to secure the object on the mount, archival tape, non-rusting pins can be used. If the textile is stable enough, then it is best to attach it using stitches. Many flat and fragile textile objects or costumes can be effectively displayed in this manner.
- **Mannequins**: The experience of any textile gallery enhanced considerable, if the costumes are displayed on mannequins or stands to depict them in 3D. A costume should be displayed on a mannequin only if the fabric and seams are stable. Adequate padding at the folds, shoulders, sleeves should be done to avoid any creases. The safe materials, which can be used for mannequins, are polyethylene foam, wax or fiberglass. The supports for accessories, like shoes, caps, bags can be carved out from polyethylene blocks, which in turn should be padded and covered with cotton fabric [9].

17.2.1.3 Points to be Considered While Planning a Textile Display

- Rare or fragile textile artifacts should not remain on display for more than three months.
- Handling of objects should be minimal during installation or de-installation.
- Textiles should not be placed directly under light fixtures, air vents and entry and exit points for visitors.
- Textiles objects should be separated from wood and other surfaces using a layer of acid-free archival tissue or polyester.
- The objects should not be in the reach of the visitors.
- Only plain white cotton or muslin should be used as a backing material colored/artificially dyed fabric should not be used.
- If the textile is to be hanged, then it should be done in the warp direction.
- Use of adhesives, staples or pins should be avoided.

17.2.1.4 Level of Interaction Between the Visitors and the Objects on Display

For the visitors to understand and appreciate the textile artifacts, it is important to display the object in context. Information kiosks, audio-video based guided tours etc. have been traditionally used to fulfill this objective. Additionally, the exhibitions have been supplemented with 2-3 minutes video footage of a traditional process focusing on a particular craft. It has been noted that this type of exhibition generates longer engagement with the visitors. With these supplementary tools in place, visitors are able to identify the functionality of the artifacts displayed and form collaborations.

Another important aspect to improve visitor experience is providing a means of tactile learning [10]. Various studies on learning styles have established that learning through touch and movement are the earliest forms of understanding for an individual. The transition to more advanced stages like reading and listening come at much later stages of life. This establishes the fact that as humans, we have an inherent desire to touch an object, especially textiles as textiles were always meant to be functional. In this perspective, exploring textile artifacts through touch can have beneficial outcomes, and can even help in understanding, and challenging the visual judgments and verbal interpretations we rely on.

However, in case of museum textiles, this cannot be allowed as there is a possibility of damage to the artifacts. This creates a contradictory argument: to touch or maintain a safe distance. Textile curators around the world have been looking for innovative solutions to create a balance between these two contradictory requirements. Replicating the look and feel of the original artifact and the process of its creation, while allowing visitors to interact with these replicas and experience the process has been a winning approach in most places. The act of making and the heritage of artisanal skills can be explained to the public through textile samples, toiles, sketches, or even lavish digital recreation of pattern making, as seen in the "Charles James" exhibition at the Metropolitan [11].

In India, many of the traditional crafts are still practiced. The museums can commission development of small samples of different weaving, printing and embroidery techniques. These can be placed near the display or can be catalogued into a book and placed at entry exit points of the gallery.

Publications including detailed information, photographs and samples of different textile ornamentation techniques can be made available for purchase at museum shops. This would help the scholars and researchers in their field of study. The Calico Museum, Ahmedabad has many such publications.

To increase the public participation, museums are also adopting a "workplace method." The visitors are engaged in the theme and context of the display by setting up workshops during temporary exhibitions. This modern approach attracts a wider range of audience and provides them with a newer method of experiencing art. The visitors can be encouraged to create something of their own but conforming to the overall theme per say. This would be a way for the visitors to "see through their hands." For instance, the artisans or craftsmen can teach the skills of embroidery or printing. Prototypes of looms can also be placed in the galleries, where a weaver can demonstrate the art of weaving a traditional sari. Such examples can be found in Crafts Museum, New Delhi. In simpler words, being in the shoes of a professional artist for a few moments, an amateur would appreciate the history and the workmanship of the craft better.

Many international curators and textile experts have emphasized upon the importance of displaying an artwork along with the whole process of manufacturing. As a beautiful example, "The Fabric of India" exhibition at the V&A (2015-2016), where Rosemary Crill and Divia Patel, paid attention to techniques and dedicated the first room to different processes of weaving, decorating, and to the materials employed [12].

17.2.2 Display Lighting

Light is a necessary evil. In museums, light is required to enrich the visitor experience by revealing the object as well as to create an ambience of the whole display with respect to the theme. The primary objective of art museums is to provide an enriched visual experience for the visitor. Thus, the function of museum lighting becomes somewhat theatrical and it helps in shaping the narrative. Hannah Crowell, an expert in illumination of artwork, says, "Lighting plays a major role in determining how people experience art, it can either create a wide-open space for self-discovery and exploration, or it can shrink the space such that visitors get a tightly scripted narrative" [13].

All forms of visible light contain at least some UV radiation, and all visible light damages fragile artwork. Moreover, lighting together with other environmental conditions (like temperature and humidity), can modify objects' properties significantly throughout their exhibition time, leading to deterioration. Textiles suffer irreparable damage from light, both visible and ultraviolet wavelengths, as well as mechanical damage caused by repeated heating and cooling (and subsequent drops in relative humidity) that occurs under strong lights. Long exposure of textiles to visible and ultraviolet light leads to photochemical degradation, which implies chemical changes at molecular level. Additionally, damage from light and heat is cumulative [14].

Therefore, there is always a dilemma to solve: *Exhibition versus Preservation.* "High standards of preservation may lead to poor conditions of exhibition. On the other hand, a stimulating ambience for exhibition may expose valuable objects to lower preservation standards" [15]. Here the role of lighting designers becomes more important. They plan the lighting in order to illuminate light-sensitive fine art as dimly and for as little time as possible. Nowadays, there are options like using lower wattages, adding metal screens or using dimmers via Bluetooth.

According to Scott Rosenfeld, a lighting designer at Smithsonian American Art Museum, "Light-sensitive objects should be illuminated just brightly enough that people can enjoy them that day, if the light is brighter than it needs to be, or if the light is too dim for people to enjoy the object, we're not using the life of that object as well as we could. This wastes energy as well!

Lighting is an ethical judgment. When we're working with precious and fragile objects, we have to make every photon count" [13].

Popular light sources used in museums are:

1. Halogen lamps including dichroic lamps
2. Fluorescent lamps
3. Metal Halide lamps
4. LEDs
5. Electroluminescent lights
6. Fiber optic systems

17.2.2.1 Hacks for Appropriate Lighting of Textile Exhibitions in Museums

- Lights should be turned off or dimmed whenever possible, and maximum lighting should be limited to light measurements and the actual exposure, for example three minutes at a time.
- The costume can be kept covered with a lightfast drape during most of the exhibition time. It implies that the drapes are removed only when there are visitors in its vicinity (can be facilitated by using sensors). This would prevent long exposure to light and subsequent damage. Curtains made of heavy velvet or cotton fabric can be used for this purpose.
- Lights should be kept at a safe distance, to avoid any temperature increase of the object.
- Flash photography should be avoided. In case, photographic documentation is necessary, the temperature and relative humidity should be regularly monitored during the photography session.
- Long continuous exposures to studio lights should be prohibited for textile objects. The safest lighting for photographing textiles is considered to be electronic flash (strobe), as tungsten and quartz lamps have very high levels of infrared light (causing heat); the high levels of UV and IR in flash lighting are of such short duration that they are not as harmful, while providing greater color accuracy and image sharpness.

An appropriate example, with respect to Indian museums with important textile collections, is *The Calico Museum of Textiles and The Sarabhai Foundations Collections*, located in Ahmedabad, Gujarat. The museum opens for a duration of 3.30 hours per day and allows only one guided tour of 20 visitors per day. The lights of each section of the gallery are switched on briefly (only when the visiting group is taken near that particular section) to provide enough time for 20 people to see all the objects and then switched off promptly. This limits the exposure time and duration of lighting to almost 10-15 minutes per day and helps in prolonging the life of delicate textile artifacts.

17.3 Exhibition Protocols Followed by the Museum

Each museum has a set of guidelines to be followed for allowing visitors in the galleries. These guidelines have been further altered in the recent pandemic situation. The focus of these guidelines is to provide access to maximum possible visitors in a day without causing any damage to the exhibition galleries. As we have already discussed, too many people would increase the temperature and affect the microenvironment of the galleries.

17.3.1 Safety Guidelines—Visitor Safety, Conduct and Access

- Footwear is mandatory for entry.
- Children under the age of 12 must be accompanied by an adult at all times.
- Still and video photography for personal noncommercial use is allowed without flash. Tripods are not allowed unless prior permission has been obtained from the competent authority.

- The museum should be wheelchair accessible.
- In case of emergency, please follow the instructions and procedures of museum staff and Security [16].

17.3.2 Safety of Museum Artifacts

- Visitors are not allowed to touch or lean on the works of art, the museum cases or pedestals. Also cases or pedestals are not to be used as writing surfaces.
- Visitors are required to maintain adequate distance from artifacts so as not to damage them by gesturing at them.
- Visitors are not allowed to bring flashlights or laser pointers to examine works of art in the museum.
- Backpacks, large bags and coats, umbrellas or portfolios must be left inside the museum's storage area during the visit.
- Sketching is welcome in the museum with dry media (pencil and erasers only). No crayons, pens, markers, or wet materials are permitted.
- Food and drinks are not permitted. Chewing gum is also not allowed.
- Smoking is prohibited inside the museum.

17.3.3 Let Us Look at Some Examples of the Protocols Followed by the Museums and the Changes that Have Taken Place After the Pandemic

V & A Museum: V & A has free entry for visitors but they have to pre-book the entry time. This ensures that there is no overcrowding of the galleries at the same time. Tickets are released monthly and are available for groups of up to six people from the same household.

Due to the pandemic, the galleries are currently closed and the museum has made provision for online exhibitions. The visitors can book online visits to special exhibitions, which can be either free or paid.

British Museum: They have now made all entries ticketed to ensure safety of the visitors. The visitors have to pre-book a time slot in advance. The museum is open on all days from 10.00 AM to 5.30 PM. On Fridays, it is open till 8.30 PM. Precautions like wearing of face shield, masks, availability of hand sanitizers have been enforced in wake of the pandemic. The visitors are not allowed to make calls from special exhibitions and keep mobiles on silent while in galleries in order to be considerate of the other visitors.

The MET: It also allows for booking a timed slot by the visitors online. This helps in keeping a check on the number of people inside the gallery. Visitor capacity is controlled and reduced in Museum buildings; access to all galleries is not guaranteed.

17.4 Photography and Memorabilia

Memorabilia & photographs have always been an inseparable part of tourist excursions. Where it allows visitors to take back fond memories of the place visited, it also widens a museum's scope of maximizing their revenues. Many museums develop replicas of the exhibits and sell them in museum shops. Many times, photographs, frames or catalogues of the museum objects are also put up for sale. This is another way of engaging the sentiments of the audience where they feel they can take away a small part of the object, especially in the case of many religious relics.

But the important aspect in this case is to safeguard the intellectual and physical integrity of the museum. The copyrights and other legalities should be resolved before replicating any object for sale. Necessary permissions should be taken from all the stakeholders.

17.5 Access Guidelines for Museum Storage

The primary function of any museum is to provide access to its collection and the related information. However, it is a well-known fact that more than 90% of museum collections rest in their storage chambers. These artifacts are taken out once every few years for display in specific exhibitions. Alternatively, museums might provide special access to these stored artifacts for students and researchers keeping in line with their primary objective of community education. Different museums have well defined guidelines for this purpose. Most of the time, it should be understood by the visitors as well as the museum professionals that physical access to some objects is not possible. There could be varied reasons for this like the fragile condition of the object or the exclusivity related to it. The museum should clearly state the reason and try to offer an alternative approach in such cases. If anyone needs access to such objects then following guidelines should be followed [17]:

- The visitors who are given access to reserved collection or storage should be briefed by the museum staff about the restrictions and precautions to be taken.
- One important aspect of providing access to non-display artifacts is ensuring their safety. Thus, all such visitors who are granted access should have proper permissions and their bonafides should be checked and recorded. Regular supervision by a museum staff or installing security cameras is essential.
- The rules for handling (allowed only if necessary) should be clearly mentioned.
  - Accession number should be easily visible to avoid unnecessary handling.
  - A clean and clutter-free space should be used to study the object.
  - Hands should be clean, free from any rings, bracelets, cufflinks or any other jewelry item that might damage the object.
  - Clean gloves should be worn while touching the object.
  - The object should be supported and lifted with both hands only when necessary.
  - One should be careful with the fastening devices like buttons, hooks & eyes or any decoration item like sequins.
  - No food, drinks or smoking should be allowed in the storage area.
  - Pins, scissors, stapler, etc. should be strictly prohibited.
  - Notes should be taken only in pencil. No pens should be permitted to avoid ink stains.

17.6 An Ideal Textile Exhibition

Let us take this discussion forward by visualizing an ideal textile exhibition. An ideal textile exhibition would be an innovative blend of ideas and display techniques that are able to develop the best possible balance between the contrasting ethos of *artifact conservation* & *visitor interaction*. It would be interesting here to note that pointers for an ideal textile exhibition would change with time, duration, location, theme etc. Interactive & open guidelines deemed appropriate for one set of artifacts might not be suitable for another exhibition or artifacts that are more fragile in nature. Let us look into some case studies to understand how innovation can go a long way in enhancing visitor interactions and thus making exhibitions more open, interactive and engaging for the visitors.

17.6.1 Case Study 1

Title of the exhibition: Sacred Traditions of the Himalayas [18]
Place: The Metropolitan Museum of Art, New York City, USA
Time: December 20, 2014 to June 14, 2015
Object in Focus: Skeleton Dance Costume [19]

There were 49 objects in this special exhibition. One of them was a traditional dance costume called the skeletal costume. It was an elaborate costume with eight different parts: a jacket, trousers, two boots, two gloves with claws, an animal apron, and a fringed belt. The materials ranged from cotton to silk to felt and even damask. The challenge in its installation and display was that the curator wanted it to be displayed in an animated and dramatic posture to place it in the context. For this purpose, the costume preparator made a custom mount using brass and wood. The mount was built in a manner that many parts were detachable to facilitate its dressing, as well as it had to be padded well, so that no stress is caused on the object. Where this facilitated the display of costume almost in action, the flexible and detachable mount ensured that no part of the precious artifact faced undue stress-strain during the exhibition. This exhibition aptly highlights the amount of research and labor-intensive work that goes into planning the display of an object in a realistic manner. These painstaking efforts improve the visitor experience and make the object memorable for them.

17.6.2 Case Study 2

Title of the exhibition: "Mode aus dem Rahmen" meaning "Fashion out of the Frame"
Place: Bayerisches National Museum, Munich, Germany
Object in Focus: Female Court Dress [20]

A small exhibition divided into two sections was created around the extraordinary dress. One part highlighted the reconstruction of the court dress so the visitors get a behind the scene look. It presented a copy of the dress in cotton fabric. The pattern diagram and life-size pattern pieces of the bodice demonstrated the different stages that the dress had gone through. Additionally, some photographs showed the conservator's work. This installation was quite popular among the visitors as they understood the construction process of the dress and the work that goes into mounting of garments.

The second part of the exhibition illustrated the professional embroideries on dress in the late eighteenth century and the craftsmanship involved. A selection of men's coats and waistcoats demonstrated the great variety of embroidery motifs and techniques that had existed in the late eighteenth century. The mounts for these objects were made of felt. To make visitors realize that all these garments had been produced on embroidery frames before they were cut and made up, some panels of embroidered, uncut waistcoats were presented in the exhibition.

There were reproductions of fashion engravings depicting these costumes as worn by the nobles of that time. The visual details were supplemented with text panels explaining the different sections of the show, like, for example, the embroidery process. It was a conventional exhibition much appreciated by the visitors. A small catalogue was also made available in the museum shop.

17.7 Conclusion

It has been duly noted in this chapter that limiting museum education to visual aspects is an aged approach that needs to evolve. Interactive and open displays are the best ways to ensure that exhibitions make a lasting impression on the minds of the audience [21, 22]. However, most textile artifacts can be too fragile to be allowed proximity and communication with visitors. Replicas and reconstructions can be effective strategies to be used for supplementing public education, while keeping the originals safe. At the same time, textile artifacts should not be governed by a standard stringent set of regulations that deem them untouchable, come what may. On the contrary, it would help if there are more detailed or stratified set of staircase guidelines that can effectively discriminate between fragile and non-fragile artifacts, making the latter available for wider public interaction. At the same time, it has been repeatedly seen that curators and exhibition designers have sometimes gone out of their way to curate experiences for their audience that have been interactive and make the audience feel involved. These exhibitions have definitely made a larger impact on museum audiences and have played a role in increasing museum footfall and return rate of visitors.

References

1. https://www.britannica.com/topic/museum-cultural-institution. [Accessed 29 December 2020].
2. Gül, S.N. and Akmehmet, K.T., Interactive spaces in art museums: A landscape of exhibition strategies. *Šolsko Polje*, 26, 5/6, 141, 2015.
3. Greenhill, E.H., *Museums and the Shaping of Knowledge*, Routledge, London & New York, 1992.
4. http://fashion.telegraph.co.uk/news-features/TMG8655222/Kate-Middletons-dress-display-is-horrid-cries-Queen.html.
5. ICOM, *Guidelines for Costume*, 1998, http://costume.mini.icom.museum/wp-content/uploads/sites/10/2018/12/guidelines_english.pdf. [Accessed 8 January 2021].

6. Zamora, M.S., O'Donell, B.M., Ajmat, R.F., Development of a methodology for materials selection to control exhibition objects in house-museums. *Int. J. Conserv. Sci.*, 11, 3, 689–702, July-September 2020.
7. https://manual.museum.wa.gov.au/book/export/html/149.
8. http://www.museumtextiles.com/uploads/7/8/9/0/7890082/displaying_textiles.pdf. [Accessed 19 February 2021].
9. Pietsch, J., The reconstruction and presentation of a french court dress, in: *Exhibitions and Interpretation: Proceedings of the ICOM Costume Committee Annual Meeting*, Toronto, ICOM, 2015.
10. Alvarez, A., *Please Touch: The Use of Tactile Learning in Art Exhibits*, 2005, https://www.getty.edu/education/museum_educators/content_play.html. [Accessed 10 January 2021].
11. https://www.metmuseum.org/art/metpublications/Charles_James_Beyond_Fashion?Tag=&title=&author=windmuller-luna,%20kristen&pt=&tc=&dept=&fmt=
12. https://www.vam.ac.uk/blog/news/getting-the-feel-of-the-fabric-of-india.
13. The Lighting Resource, The art of museum lighting, June 9 2017. https://www.soraa.com/learn/beautiful/art-museum-lighting.php. [Accessed 13 February 2021].
14. Sachdeva, K., Suri, M., Bhagat, S., Assaying changes in museum textiles on ageing-a short note. *Conserv. Cultural Property India*, 39, 119–122, 2011.
15. Ajmat, R., Sandoval, J., Sema, F.A., O'Donell, B., Gor, S., Alonso, H., Lighting design in museums: Exhibition vs. preservation. *WIT Trans. Built Environ.*, 118, 195–206, 2011.
16. John Hopkins Archeological Museum, *Visitor Guidelines*, Jhon Hopkins University, Maryland, USA, https://archaeologicalmuseum.jhu.edu/visit/museum-visitor-guidelines/.
17. Museum & Galleries Commission, *Standards in the Museum Care of Costume & Textile Collections*, Museums & Galleries Commission, London, 1998.
18. https://www.metmuseum.org/art/collection/search/69054?&exhibitionId=%7b6ca4d3d3-0c0d-453d-81b2-6e5d8d35b73e%7d&oid=69054&pkgids=281&pg=1&rpp=4&pos=2&ft=*
19. https://www.metmuseum.org/blogs/now-at-the-met/2015/installing-sacred-traditions
20. http://costume.mini.icom.museum/wp-content/uploads/sites/10/2018/12/Johannes_Pietsch_Toronto_article_PDF.pdf
21. Vom Lehn, D., Lehn, C., Hindmarsh, J., Rethinking interactivity: Design for participation in museums and galleries, Work, Interaction & Technology Research Group, Kings College, London, 2005.
22. Bedno, J. and Bedno, E., Museum exhibitions: Past imperfect, future tense. *Mus. News*, 1–11, October 1999.

18

Educational Value of Clothing and Textile Museums

Sara Marcketti* and Jennifer Gordon

Apparel, Events, and Hospitality Management Department, Iowa State University, Ames, Iowa, United States of America

Abstract

Textiles and clothing protect us from the environment and provide a means of modesty, as well as communicating valuable information about our beliefs and identities to the world around us. Once accessioned into a university or college collection, our garments take on a second life as historical evidence of a specific time and place. The conservation of garments can be costly. However, their use in college and university classrooms can provide significant learning experiences for students, who are often more familiar with analyzing text and images. In this chapter, we discuss factors when conserving historic textiles and clothing within the university environment, frameworks for material culture analysis, and the value of collections to students' higher education. We then apply the Taxonomy of Significant Learning Experiences to the many possible uses of textiles and clothing collections within classroom settings.

***Keywords*:** Conservation, textiles, collections, object-based learning, teaching and learning

18.1 Introduction

Different from paintings or sculpture, textiles and clothing in museum collections are "touched" by the owners. Rare is the garment that has not lived a life, whether on the runway or in a wardrobe, before entering a collection's holdings. There are often signs of use, adaptations due to personal reasons of comfort, and alterations made for style or functional considerations. As conservators Sarah Scaturro and Glenn Petersen note (2004, p. 235), clothing ".... is damaged every time it is put on and worn. Seams are stressed and strained, and the materials are exposed to environmental factors like sunlight, dirt, stains, and sweat" [1]. Garments provide protection, modesty, and adornment, as well as reflecting societal beliefs and attitudes [2]. Within the costume, fashion, and apparel discipline, these objects, or material culture, provide evidence of past lives, often raising new questions, or bringing new insights into our beliefs regarding a previous time. While some collections accession only rare, valuable, and pristine items, most university and college collections represent the everyday objects of the individuals living in the region or affiliated with the institution, providing a lens into the prevailing trends, attitudes, and perspectives of a period and a person.

**Corresponding author*: sbb@iastate.edu

Seiko Jose, Sabu Thomas, Pintu Pandit, Ritu Pandey and Vandana Gupta (eds.) Handbook of Museum Textiles: Volume I Conservation and Cultural Research, (341–354) © 2023 Scrivener Publishing LLC

According to Prown (2000, p. 2-3), "Artifacts constitute the only class of historical events that occurred in the past but survive into the present...they are authentic, primary historical material available for first-hand study" [3]. Primary sources provide opportunities for students' to critically analyze and make meaning of past events. As stated by Osborne (1986), "new history" in the 1960s and 1970s led to an increase in student-centered, activity-based use of historical documents and the need for archivists to act as educators [4]. Used within educational settings, primary documents, including first-person accounts, photographs, images, performances such as songs and dance, ephemera, garments, and textiles breathe new life into secondary sources such as journal articles and textbooks.

Textile and clothing collections, as repositories of primary sources, have significant value in learning and teaching the many facets within the fashion studies curriculum. Within art museums, objects may need to be paradigms in design and condition to be useful in mounting exhibitions [5]. This is not necessarily the case in academic collections, which support learning, whether through exhibition or hands-on access. Most often documented in the literature are the benefits of extant garments in the teaching of fashion history [6], teaching topics such as construction and embellishment [7], and technical drawing to Optitex CAD reconstructions [8].

Even with the empirical and scholarly evidence of the numerous benefits of object use within classrooms, there remains opposition to the costs and perceived niche interest of maintaining historic textile and clothing collections [9]. The purpose of this paper is to provide the foundational literature and ideas discussing the educational value of museums. Sections within this chapter include conserving historic textiles and clothing, frameworks for material culture analysis, and the value of collections to students' learning within the college and university setting. We then apply the Taxonomy of Significant Learning Experiences to the possible uses of textiles and clothing collections within various types of classes, from textile science to design, to history. With this framework, instructors and administrators can gain evidence-based strategies to deepen students' knowledge, application, and connection to the subject and world around them.

18.2 Importance of Conservation in Textiles and Clothing Collections

Many university collections have in their mission statement a commitment to educational opportunities for students, and by extension, support the curriculum of the department or program in which they are housed [10]. Once a garment enters a historic textile and clothing collection, these objects are no longer "touched" by wearers, but handled by learners as part of their study in higher education classrooms. Indeed, this use was the origin of many collections. Within university collections, there are typically two tiers of objects: those that are rarely touched due to their fragile nature, rarity, or importance and the study collection brought into the classroom for observation and analysis. Many collections also include nonaccessioned examples with the purpose to be deconstructed by students to better understand construction methods and pattern-making techniques.

Textile-based materials pose unique challenges related to proper care and storage. Beyond the conservation and preservation of the materials damaged from previous use, garments are subject to a litany of factors that may contribute to deterioration. Modern conservators are engaged in the hands-on treatment of objects, as well as assessing and managing risk from the attraction of pests, mold, susceptibly to fading, and other environmental factors [11]. Even the most careful plan and dedicated care cannot maintain an object in perpetuity [12].

Certain clothing objects are subject to "inherent vice," a general conservation term pertaining to a characteristic that can cause damage and ruin. Inherent vice can stem from the fiber, fabric, or construction, as well as the maintenance or lack thereof of the artifact [1]. Inherent vice is a factor in the aging and deteriorating historic silks that commonly vex caretakers of collections [13]. Yet modern fibers and fabrics can also challenge collection managers and conservators in their ability to deteriorate. For example, 1960s knitted dresses with crumbling polyurethane foam interlinings [14] or plastic materials and polyvinyl chloride (PVC) can become sticky and frangible as they age [15].

Many academic collections cope with the duality of the artifact's original purpose as a worn garment and its present educational functions—and the gradual degradation compounded by both of those uses. All of these considerations factor into the potential need for conservation of the artifacts. Collections may lack the financial resources, facilities, equipment, or trained staff for a rigorous conservation program. In university collections, and public and private institutions alike, the laborious and expensive nature of textile and garment conservation means that items defined as having greater historical significance or value or those intended for exhibition receive priority attention [16]. Objects earmarked for study may not be an institutional priority [17] and many museums and collections have more objects in need of stabilization or repair than is feasible. The result is that some academic collections may manage little to no conservation work [18].

It is a difficult balance. As DeiTos notes (2016, p. 65): "Caring for these materials is an ever-growing and costly challenge, as standards of conservation and care have grown and changed over the last several decades" [19]. Visitors and students often do not see the hidden work that facilitates the access to collection objects, the record keeping and inventories and maintenance done by collection managers and registrars or the careful consideration of condition and treatment by conservators. Conservation work is so often concealed and unrecognized, that conservator-scholar Sarah Scaturro termed it "ghost labor." It is important work, however, as a conservator's knowledge and skillset can make "objects on the brink of 'death' viable for display yet again" [20]. While Scaturro [20] refers to the primary function for exhibition, the same holds true for objects utilized in educational spaces. Conservators create possibility of access to objects, both through interventions to repair objects and preventative conservation, which may encompass monitoring an object's storage environment [21].

The question of access is one curators and collection managers often face. In one regard, touching and active examination is encouraged as providing a sensory experience with the garments. However, touching threatens the very life of the garment. Peters and Romanek (2008, p. 2) pose an interesting point-of-view about access and touch: "..... [the] idea is that access always means objects will suffer... what does it mean for an object to suffer, and is this always related to its physical condition—can an object prosper, even as it suffers physical degradation?" [22]. In the case of touch in academic teaching collections, the question may be, do our collections thrive by use in service of education, rather than tucked away, undamaged but unseen, even if it means they may be eventually beyond repair? Does the benefit of learning outweigh these consequences? Malkogeorgou contends (2011, p. 442), "conservation is driven by the paradox of treating objects with a view to transmitting them into the future and a search for truth in authentic objects that may link original people and practices to us today" [23]. Conservation is philosophically linked to education, and if we prioritize the conservation of our teaching collections, we ensure that generations of students have an unparalleled opportunity to learn directly from primary source objects.

18.3 Frameworks for Material Culture Analysis in the Learning Process

Despite the commonness of many of the objects within our collections, most students have little experience interpreting garments and other material culture objects as source material. When students enter our university and college classrooms, they have often practiced making meaning out of textual and visual media, but are less adept at investigating tangible things. Because of this, Lafrenz and Mida state (2013, p 20), "many students, and even some faculty, demonstrate initial reluctance to engage with actual objects" [24]. Students are accustomed to taking a passive view of collections as many museum objects serve as static props or the background illustrations to a story. As Cain points out (2010, p. 198), object-based learning is "inextricably linked to… the promotion of active, open-ended, student-centered learning" [25].

As historians and educators, we must develop skills and help our students develop the ability to "read" and analyze the cultural objects that surrounds us. Students often need guidance in the act of proper object handling, as well as skills in making meaningful connections between that object and the producing culture. Even with provenance, the full meaning and meaningfulness of garments and textiles can remain concealed, as we bring our own 21st century perspectives, attitudes, and experiences to the viewing and meaning making of the objects.

Depending on the course goals, objectives, and discipline, different frameworks can be used to examine material culture objects. Each has in common the systematic examination of the artifact. E. McClung Fleming outlined the following steps: identification, evaluation, cultural analysis, and interpretation [26], while art historian Jules Prown suggested engaging with an object through description, deduction, and speculation [27]. In the latter, the researcher begins by looking at the artifact, maintaining as much objectivity as possible, to describe what is observable about the object in its present state; for instance, its dimensions, materials, and visual elements. As the methodology advances to the next stages, the researcher may begin to relate physically and mentally to the object, imaging and making suppositions about its use. At the point of speculation, the researcher generates ideas, theories, and develops a research plan for supporting evidence. Scholar and curator Valerie Steele applied Prown's approach to the study of fashion objects [28]. Instructors continue to adopt and modify Steele's approach in the careful reading and analysis required for studying historic objects.

18.4 The Value of Collections to Students' Education

Scholars have written of the benefits of collections to enriching students' education. As emphasized by Riello (2011), there is an overabundance of evidence that material objects enrich interpretation, furthering our knowledge of individuals, cultures, and societies [29]. For students, the physical touch and sensory engagement brings a dimension of learning that can deepen enjoyment and retention of a knowledge about a historical period [30]. The hands-on skills associated with the making of clothing, flat patternmaking, technical design, and creative design are enriched through close examination of garments. Further, the observation and combined drawing of historic objects can strengthen students' abilities in terms of visual literacy, resulting in improved written communication in more thoughtful artifact descriptions [31].

In addition to strengthening discipline specific knowledge, the thoughtful study of garments can further twenty-first-century skills. As brought forth by Hodge (2018), these skills can include cultural competency (via multiplicity of meaning in past and present), critical thinking (via evaluation of multiple forms of evidence), communication (via consideration of multiple standpoints in the classroom), and the application of theory to practice [32].

18.4.1 The Collection at Iowa State University

Many historic textile and clothing collections within the United States began as part of Home Economics or Family and Consumer Sciences programs, departments, and colleges. Iowa State University in Ames, Iowa, was the first land-grant institution in the United States to teach home economics (then referred to as domestic science). From 1871 to the present, coursework in nutrition, personal finance, childcare and parenting, design and merchandising formed the core of the ISU College of Human Sciences. Within ISU, Professor Olive Settles collected the original contents comprising what is now known under the umbrella term: The Textiles and Clothing Museum. Today, the Museum facilities include the Mary Alice Gallery, a 504 square foot exhibit space, the Donna Rae Danielson Conservation Lab and Classroom, and the Bertha and Edward Waldee Storage Facility.

Settles' pieces, acquired during trips to Russia, India, and South America (particularly Guatemala), represented changes in technology and consumer interests from the 1920s to the 1960s. Faculty, staff, alumni, and others continue the tradition of collecting and donating pieces from their travels and their personal collections, as well as donating financially to the Museum. While a bulk of the 9,500 object collection includes garments to signify rites of passage such as baptismal gowns, wedding dresses and suits, high fashion designer pieces and rich examples of textiles from around the globe are used across the curriculum, from history, to patternmaking, to illustration. Pieces often serve multiple functions. For example, the 100 plus collection from Anne Eggenberger Zimmerman (b. 1942-d. 2007) includes many notable 20th century fashion designers including Rudi Gernreich, Adrian, Claire McCardell, Kenzo, and Schiaparelli and formed the basis of an exhibit, a scholarly presentation to a disciplinary organization, and as classroom examples when discussing 20th century fashion designers.

18.5 Taxonomy of Significant Learning and Collections

Within the literature, the connection between the use of collection objects and increased student interest, engagement, and learning, is abundant. Within this paper, we sought to apply a framework or taxonomy of significant learning experiences to the many ways collection objects are invaluable to teaching today's students. Benjamin Bloom and associates created one of the most well-known educational taxonomies in 1956, consisting of six kinds of learning arranged in a hierarchical sequence: knowledge, comprehension, application, analysis, synthesis, and evaluation. A revised taxonomy for teaching, learning and assessment was proposed in 2001, to demonstrate a more dynamic concept of the classification and renamed the classifications: remember, understand, apply, analyze, evaluate, and create. Educational scholar, researcher, and author L. Dee Fink proposed a taxonomy of learning that went beyond the cognitive domains of Bloom's 1956 and 2001 models. Fink suggested that for learning to occur, there has to be some kind of lasting, life-long change in the learner. Fink created a taxonomy that includes six kinds of significant learning: foundational knowledge, application, integration, human dimension, caring, and learning how to learn. Rather than the hierarchical portrayal (and sometimes conception) of Bloom's Taxonomy, Fink's categories are interconnected and can occur at the same time [33]. Within the next section, each of Fink's dimension is presented with examples drawn from the literature and the Textiles and Clothing Museum, Ames, Iowa.

18.5.1 Foundational Knowledge

Foundational knowledge is the base of every discipline, from science to history to mathematics and literature to conservation. Students need to know key facts, formulations, concepts, principles, and relationships. Further, students must be able to know the central ideas or perspectives underlying the field. For example, most apparel curricula begin with an introduction to textile science, as students must first remember fiber classification, yarn identification, and fabric construction before enrolling into retailing or design coursework. Foundational knowledge, "provides the *basic understanding* that is necessary for other kinds of learning" (Fink, 2003, p. 31).

At ISU, within the second-year Introduction to Textile Science class, students are expected to thoroughly understand the composition, properties, and characteristics of fibers, yarns, fabrication processes, and dyeing, printing, and other finishing techniques. The use of PowerPoints and textile swatches help students understand the material. However, this often occurs discretely. For example, a student may be able to identify a rib-knit, but has challenges identifying the rib-knit portion of a sweater that includes both jersey and rib-knit sections. By utilizing examples from the collection, students can better make sense of a textile property on a garment. Depending upon the specific lesson's objectives, instructors can select a specific example to illustrate a point, such as the use of jersey on the body of a sweater and rib-knit on the cuffs, or provide examples that demonstrate the richness of the combination of fiber, yarn, fabric, and finishing detail. An exhibit curated by then-graduate student, Sonali Diddi, entitled, *Tana-Bana: A Celebration of Indian Textiles* provided ISU textile science classes a rich opportunity to see the visual differences created through fiber and fabrication choice. For example, a *Kanchipuram* sari made of silk featuring jacquard designs woven in gold thread reflected a great deal of light and looked smooth to the touch. In contrast, a sari made of cotton with silk embroidery, was matte looking with the silk embroidered flowers "popping" off the green georgette fabric. Students who viewed the exhibit overwhelmingly stated how the garments reinforced, brought to life, and expanded concepts they learned in the class to a global context [34].

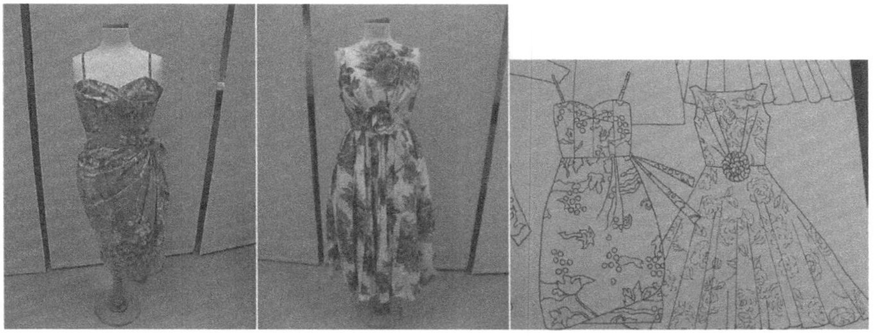

Figure 18.1 On the left, two garments from the Textiles and Clothing Museum, Ames, Iowa. On the right, drawings from a second-year fashion illustration course, 2013. Courtesy of Michael Mamp.

Another example of building students foundational knowledge through collections is from a fashion illustration course. Students closely examined two garments from the collection and then drew the garments in 1.5" scale (Figure 18.1). As is evident in the images below, the future designers practiced and sharpened their technical acumen from examination and drawing of these artifacts.

18.5.2 Application

Students must move beyond mere comprehension to apply what they have learned. With application, students use their knowledge in new situations or in increasingly complex ways. This category of significant learning also includes developing certain skills, such as communication, critical, and creative thinking. According to Fink, "application learning allows other kinds of learning to become *useful*" (p. 31) [33].

Conservation can be a critical part of the life cycle of the object, preserving it for future use. The ways in which conservation connects to learning are manifold. Conservation practice, according to Malkogeorgou (2011, p. 451), provides an intimate way of learning about a material object that other methods do not, particularly as relates to craft:

> The way that information about the object is revealed means that it is specific to the textile conservators themselves and it is they who ultimately have access to it. Neither connoisseurship, study nor erudition offer helpful ways of access into the practical knowledge of the object's construction, and when the object is preserved a whole world of craftsmanship is preserved with it [23].

Students in conservation courses often benefit from the opportunity to apply fundamental knowledge and skills in the care of objects; in turn, this benefits collections that may not have the funds to pay for expensive treatments. Further, objects, even those in a state of deterioration, can have value as examples of condition problems that conservators must learn to recognize, understand, and treat (Figure 18.2).

Figure 18.2 A former ISU graduate student, Kate Greder, conducting a conservation project as part of her graduate assistantship. Courtesy of the Textiles and Clothing Museum, Ames, Iowa.

18.5.3 Integration

When students are able to make connections, view the similarities, differences, and interactions between things, an important kind of learning has occurred. Connections can be made between specific ideas, people, courses within a discipline, as well as with integration with a student's personal, social, and/or work life. Fink states, the "act of making new connections gives learners a new form of *power*, especially intellectual power" (p. 31) [33].

The comparison of one category of objects collected in different periods can be illuminating for both students and researchers. At Iowa State University, three individuals donated examples of adinkra stamps and cloth collected during travels to the Ghanaian village of Ntonso, West Africa. Developed by the Ashanti people, Adinkra means "saying good bye" and was traditionally used in mourning and funerary ceremonies. The process begins by sewing together several lengths of hand-woven or imported cotton or linen cloth. The patterns are printed with a vegetable-based dye using a stamp onto a cloth. Each pattern has an individual meaning (often multiple meanings) and the combination of the patterns can tell rich stories. The quality and technique of the stamps and cloth demonstrate change in the twenty-year period of collection. For example, the 1990s cloths were constructed of cotton with yellow, red, and green thread embroidered over the seams combining multiple cloths. The cotton cloth collected in the early 2000s, while thinner than that collected in the 1990s, had a damask woven pattern that provided additional glean and decoration to the fabric. The cloth from the mid-2010s was off-white, with no discernable construction pattern, and the stamps were made of balsa wood, rather than the traditional calabash gourd. The instructor of the Cultural Perspectives of Dress courses routinely instructed students to observe and note the similarities and differences of the material and the technique, to contemplate the ways technology, economic, and social factors influence societies, cultures, and traditions.

Researchers from the University Archives Textile Collection at The University of North Carolina at Greensboro used gym suits from various decades of the 20th century within a course on sports and physical education history. The contrast between the constricting nature of some gym suits as compared to looser styles, allowed students to see the evolution in design, but also, how the changes related to attitudes about physical exertion for girls during the twentieth century [35].

18.5.4 Human Dimensions

When students learn about themselves or others, empathy can grow, with interactions becoming more sincere and effective. Students discover the larger implications of what they have learned, and place this knowledge into greater understanding of their goals personal and career goals. They can also acquire a deeper understanding of the ways others behave, how they function under different stressors and circumstances. This understanding can deepen students' empathy and positively affect future interactions with others. According to Fink, "this kind of learning informs students about *human significance* of what they are learning" (p. 32) [33].

By nature of their being, objects provide a way for students to interact on a physical, sensory, and emotional level. Dudley (2010, p. 6) argues that "the very materiality of the material" allows a going beyond the narrow aesthetic, formal qualities, or technical analysis of specimens [36]. Rather, interacting with objects provides a communicative dimension involving physical sensation. The inherent multisensory levels of objects, their smoothness or roughness, weight, even the sound the fabric or embellishments makes, provides a stark contrast to the dependence on written text and visual images alone.

In garment analyses, the sensory touch of the objects, the intellectual work of determining an artifact's features, purpose, and date, and the emotional involvement of sharing and exchanging ideas with fellow students provide opportunities to connect with one another and past people and histories [37]. In addition to individual learning, the use of guided questions regarding an artifact often facilitates oral communication as student teams work together to form a cogent and logical date or explanation of the use of the garment. For example, students may be placed into small groups and asked to work together in discussing and answering the following questions.

- Describe the silhouette.
- Describe the fabrics used and identify possible fibers.
- Examine evidence of hand (typically irregular) or machine sewing (very regular stitches).
- Describe the fastenings used, boning, padding, hem finishes, and any other features related to the cut and construction of the garment.
- Describe any evidence of alterations and/or repairs. (This is sometimes very difficult to discern, but look for different fabrics in one section of the item, different shadings of the same fabric, places where the hem or seams were taken out and redone, movement of buttons and other closures).
- Sketch the front and the back of the garment in the space below.
- How was the silhouette supported or enhanced?
- What sort of accessories might have been worn with this garment to complete the desired "look?"
- Establish a date range of no more than ten years for the garment. Justify this time span.

Figure 18.3 Later twentieth century garments are sturdy enough to be examined while on hangers. In addition to their use in classes, artifacts from the Textiles and Clothing Museum are often incorporated into public programs highlighting the academic program that houses the facility.

- Imagine the artifact being worn on a specific individual. What is the visual effect from head to toe? How does the garment relate to the wearer? What message or meaning does the garment express?
- How does the artifact "fit" into the time period?
- What questions developed about the garment or period after viewing this garment?

By nature of using student groups to discuss these questions in the analysis of a garment, rich conversation, engagement, and thinking occurs (Figure 18.3). The importance of this type of learning cannot be overestimated, as Chatterjee contended (2010, p. 179): "objects can... be used to inspire discussion, group work and lateral thinking—all essential key, transferable, skills in higher education" [38].

18.5.5 Caring

Sometimes a learning experience changes the degree to which students care about something. This may be reflected in new feelings, interests, ideas, or values. According to Fink, "when students care about something, they have the *energy* they need for learning more about it and making it a part of their lives. Without the energy for learning, nothing significant happens" (p. 32) [33].

The link between learning and emotion is powerful. When we have learning experiences that involve emotion, they are more likely to be remembered. As the brain detects emotions, neurons release the chemical dopamine, which ensures the experience will be processed more deeply [39]. Historical information can take in new life when students envision themselves in the garments they are observing and analyzing. For example, in a History of Dress course at Iowa State University, an 18th century living history interpreter attends class and "dresses" from her under garments to her outer dress. Throughout the experience, the interpreter fielded questions from the students, bringing to life not just the aesthetics of the

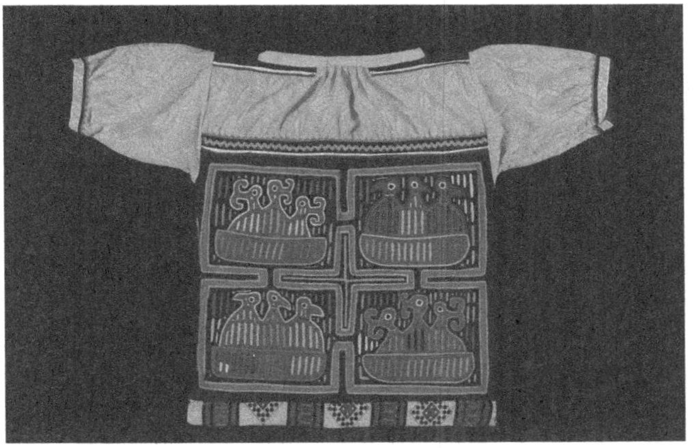

Figure 18.4 Mola, circa 1970. Panama. Gift of Dr. Lillian Matthews. 3668. Textiles and Clothing Museum, Ames, Iowa.

garments, but also their function and placement within 18th century life. In one representative comment on the end-of-course evaluation, a student stated, "I tried and put myself in the period, imagined who I would be, what I would be doing in these garments. It inspired interest in museum work."

Garments and textiles can be powerfully tied to the use of storytelling, another attention grabbing teaching technique that can evoke emotion. Storytelling through cloth is an ancient tradition with evidence in many cultures, regions, and periods, including the *paj ntaub* or flower cloth created by the Hmong people, West African wax print cloths, and the *huipil* of Mexico and Central America. The term mola refers to the individual stitched panels as well as the blouses constructed from them by the Kuna women of the San Blas archipelago of Panama. The mola communicates the maker's creativity and fashion choices, with the inspiration for mola designs coming from many different sources, such as local politics, history, Kuna myths and religion, and outside influences such as tourists (Figure 18.4). Students can utilize their drawing and creative thinking skills by examining mola patterns or other story cloths and create their own designs inspired by events surrounding them. Mental health professionals, including art therapists use textiles and fabric to help individuals tell their stories to cope with traumatic events [40].

18.5.6 Learning How to Learn

Learning how to learn is an important broader goal of teaching students. Within their studies, students gain opportunities to learn how to be a better student, how to engage in a particular kind of inquiry, and how to become a self-directed learning. This can translate to increased efficacy in other courses and other dimensions of a students' life. According to Fink, "this kind of learning enables students to *continue* learning in the future and to do so with *greater effectiveness*" (p. 32) [33].

Artifacts are often used in tandem with lectures, readings, and additional primary source evidence, such as first-hand accounts, letters, or advertisements. Instructors can use this integrated approach to provide depth via context and triangulation. Instructors can guide students through the iterative process of scholarly research in which interpretations are built via multiple, and sometimes contradictory, forms of evidence. It is not enough to examine a garment, but a student must continuously collect additional information. For example, students can consult social media, newspaper articles, trade publications, films, and conduct oral histories to analyze, critique, and create hypotheses, arguments spurred through initial examination of a garment. Learning through objects requires exploration and continual reevaluation. According to Hendrickson (2016, p. 140), "students who are engaged with artifacts, and who are invited to develop complex historical thinking, emerge with greater competence" [12].

Learning how to learn with historic textile and clothing collections can include infusing material culture as a primary source in student research projects. In advanced undergraduate and graduate courses, students can contribute to an in-person or digital exhibitions. The experience of curation pushes students to not only apply their disciplinary knowledge, but also reach for, learn, and apply communication skills, promotional strategies, and teamwork skills in exhibit creation. The ISU Textiles and Clothing Museum has served as an invaluable opportunity for student learning and applying museum-based skills, such as object research, cataloguing, exhibit conception, and exhibition installation and deinstallation. Undergraduate students have curated exhibits as part of their honors projects, as part of a college-wide Rosenfeld Internship program, and through research assistantships. These students are in diverse majors, from history, journalism, and sociology, with these experiences providing a springboard for internships and employment.

18.6 Conclusion

Opportunities provided by instructors through the use of textile and clothing collections offer deep, significant, and often life-long learning. This learning can be viewed as a "change in a person's way of seeing, experiencing, understanding, conceptualizing something in the real world" (Ramsden, 1988, p. 27) [41]. Indeed, the very purpose of higher education is for students to learn how to think, developing the skills to identify problems, gather and critically analyze existing information, develop, and then evaluate these ideas.

Educators, curators, and collection managers understand the vital role that objects play in teaching and training future professionals. Decisions about what to conserve, or even whether to conserve, as well as the availability of funding, time, materials, and staff, are significant challenges faced by caretakers of university teaching collections. Conservation can have a crucial role in ensuring that these types of educational opportunities are available to subsequent generations of students. Ultimately, we are preparing our students to help solve the problems, challenges, and opportunities identified and awaiting the remainder of the 21st century. Objects, while ephemeral and from the past, provide one way for students to develop, practice, and apply skills needed for our future.

References

1. Scaturro, S. and Petersen, G., Inherent vice: challenges and conservation, in: *Charles James: Beyond Fashion*, H. Koda and J. Glier Reeder (Eds.), pp. 233–249, Yale University Press, New Haven, CT, 2014.

2. Bolin, P.E. and Blandy, D., Beyond visual culture: Seven statements of support for material culture studies in art education. *Stud. Art Educ.*, 44, 3, 246–263, 2003.
3. Prown, J.D., The truth of material culture: history or fiction?, in: *American Artifacts: Essays in Material Culture*, J.D. Prown and K. Haltman (Eds.), Michigan State University Press, East Lansing, MI, 2000.
4. Osborne, K., Archives in the classroom. *Archivaria*, 1, 23, 16–21, 1986.
5. Scaturro, S. and Fung, J., A delicate balance: ethics and aesthetics at the costume institute, metropolitan museum of art, New York, in: *Refashioning and Redress: Conserving and Displaying Dress*, M.M. Brooks and D.D. Eastop (Eds.), pp. 159–172, Getty Conservation Institute, Los Angeles, CA, 2016.
6. Gam, H.J. and Banning, J., A collaboration to teach students to utilize historic dress as inspiration for apparel design. *Fam. Consum. Sci. Res. J.*, 41, 1, 56–68, 2012.
7. Sauro, C., Digitized historic costume collections: inspiring the future while preserving the past. *J. Am. Soc. Inf. Sci. Technol.*, 60, 9, 1939–1941, 2009.
8. Cobb, K., Orzada, B., Lopez-Gydosh, D., History is always in fashion: the practice of artifact-based dress history in the academic collection. *J. Text. Des. Res. Pract.*, 8, 1, 4–23, 2020.
9. Marcketti, S.B. and Gordon, J., "I should probably know more:" reasons for and roadblocks to the use of historic collections. *J. Conserv. Mus. Stud.*, 17, 1, 2019.
10. Welters, L. and Ordoñez, M., *Historic Textile and Costume Collection in the Academic Setting. ITAA Monograph 11*, International Textile and Apparel Association, Cary, North Carolina, 2011.
11. Halahan, F., Changes in attitudes and training in conservation. *J. Inst. Conserv.*, 41, 1, 79–84, 2018.
12. Hendrickson, L., Teaching with artifacts and special collections. *Bull. Hist. Med.*, 90, 1, 136–140, 2016.
13. Garside, P., Wyeth, P., Zhang, X., Understanding the ageing behavior of nineteenth and twentieth century tin-weighted silks. *J. Inst. Conserv.*, 33, 2, 179–193, 2010.
14. Lovett, D. and Eastop, D., The degradation of polyester polyurethane: preliminary study of 1960s foam-laminated dresses. *Stud. Conserv.*, 49, 2, 100–104, 2004.
15. Tonkin, L., Visionary design practices: preserving plastics and designers' intent in collections of modern and contemporary fashion. *J. Inst. Conserv.*, 40, 2, 152–167, 2017.
16. Battistini, S. and Benevolo, G., The fashion and sartorial documentation section of the Davia Bargellini Museum of Bologna: history, conservation and exhibition. *Conserv. Sci. Cult. Herit.*, 15, 1, 191–233, 2015.
17. Derbyshire, A., An approach to conserving study collections. V&A *Conserv. J.*, 59, Spring 2011. http://www.vam.ac.uk/content/journals/conservation-journal/spring-2011-issue-59/an-approach-to-conserving-study-collections/
18. Boyd, S.L., *Resources Needed to Support an Academic Historic Costume and Textile Collection in an American College or University* (UMI 1406665), [Master's thesis, California State University, Fresno]. ProQuest Dissertations Publishing, CA, 2001.
19. DeiTos, K.M., *The Art of Removal: An Analysis of Deaccessioning in Clothing and Textile Teaching Collections*, p. 843, Open Access Master's Theses Paper, 2016.
20. Scaturro, S., Confronting fashion's death drive: Conservation, ghost labor, and the material turn within fashion curation, in: *Fashion Curating: Critical Practice in the Museum and Beyond*, A. Vanska and H. Clark (Eds.), pp. 21–38, Bloomsbury, London, England, 2017.
21. Ashley-Smith, J., The ethics of doing nothing. *J. Inst. Conserv.*, 41, 1, 6–15, 2018.
22. Peters, R. and Romanek, D., Approaches to access: Factors and variables. *Stud. Conserv.*, 53, Sup 1, 1–6, 2008.

23. Malkogeorgou, T., Folding, stitching, turning: putting conservation into perspective. *J. Mat. Cult.*, 16, 4, 441–455, 2011.
24. Lafrenz, L. and Mida, I., Engaging students with objects: Preliminary experiments in reviving a dormant fashion research collection. *International Textile and Apparel Association Annual Conference Proceedings*, 2013.
25. Cain, J., Practical concerns when implementing object-based teaching in higher education. *Univ. Mus. Coll. J.*, 3, 197–201, 2010.
26. Fleming, E.M., Artifact study: A proposed model. *Winterthur Portf.*, 9, 153–173, 1974.
27. Prown, J.D., Mind in matter: An introduction to material culture theory and method. *Winterthur Portf.*, 17, 1, 1–19, 1982.
28. Steele, V., A museum of fashion is more than a clothes-bag. *Fash. Theory*, 2, 4, 327–336, 1998.
29. Riello, G., The object of fashion: Methodological approaches to the history of fashion. *J. Aesthet. Cult.*, 3, 1, 1–9, 2011.
30. Duhs, R., Learning from university museums and collections in higher education: University College London (UCL). *Univ. Mus. Coll. J.*, 3, 183–186, 2010.
31. Mida, I. and Kim, A., *The Dress Detective*, Bloomsbury, UK, 2015.
32. Hodge, C., Decolonizing collections-based learning: Experiential observation as an interdisciplinary framework for object study. *Mus. Anthropol.*, 41, 2, 142–58, 2018.
33. Fink, L.D., *Creating Significant Learning Experiences: An Integrated Approach to Designing College Courses*, Wiley, Hoboken, NJ, 2013.
34. Diddi, S. and Marcketti, S.B., The use of museum exhibits in teaching textile science. *J. Fam. Consum. Sci.*, 106, 1, 47–49, 2014.
35. Koelsch, B., McCarty Smith, K., Motszko, J., Collecting textiles: Is it worth it? *J-SNACA*, 10, 2, 2–17, 2013.
36. Dudley, S., *Museum Materialities: Objects, Engagements, Interpretations*, Routledge, New York, 2010.
37. Csikszentmihalyi, M. and Hermanson, K., Intrinsic motivation in museums: why does one want to learn?, in: *The Educational Role of the Museum*, E. Hooper-Greenhill (Ed.), pp. 146–160, Routledge, NY, NY, 1999.
38. Chatterjee, H.J., Object-based learning in higher education: The pedagogical power of museums. *Univ. Mus. Coll. J.*, 3, 179–182, 2010.
39. Persellin, D.C. and Daniels, M.B., *A Concise Guide to Improving Student Learning: Six Evidence-Based Principles and How to Apply Them*, Stylus, Sterling, VA, 2014.
40. Garlock, L.R., Stories in the cloth: Art therapy and narrative textiles. *Art Ther.*, 33, 2, 58–66, 2016.
41. Ramsden, P., Studying learning: improving teaching, in: *Improving Learning: New Perspectives*, P. Ramsden (Ed.), Kogan Page, London, 1988.

19

Career in Textile Museum

Maanasaa Sethuraman[1], Suruchi Pandey[1]* and Ritu Pandey[2]

[1]Symbiosis Institute of Management Studies, SIU, Pune, India
[2]Chandra Shekhar Azad University of Agriculture & Technology, Kanpur, India

Abstract
Museums are often considered the reserve houses where the history and legacies of the past are preserved. They play a critical role in acquainting the people of today with the history and lore of yesterday and yesteryears. Historic collections of textile and clothing have inspired many modern designers in curating trends, helped academia in teaching students about the history of textile and clothing, have enhanced the understanding of the history of various cultures across the globe and across timelines. Aside from the various educational opportunities offered by these museums, they also offer career opportunities less known to the world. Like any other industry or segment, museums need skilled and qualified resources and offer a niche career to few. The chapter discusses the various career opportunities available in this segment for the students in India and abroad while also exploring the skills required for capitalizing on these opportunities. The chapter also details the various sourcing techniques used by employers in this industry that provide students with the aforementioned opportunities along with a list of sourcing techniques for museum administrators and managers to do staffing effectively. The chapter also provides a detailed view of the various sources of museum textiles and their significance. The chapter also allows the user a glimpse of the limitations and challenges in the industry to help the reader make informed opinions and choices regarding the opportunities in this field.

Keywords: Career, textile museum, job profile, competencies, textile conservators

19.1 Introduction

Museums across the world display various figments of art and history and work tirelessly to preserve these artifacts. One of the most important objects that are part of any display in a museum or heritage exhibition is textile. Across the globe, museums display and preserve garments aged thousands of years with most of these garments needing special care and conservation. The encyclopedic collection includes tapestries, carpets, embroideries, costumes, and three-dimensional accessories from all periods and cultures. The significance of these collections are such that up until 2012, non-departmental public bodies like the Museums and Galleries Commission in England were very actively involved in laying down

**Corresponding author*: suruchi.p@sims.edu

Seiko Jose, Sabu Thomas, Pintu Pandit, Ritu Pandey and Vandana Gupta (eds.) Handbook of Museum Textiles: Volume I Conservation and Cultural Research, (355–382) © 2023 Scrivener Publishing LLC

policies, standards and promoting best practices to ensure proper care of these heritage artifacts and garments by the museums [1–3].

While potential and scope for work are immense in this area, it is imperative to understand the aspects of pursuing a career in museum textile curation and conservation. The prospects of pursuing a flourishing career in this field have increased with the turn of the 20th century with collections of textiles and historic clothing inspiring new-age designers, and these have been featured as the highlights of major fashion and textile museums.

Textiles in museums vary enormously with regard to collections. It has historic value, aesthetic appeal and is culturally significant. Along with projecting and taking care of the heritage, textile museums have made in news off late for unconventional display. London's Fashion and Textile Museum announced that its featured exhibition in 2020 would be a celebration of Designers Guild's 50th anniversary, an influential design company whose products have changed the way consumers view "color, pattern, and texture" in their homes, inspired from ancient Indian textiles, Renaissance-style velvet, or Swedish Gustavian wall treatment, but the resulting interiors are never based solely on one traveller's story or a reworking of an old document. Instead, each is an eclectic amalgam in which there is harmony between East and West, present and past. Thus, it is evident that these textiles and artifacts provide a deeper understanding of the culture and history of the times they represent and storing, preserving, maintaining, and documenting them are the prospective arenas that a career in this field offers to aspirants worldwide [4].

This chapter seeks to explore the skills, knowledge, professional background, career path, pay scale, and the roles that a career in the field warrants. The objective of this chapter is to highlight the options available for a career in textile museums. The scope and validity of the discussion are constrained and limited to the information available and the information compiled by the authors in the time frame whence the paper is set.

19.2 Sources of Textile Museum Collections

Textile museum, in layman terms, is a place displaying the history and art of textiles. The list could be exhaustive, ranging from production to utility, and agriculture to decoration, for example, yarn, carpet, cloth, bedding, dresses, types of silk, embroidery, quilts, etc. These museums are visited by students, teachers, travellers, researchers, anthropologists, historians, and developers. The reasons and views that attract each of them to these museums are distinct. These museums are owned by the government, private owners, trusts or under public-private partnerships. They are doing a great job in preserving the heritage and culture.

Museum textiles, no doubt, are extremely sensitive and fragile owing to their age. This implies that the work of a museum textile conservator is far more tedious and comparatively slow-paced unlike other professional assignments in the clothing and textile industry. Like any other kind of restoration and conservation work in a museum, textile curation and conservation also warrant immense patience with zeal and passion unparalleled to any other, to bring out the true lustre of the historic value and beauty held by these artifacts. These historic collections in the form of textiles exemplify a sense of belongingness to an individual's identity, sheds light onto one's personal history, cultural heritage and a family link to the historic events in the past. These collections are preserved and cared for in glorious museums, as part of stand-alone museum collections, a sub-group of a larger museum collection or as single pieces of historic value owned by museums, historic societies, or individuals alike [1, 5].

Many European museums (and Indian museums alike) hold objects of everyday life that have become obsolete for everyday use, often termed 'bygones' which also inspire modern fashion in many forms. Local museums also hold collections of the garments that have seen large scale production in the area, such as Kancheepuram silk, handwoven garments, etc being displayed in places like Dakshin Chitra, Chennai. These museum textiles represent the industrial efficacy of the people in the area from a bygone era and hold immense value and hence also form interesting objects for textile conservators and curators alike. Many institutions use such collections or individual pieces as a teaching aid for their textile design, history, or museology courses and hence possess collections of their own in this regard. Table 19.1 is a list of textile museums and other museums that hold textile collections in India and internationally. Museums across the world (such as The Textile Museum of Canada) work with the sole aim of facilitating the understanding of the ancient human experience through textiles and textile arts. International museums often present and preserve local as well as international textile collections to help recreate history as well as to ignite creativity inspired by the past amongst the patrons and students of fashion and art. The Boards of many of these museums constitute professionals from diverse backgrounds ranging from academicians, writers, lawyers, historians, educators, and more, thus allowing us to conclude that careers in museums include various other opportunities apart from conservation and curation of museum textiles [6].

Table 19.1 Ten Indian and International museums with textile collections.

List of museums with textile collections in India	**List of museums with textile collections abroad**
Calico Museum of Textiles, Ahmedabad	Fashion and Textile Museum, London
Sanskriti Kendra, New Delhi	Royal Textile Academy of Bhutan, Thimpu
National Crafts Museum & Hastkala Academy, New Delhi	Victoria and Albert Museum, London
Kasthuri Sreenivasan Art Gallery & Textile Museum, Coimbatore	George Washington University Museum, Washington D.C.
Indian Museum, Kolkata	National Textiles Museum, Kuala Lumpur
Vimor Museum of Living Textiles	Museum of Textile, West Jakarta City
National Museum, New Delhi	Museum for Textiles, Toronto
Textile Museum of Ladakh, Ladakh	Metropolitan Museum of Art, New York
Dakshin Chitra, Chennai	Textile Museum of Decorative Arts, Lyons
Traditional textile museum, Jaipur	Yokohama Silk Museum, Japan

While some of these museums are nationalized and funded by the respective governments (Ministry of Culture, in the case of India), there also exist a few privately owned museums or collections. A major portion of the career opportunities in this field is provided by privately funded museums or private collectors seeking to restore family heirlooms from the past. These museums are not limited just to providing information on the textile choices, popular clothing materials and styles, but they also aim to showcase a glimpse of the lifestyle that was inspired by these heirlooms, driving men, women and animals to look the way they looked and carry themselves the way they did. Some museums (like the Sanskriti Kendra, New Delhi), serve as a resource for the study of textile traditions rather than harvesting the "show value" of these heritage textiles, aiming to generate as much information as possible around the various genres of textiles available in India that would eventually serve as academic resource for study and research. Heritage exhibitions, private museum tours, endorsements and fashion inspired by the past are a few key channels that have kept the passion for conservation alive amongst our museum curators even to this day [7–9].

Museums have also begun digitizing literature and records related to their textile collections since the current audience of these museums belong to the "Generation Y" who have grown up enjoying the fruits of the Internet, Google, Wikipedia and other sources of instantaneous knowledge and expert advice [5, 10].

19.3 Scope of Careers in Textile Museum

A career is the set of jobs and positions a person holds during one's lifetime. A career is associated with advancement unlike doing a job that may be done merely as an occupation to earn a living. The career decisions of individuals are usually influenced by intrinsic, extrinsic and interpersonal factors. Textiles and museums give a promising career. At times lack of awareness becomes a boundary between the job seeker and the job [11].

Careers in Museum textiles, although driven by passion, are very subjective, with the current trends showing a minority of takers. This trend is observed to be prevalent due to a lack of awareness amongst students about the scope and opportunities of pursuing a career in this field. Given the supposed unconventionality of this field, more and more enthusiasts in the form of students, academicians and professionals currently pursuing a career in museum textiles are advocating students and aspiring professionals to commit to an informed decision in this regard. Most job opportunities in this field are dependent on the fund available to them. Job duties are often rewritten when budgets contract or expand to include or exclude traditionally distinct roles [5].

A successful career requires a level of competencies. Competencies are a set of knowledge, skills and attributes that a candidate must possess to perform the job. Knowledge can be gained through degrees and certifications offered by specialized universities. Skills are gained by internship, project work or on the job training provided by the organizations. The right attribute is the key and is intrinsic to making sound career decisions.

19.3.1 Job Opportunities

Museum textile in itself is often a subset of the broader career choice of the field of museum conservation and art restoration. Like any other career chart, the growth and hierarchy of roles for both government employed professionals as well as privately employed professionals are efficiently demarcated over the years. The most likely employers for students and professionals in India in this domain range from government museums to private organizations consulting in conservation and ultimately the private or free-lance conservators who operate as smaller units with a singular head (the conservator) teamed with their assistants.

Job opportunities in India for textile conservation in specific are minimal owing to the fact the government museums that employ a majority of the professionals in this domain, do not specialize in textile conservation. Most of the textile pieces and collections available in these museums are decorative. Owing to the lack of demand for the job role in general, Government museums refrain from hiring for positions specializing in textile conservation and more so at the entry-level. The number of museums specializing in textiles are but few and employ the majority of those being hired [12].

19.3.1.1 Public Sector

The public sector is referred to as a government-owned organizational setup. In the Government sector, the job opportunities in the initial/entry stage, as Technical Assistants (TA) are far more generic and they do not require specialized education in any field related to conservation. Their role is that of a keeper who handles the storage and movement of the objects. In the hierarchy of roles in government museums, two parallel career paths span out. Technical Assistants can choose the path of a curator or that of a conservator in the museum based on their competencies and qualifications. In the scenario that TAs choose the curatorial career path, the next position in the hierarchy is that of a Senior Assistant. Higher up in the hierarchy is that of the Deputy Curator and subsequently the Curator. With regard to the nature of the occupation, the curatorial career path defines job roles that deal with the artistical and historical representation of the object (textile in this case), while the conservatorial career path deals with the storage, restoration and conservation of the piece to preserve the quality and characteristics of the material. The hierarchy of roles for the conservatorial path is also similar to that of the curatorial path, however, with job responsibilities specific to conservation. Most of the projects assigned to these curators and conservators are long term and require detailed and effective planning. While the technical assistants in the government museums draw salaries in the reasonable pay scale (as per set grade), the senior positions are also progressively paid well. As you grow and reach out to higher positions, these jobs attract high compensation. Starting or owning your venture in form of a consultant is another rewarding option. However, the number of positions for which the government museums hire candidates are limited to a sparing minority [1]. Figure 19.1, maps the hierarchy of roles offered in the government museums for an aspiring Technical Assistant.

1. Technical Assistant

As described above, the role of the TA is more generic and involves storage, movement and handling of the collection or piece in both cases- curator and conservator. The educational qualification as mandated by the Ministry of Culture requires a TA to hold an MA in Ancient History or Medieval History or a similar degree. Many museums, including the National Museum, refrain from hiring candidates with a Masters in Modern History for this role. The basic skill for the job includes agility. The TAs are expected to handle the objects with immense care and are tasked with the handling of the object while the curator/ conservator works on them.

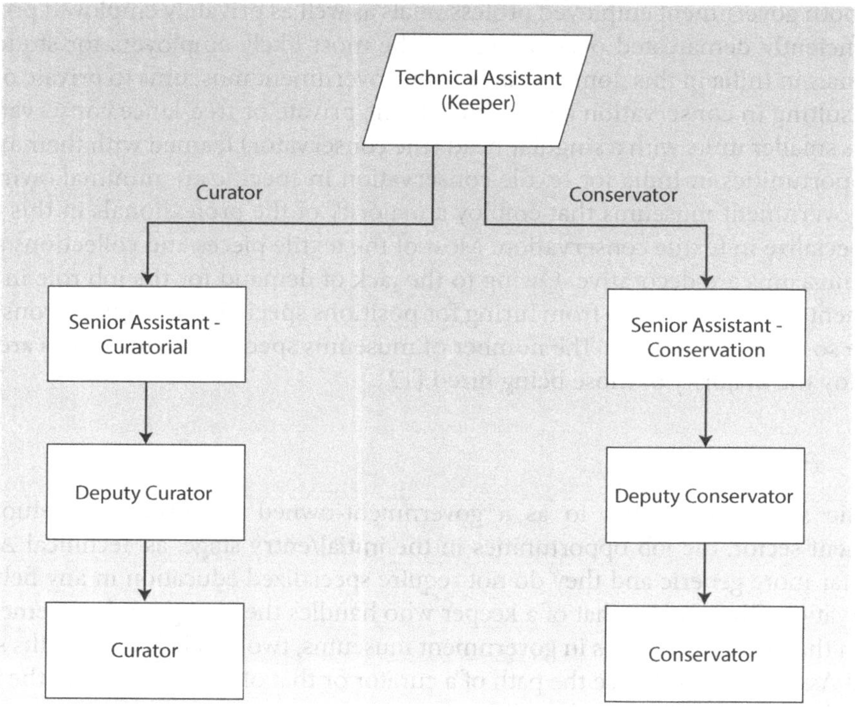

Figure 19.1 Career path in Indian Government museums.

2. Senior Assistant Curator, Deputy Curator, and Curator

All roles in this career path mandate the candidate to qualify for the UPSC exams and the interview while holding a Master's degree in Ancient or Medieval History. Senior Assistants are expected to hold better knowledge of the object, its historic and artistic value and aid the Deputy Curator and the Curator with the conceptualization of the exhibit. The Deputy Curator and the Curator are tasked with the procurement of objects to curate the collection and conceptualize the exhibitions and displays in the museums. All these roles require candidates to possess leadership, decision-making skills and creativity along with the ability to recreate historical moments through their display of the collections. These positions also require skills such as fundraising, management and marketing. The curatorial positions do not mandate any preliminary background in the school or undergraduate level. This line of work is less lucrative compared to that of the conservator and hence have fewer opportunities as well [5].

3. Senior Assistant Conservator, Deputy Conservator, and Conservator

Similar to the career path for curators, this career path also mandates the candidates to qualify for the UPSC exam (in India) and the subsequent interview to be shortlisted. While Senior

Assistants and Deputy Conservators are expected to hold a Master's degree in Chemistry and preferably a background in material sciences. The preferred education qualification for Conservators is a Masters in Conservation and more specifically in textile conservation, in the absence of which a Masters in Chemistry is also expected. Candidates in these senior conservatorial positions are expected to possess in-depth knowledge of chemistry, physics, history and various restoration and conservation techniques. They are also expected to possess competencies that help them in documenting the repair and restoration done on these garments for future conservators to be able to retrieve a complete history of restoration performed on the material. They must also possess a good idea of the cultural, historical and artistic significance of the garment they are working on to differentiate wear and tear that hold historic significance from those that are a function of the age or mishandling of the garment. Only damages of the latter type are repaired by these conservators [1, 12].

Most aspiring conservators are recommended to complete one or more internships under established and experienced conservators to hone the skills and competencies required by the job. This career path is the most lucrative, with conservators continuing to stay in demand even post-retirement.

Basic skills required for the role of textile conservator [13]:

- They are required to possess decision-making skills
- Agility and aptitude to empower them to work on fragile and sensitive pieces of heritage textiles.
- Long-term planning and creative problem-solving skills.
- Excellent sewing skills and perfect vision and eyesight
- A basic understanding of organic chemistry
- Clear, descriptive, writing skills
- Patience, concentration and curiosity.

19.3.1.2 Private Sector

Careers in museum textile in the private sector have two varied paths. The first is that of a curator or a conservator in a private collection/museum and the second is that of freelance conservators. Freelance curator, as a job role, is almost non-existent. The hierarchy of roles and the skills required by private museums are similar to those in government museums. The major difference in the career progression in the sectors is directed towards faster career progression in the private sector compared to the government sector. The private organizations mostly hire interns with a stipend and proceed to promote them up the hierarchy. However, employees in the government museums possess a sense of job security that private museums do not, as the former is funded by the government while the latter fund themselves or depend on their benefactors for funding. In the lack of footfall into the museums, or the inability of shows and exhibitions to attract the audience, the private museums are forced to let go of a lot of their employees. A lot of these museums also deploy their workforce to the government museums when they take up restoration projects for collections held by these government museums. This cross-sector exposure is a definite advantage of this field of work. Indian National Trust for Art and Cultural Heritage (INTACH), and Indira Gandhi National Centre for Arts (IGNCA) are among the many private organizations in India that are involved in heritage textile conservation. Private organizations act as a learning area for a lot of conservators who eventually move out to become freelancers owing to the experience, exposure and network they are able to build in their tenure. Curators can also reach out to other curators with a specialization in textiles when they are planning a textile-specific exhibition, who will then be brought in as a textile expert for curating this exhibition. The general model of hiring in these museums is "word of mouth."

In the case of private conservators, who work as freelance consultants, they hire assistant conservators on a need-basis based on the projects they sign and the funds they are provided. While freelance conservators in India are very few, they are extremely popular and established for their quality of work. In most cases, the assistant conservators are apprentices and freshers and hence lack the expertise and knowledge required for effective decision making concerning the conservation methods. Most private conservators hire assistants at a pay scale while their boarding and lodging are taken care of at the project site. More often than not, the restoration projects in heritage palaces and their adjunct museums are signed by these freelance conservators. Assistant conservators are delegated basic tasks like documentation, mechanical cleaning, vacuum cleaning and other nonsolvent cleaning methods, minor repairs, etc, while a major part of the restoration is personally undertaken by the project head or the freelance conservator. With experience, these assistant conservators either branch out into freelancers themselves or join other organizations as conservators. Some conservators with a zeal to create awareness about the field, also take up part-time or full-time teaching and/or conducting workshops for students seeking to pursue a career in this domain. In the words of an established freelance textile conservator "Textile conservation is analogous to doctors. It is a practical field where seniors and juniors work together but juniors do the follow-up treatments suggested and mostly serious treatments are executed by the head."

Aside from the positions opened up by the museums, other freelance positions for museums also include conservation or collection care specialists who are hired to conduct curatorial and conservation audits in museums.

19.3.1.3 Opportunities Offshores

Museums abroad often collaborate with museums in India in conducting shows and exhibitions in India as well as abroad which act as very opportune platforms for aspiring conservators. Aside from these collaborations, museums abroad also offer internships and other contractual employment opportunities for various positions for students of textile conservation, museology and other nondegree courses related to museums. The myriad opportunities off-shores is depicted through the Figure 19.2, highlighting the organization structure of the international private museums. Figure 19.3 describes a sample job posting for the position of Associate Conservator, in the Art Institute of Chicago while Figure 19.4 details the job description for the position of the Curator, Textile Arts of Americas.

19.3.2 Changing Hiring Trends

Synchronous with any other industry in the country, the general hiring trend in museums across sectors have been shifting towards contractual engagement of curators and conservators. While progression from a contractual engagement to a permanent engagement

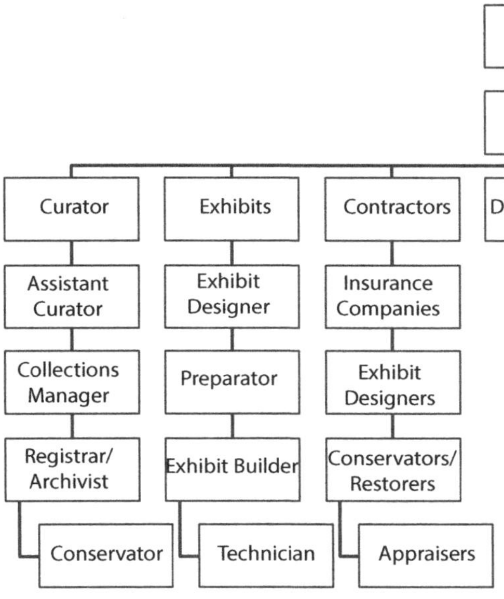

Figure 19.2 The organizational structure of international private museums [5].

with the museum is still possible, museums prefer to retain contractual relationships while promptly renewing the contract in adherence to the duration of the corresponding project associated with the Conservator. While contractual engagements offer an open avenue of career paths to the employee, it is even more advantageous and cost-efficient for the employer and is hence preferred as a growing trend.

19.3.2.1 Work-Life Balance in Careers in Museum Textile

As in the case of any creative line of work, policies for work-life balance in careers in museum textiles are nonexistent in the case of employees of private museums and freelancers. Public sector museums follow guidelines and work-timings as established by the government and thus offer better work-life balance. The amount of work-related travel, long working hours, especially for on-site conservation are deterrents that tilt the balance between work life and personal life ominously. Nonetheless, many modern museums and legacy museums alike have established some basic policies to ensure equality, inclusive work culture and a semblance of work-life balance for their employees. These policies include Flexible Working Hours Scheme, Adoption and Parental Leave, Flexible Working Arrangements, Shared Parental Leave, Part-Time Working, Employee Assistance, Special Leave, Child Care Voucher scheme, Dependent Leave, Gender Reassignment, Work-Life Balance: Career Breaks and Flexible Working, Maternity and Paternity Leave etc. [14, 38].

19.3.2.2 Job Description

Job Description is a written internal document maintained by organizations of all kinds. This document is referred to as the source of advertising for the job, hiring candidates,

Associate Conservator, Textiles
The Art Institute of Chicago

Job Type: Full-time

Job Description

Introduction
The Art Institute of Chicago shares its singular collections with our city and the world. We collect, care for, and interpret works of art across time, cultures, geographies, and identities, centring the vision of artists and makers. We recognize that all art is made in a particular context, demanding continual, dynamic reconsideration in the present. We are a place of gathering; we foster the exchange of ideas and inspire an expansive, inclusive understanding of human creativity.

Position Summary
Reporting to the Director of Objects and Textiles Conservation, the Associate Conservator, Textiles, is involved in all aspects of the care, treatment, preservation, preparation, installation, and research of all textile objects in the museum's collections. In consultation with the Director of Objects and Textiles Conservation, the Associate Conservator carries out conservation-related activities to support exhibitions, permanent collection rotations, and other special needs regarding shipping and receiving of permanent collection, new acquisitions, gifts, loans, imaging, technical examination, and researcher and class visits. Provides basic care for the collection including maintenance, proper storage, and monitoring environmental conditions in storage and exhibition. Collaborates in the planning, maintenance and day-to-day workflow of the textile conservation laboratory. Supervises fellows, interns and volunteers as needed; is actively engaged in fostering a diverse, inclusive, and equitable work environment.

PRIMARY DUTIES AND RESPONSIBILITIES:
- Determines and documents an object's condition. Prepares treatment proposals in consultation with the Director of Objects and Textiles Conservation. Executes conservation treatments and maintains full documentation of treatment. Supervises the execution of minor treatments by other textile conservation staff and outside consultants as needed.
- Prepares full conservation records and structural analyses for textile objects, including microscopic examination of structure and materials analysis as well as the physical examination of the condition of objects. Maintains digital documentation and hard-copy conservation records.
- Working with the Assistant Conservator, Preparation and Mounting, and the Conservation Technician, assists with all aspects of preparation, mounting, and installation of exhibitions, including examination, documentation, research and treatment of artworks.
- Carries out examinations on textiles as they enter and leave the museum for loans, and prepares and shares related documentation.
- With varying levels of supervision is responsible for review of works under consideration for purchase and preparation of relevant documentation.
- Trains, and supervises, art handlers, photographers and other specialists in all aspects of their work with objects and equipment.
- Assists with and oversees monitoring temperature, relative humidity and air quality in the galleries, exhibitions and storage. Advises on environmental standards for textiles in the permanent collection and for works that are proposed for loan.
- Advises on and assists with gallery and storage maintenance, including regular and preventative care of textiles artworks.
- Participates in daily activities including exhibition maintenance, lab maintenance and safety, ordering supplies in consultation with the Administrative Coordinator for Conservation and Science, and using/maintaining equipment.
- Serves on committees and undertakes special projects and responsibilities as appointed by the Director of Objects and Textiles Conservation.
- Corresponds with professionals with regard to research, answers inquiries of a technical nature, in-house and outside the institution. Answers public inquiries by phone and mail as necessary. Conducts studio tours and lectures.
- Responds to emergencies that occur in galleries or storage areas.
- Conducts research and publishes on textiles technique and on methods and materials.
- Publishes results of research projects in Museum publications, social media, and journals aimed at both professional and general audiences.
- Travels as a courier for artwork if needed.
- Supervises fellows, interns, volunteers and visiting scholars as needed.
- Supports the museum in its commitment to creating a diverse, inclusive, and equitable institution; treats colleagues and visitors fairly and respectfully, recognizing the value that diverse backgrounds and perspectives bring to our organization.

Qualifications
Master's degree from a recognized conservation training program with a minimum of five years of post-graduate experience or equivalent training and experience. Proven expertise in the treatment and preservation of textiles. Must possess knowledge of sewing techniques, textile structures and garment construction. Must have excellent verbal and written communication skills as demonstrated by a track record of publications and presentations. Proficiency with project prioritization and demonstrated project management experience. Embraces the museum's commitment to creating an inclusive and equitable institution.

Benefits
We offer a competitive, comprehensive benefits package:
- 5+ weeks of paid time off
- Paid holidays (up to 14 days)
- Medical, dental, and vision insurance
- Generous parental, caregiver, and medical leave benefits
- Competitive retirement plan (up to 9% matching contribution)
- Tuition remission and Tuition Exchange Program
- Great employee discounts
- Complimentary general admission to participating cultural institutions

Want to stay abreast of the latest job openings at the Art Institute? Sign up for our Employment Opportunities E-announcements.

Application Time Out Warning
The Art Institute of Chicago requires all applicants to complete an online application to be considered for employment. 60 minutes are allotted for applicants to complete the employment application. The application will automatically time out after 60 minutes. If you are unable to submit the completed application at this time, please choose the "Save for Later" option to avoid losing the information you have entered so far.

Closing Statement

The Art Institute of Chicago is an equal opportunity, equal access employer fully committed to achieving a diverse and inclusive workplace.

Figure 19.3 Sample JD for associate conservator, textiles [15].

interviewing and training. Understanding the job description can help career seekers in knowing what is expected out of them. The job description helps in knowing the competencies required for the job, which will help in making a well-informed career decision.

- **Sample Job Descriptions**

Curator, Textile Arts of Americas

Employer: The George Washington University Museum and The Textile Museum

Contract Type: Full-time

The George Washington University Museum and The Textile Museum seeks a creative, highly motivated individual with strong organizational skills, professional demeaner, and great attention to detail for the endowed curatorship position dedicated to the textile arts of the Americas.

Position overview

This position will perform responsibilities related to strengthening the museum's collections of textiles of the indigenous cultures from Central and South Americas, particularly from pre-Hispanic period; advancing knowledge through the critical inquiry and study of the collections; disseminating knowledge through exhibitions, publications, and lectures that build academic and popular interest; expanding participation by GW faculty and students, and build collaborations with peer institutions. Curator is a full-time position and reports to the senior curator. Title and salary will be commensurate with experience.

Major Responsibilities

- **Exhibitions:** The curator will identify and develop concepts for future research and exhibitions through collection assessment, and primary and secondary research. The curator will assist with the development and implementation of exhibitions and their catalogues and will conceptualize original ideas for audiences that include the university community and the general public. Strong dialogue with GW faculty and students, and wider scholarly community in this process is a must. Identify areas for future research and exhibition through collection assessment and research; Develop new exhibition concepts and proposals to present to the staff Exhibitions Task Force and occasionally to the Board, Assess and recommend loan exhibitions related to the textile arts of Americas; Assist with institutional exhibition planning/schedule; Curate and/or coordinate exhibitions; Provide curatorial support as in-house curator to loan exhibitions related to the textile arts of Americas
- **Collections:** Working with museum colleagues and stakeholders, and experts from the art world, the curator will strengthen, through accessioning and deaccessioning, the museum's holdings of textiles from pre-Hispanic Central and South America as well as textiles of the indigenous cultures from the region. The curator will develop and maintain relationships with collectors, donors and prospective donors, and dealers, including auction houses, to build the collections and acquisition funds, as well as solicit work from collectors. Contribute to collections documentation by identifying areas of collection for research and conducting through documentation and research in those areas; Strengthen Collections Database (Museum+) and Collections online project as an institutional resource by inputting the new data developed through collections documentation as well as retrospective data-entry; Assess collections for future areas of accession and deaccession; Research offers-of-donations and other prospective acquisitions by evaluating private collections for their suitability for The Textile Museum Collections; Present results of research to staff Collections Management Task Force and Collections Committee of GW/TM Board for approval of new accessions; Document new accessions; Develop and maintain relationships with collectors, donors and prospective donors, and dealers, including auction houses; Support continuing efforts to provide research access to textile arts of Americas in the collections for qualified researchers.
- **Research & Publications:** The curator will undertake primary and secondary collections-based research, and disseminate knowledge through exhibitions, publications, and lectures that build academic and popular interest. Through engaging with the scholarly community and contributing other museum-generated scholarly initiatives, the curator will contribute to strengthening the museum's scholarly role and visibility. Research and write articles, reviews, and book-length manuscripts for publication; Represent the GW/TM in the scholarly community and maintain an active professional role and visibility by engaging in collegial exchange and dialogue as well as attending and participating in pertinent symposia, colloquia, and conferences and giving presentations and lectures as appropriate; Support and contribute to the efforts to develop strategies for dissemination of scholarly information, including electronic means; Support and contribute to development and execution of strategies to strengthen institution's scholarly role and visibility; Provide scholarly input for the GW/TM projects, such as the Collections Online
- **Educational, Professional, and Public service Collaborating with GW academic departments and beyond:** The curator will assist the museum's education team in developing lectures, discussions, workshops, screenings, and other special events to build an audience for the museum's collections of textiles from central and south America within the university and the general public. Sharing his/her specialized knowledge on Indigenous American art and textiles through public and scholarly lectures, publications, and other means with diverse museum audiences is a must. Share specialized knowledge with diverse GW/TM audiences; Prepare and present public lectures and exhibition tours; Develop teaching resources and train docents; Conduct monthly public curatorial consultations (ask-a-curator, ask-a-conservator); Address public inquiries (mail/phone/email requests); Share knowledge as appropriate with Board, The Textile Museum Council, docents, and staff;

Figure 19.4 Sample JD for curator, Textile Arts of Americas [16]. (*Continued*)

Work with Education Department on programming related to the textile arts of Americas; Contribute to planning/implementation of the GW/TM development events as appropriate

- **GWU Engagement and Administrative:** A portion of the curator's duties includes developing opportunities for student and faculty research, especially through museum career training, collaborative academic projects, and teaching. The curator will also contribute and maintain the departmental activities and archives. Assist in training, utilization, and supervision of curatorial staff, interns, and volunteers; Contribute to the GW/TM planning and budgeting process by preparing and monitoring pertinent projects and corresponding budgets; Support GW/TM long-term planning and fund-raising efforts Participate in staff Research, Publication, Library and Education (RPLE), Exhibitions, Collections Management Task Forces, as well as other ad-hoc committees established for specific projects; Attend and report during the Board committee meetings

Qualifications

- **Minimum qualifications:** Bachelor's degree in an appropriate area of specialization plus 2 years of relevant professional experience, or, a Master's degree or higher in a relevant area of study. Degree must be conferred by the start date of the position. Degree requirements may be substituted with an equivalent combination of education, training and experience.
- **Preferred qualifications:**
  - MA and knowledge of textiles and textile structures from pre-Hispanic American textiles and/or textiles of the indigenous cultures from Central and South Americas required;
  - PhD preferred in Art History or related field.
  - Two or more years of museum experience or experience working directly with art objects desired.
  - Excellent interpersonal and communication skills (written, oral, and visual); organized and detail oriented; ability to work collaboratively with staff, students, and faculty, as well as diverse museum audiences.
  - On-site time includes days/evenings/weekends as appropriate. Some travel required.

Knowledge, Skills and Abilities

- Knowledge of pre-Hispanic American textiles and textiles of the indigenous cultures from Central and South Americas including the means of textile creation;
- Rich understanding of art history; interest in foreign cultures;
- Excellent research and writing skills for both scholarly and public audiences, and verbal and interpersonal communications;
- A self-starter with the ability to operate with purpose, urgency and accuracy both independently and as part of a team in a dynamic environment with concurrent deadlines and rapidly changing priorities;
- Ability to share projects and form alliances with institutions;
- Comfortable communicating with Board Members, faculty, university executives, and VIPs; professional demeanor and ability to interact with all levels within the organization;
- Exceptional time management, project management and organizational skills;
- Detail oriented with ability to prioritize, manage and multitask; ability to coordinate complex schedules and analyze and solve problems;
- Experience in nonprofit or higher education setting a plus, as well as willingness to work non-traditional business hours on occasion.

Typical Hiring Range: $52,959.04 - $72,845.69
Campus Location: Foggy Bottom, Washington, D.C.
Family: Museum & Performing Arts
Sub-Family: Museum and Arts
Stream: Individual Contributor
Level: Level 2
Hours Per Week: 40
Work Schedule: Monday - Friday: 8:30am - 5:30pm

Figure 19.4 (Continued) Sample JD for curator, Textile Arts of Americas [16].

19.4 Glimpses of Work in Progress on Museum Textile

Figures 19.5–19.11 presented here showcase the conservator working on the various artifacts. This is a good depiction of dedication, involvement, physical agility, fine skills and knowledge required for a successful career in a textile museum.

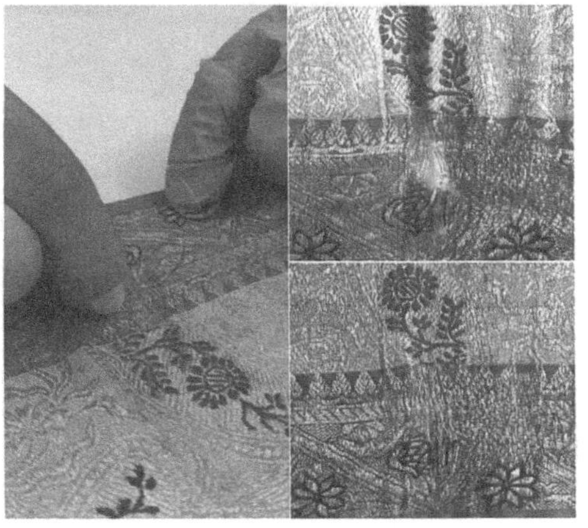

Figure 19.5 Stitch repair of a hole using curve needle by giving a backing of a support material and using laid couching stitches. Textile Conservator: Radhana Raheja. Object courtesy: wedding trousseau veil from the private collection of Seema Khandlwal. Image courtesy: Ms. Radhana Raheja, Textile Conservator.

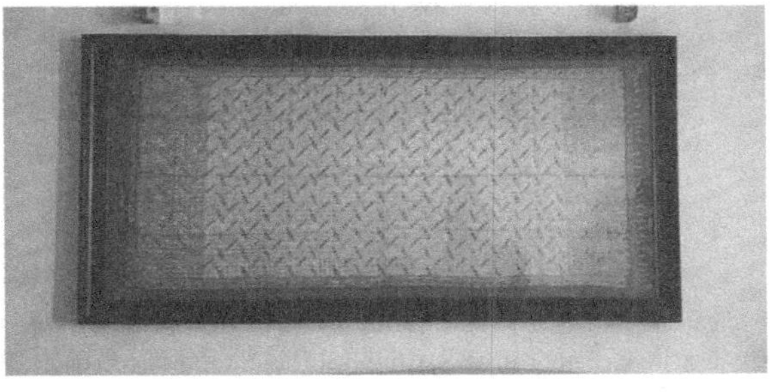

Figure 19.6 Display of the conserved wedding trousseau veil. Object courtesy: from the private collection of Seema Khandelwal. Image courtesy: Ms Radhana Raheja, Textile Conservator.

Figure 19.7 Lining given to the wedding trousseau veil for display. Textile Conservator: Radhana Raheja. Object courtesy: from the private collection of Seema Khandelwal. Image courtesy: Ms Radhana Raheja, Textile Conservator.

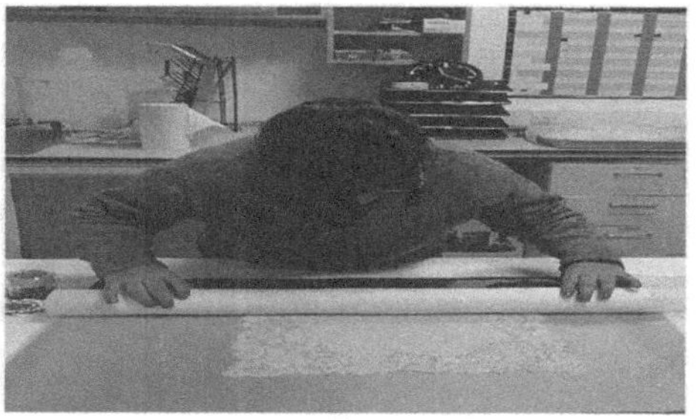

Figure 19.8 Careful rolling for safekeeping of lace handkerchief after conservation. Textile Conservator: Radhana Raheja. Object courtesy: Royal Albert Memorial Museum (RAMM), Exeter, United Kingdom. Image courtesy: Ms Radhana Raheja, Textile Conservator.

Figure 19.9 Flattening the creases of a historic lace handkerchief by providing moisture through the layer of sympatex and overlay of net. Textile Conservator: Radhana Raheja. Object courtesy: Royal Albert Memorial Museum (RAMM), Exeter, United Kingdom. Image courtesy: Ms Radhana Raheja, Textile Conservator.

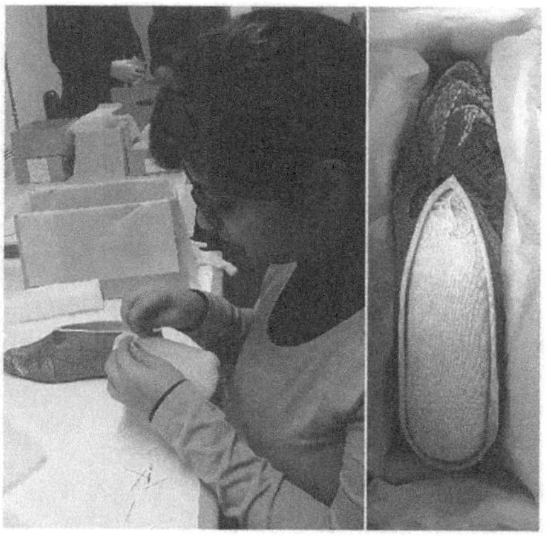

Figure 19.10 Preparation of padding for storage of the shoe collection using cotton stockinette and poly-wadding. Textile Conservator: Radhana Raheja. Object courtesy: Royal Albert Memorial Museum (RAMM), Exeter, United Kingdom. Image courtesy: Ms Radhana Raheja, Textile Conservator.

Figure 19.11 Experimentation with varying concentrations of 'Deca Silk' and preparation of overlay using the varying concentrations on silk crepe line for an object to be conserved. Textile Conservator: Radhana Raheja. Location courtesy: Royal Albert Memorial Museum (RAMM) Laboratory, Exeter, United Kingdom. Image courtesy: Ms Radhana Raheja, Textile Conservator.

19.5 Sourcing for Talent at Textile Museums

The aspirants or prospective employees are referred to as Talent by organizations. There are various ways by which organizations conduct sourcing for talent. The objective is to make the position or vacancy known to the right and deserving candidates. Organizations have specific sourcing methods that are preferable to the type of positions being filled, while some organizations use a combination of these sourcing methods to increase the talent pool sourced. Some popular sourcing techniques used by museums are mentioned below.

19.5.1 Private Job Sites

Strenuous tasks like sourcing talent can become comparatively easier thanks to the multitude of private job sites, the like of Naurki.com, Indeed.com and many more who allow museums to open up a job posting, inviting aspirants to apply for the positions. While this is not the most preferred sourcing method for entry positions in museums, these sites

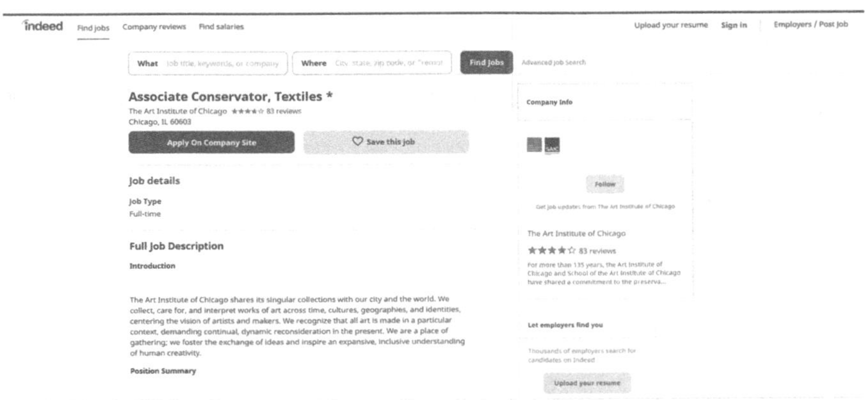

Figure 19.12 Job posting on Indeed.com for associate conservator, textiles, The Art Institute of Chicago [15].

are often used for advertising and inviting applications for senior-level positions in private museums. Most government museums, especially in India do not use this model of sourcing while many international museums rely, significantly, on this method of sourcing talent. Figure 19.12 showcases a job posting by the The Art Institute of Chicago on Indeed.com for the position of an Associate Conservator, Textiles [15].

19.5.2 Museum Websites

Museums across the world use the careers sections of their websites to invite applicants as a sourcing model. Most museums in India, both private and public funded use this mode of talent sourcing frequently and more so in the case of vacant administrative positions in their organizations. Organizations use this method of sourcing often in combination with social media hiring, especially for entry positions in the museums. Figures 19.13 and 19.14 offer a glimpse into this often-used sourcing model thorough application invites on the websites of museums such as Dakshin Chitra and Chhatrapati Shivaji Maharaj Vastu Sangrahalaya.

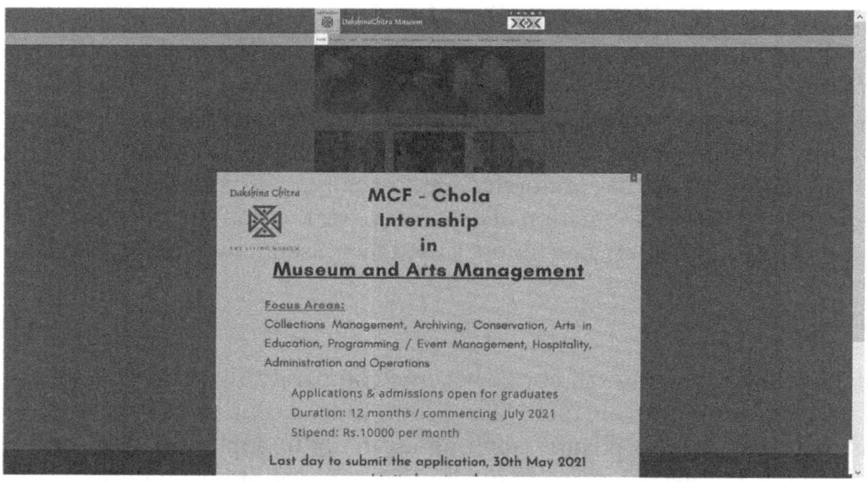

Figure 19.13 Internship posting in Dakshina Chitra website [17].

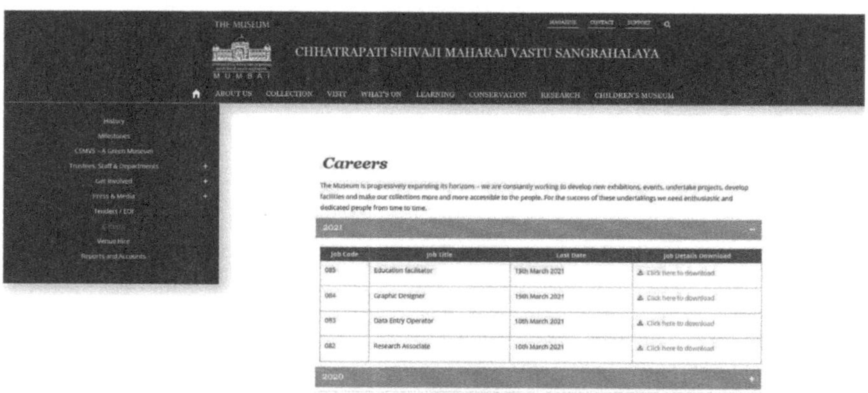

Figure 19.14 Job postings on the website of Chhatrapati Shivaji Maharaj Vastu Sangrahalaya [18].

19.5.3 Consultant Hiring

While consultant hiring or headhunting are methods of sourcing talent for niche and senior positions in most organizations, in the case of museums and career opportunities in this field, consultant hiring is a rarity. While freelancers recruit assistants on a project-project basis, they occasionally prefer to reach out to consultants who help them attract the right talent. Owing to the speciality, uniqueness and lack of awareness about the field in itself, consultant hiring is not a prevalent trend yet in this arena. While senior positions like Board Members, Legal Counsel, Trustees etc, for these museums might still be filled by this mode of talent sourcing, museums prefer other modes to source talents for the more common positions or vacancies in the organizations.

19.5.4 Social Media

This is a mode of talent sourcing used by freelancers and private museums alike. Most freelance conservators, museums and private organizations post job openings on social media websites including LinkedIn, Facebook, and Instagram. Organizations use their social media networks and contacts to advertise their job vacancies through posts in the job section in LinkedIn or posters and videos announcing vacancies in the organization. Most of these posts contain links redirecting applicants to key in their details and upload applications within a set deadline. Social media is here to stay and can influence marketing and publicity [34]. Once organizations are satisfied with the talent pool sourced, the links are disabled and are no longer visible on the page or account associated with the organization. Figures 19.15 and 19.16 allow an understanding of the various kinds of social media posts that organizations use to source talent.

19.5.5 Newspaper Advertisement

Newspapers serve as the most popular choice of talent sourcing method amongst a majority of the government-funded or government-aided museums in India. These advertisements include notices on the vacancy of both entry level, as well as senior positions in the job hierarchy of museums. Candidates are sourced through this method to fill all modes of employment including full-time, part-time, temporary, and contractual engagement. As depicted

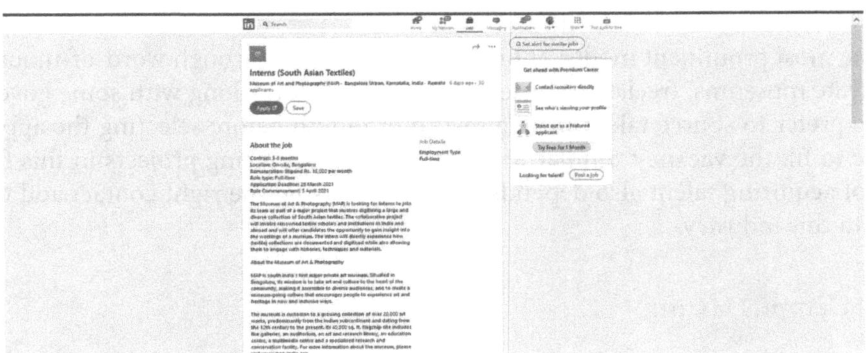

Figure 19.15 Job posting intern at museum of art and photography in LinkedIn [19].

Figure 19.16 Job posting for vacancy in customer experience team at The Durham Museum on their official facebook page [20].

by Figures 19.17 and 19.18, advertisements are aired in both national as well as local or regional newspapers to aid far-reaching coverage. Some museums also opt for advertising job openings through radio channels, especially nationalized radio channels, like the AIR, to improve the coverage of messages in a bid to attract talent from across the country.

19.5.6 Word of Mouth

One of the most prominent modes of hiring in museums is through word-of-mouth. Most small private museums, freelance conservators and curators along with some government museums prefer to source talent through word of mouth before selecting the appropriate candidate to fill the vacancy. Similar to the method of acquiring projects in this field, the method of acquiring talent also depends heavily on having the right contact and the right network in this industry.

19.5.7 Campus Hiring

While campus hiring is a rage in almost every industry, the same does not apply to the Museums in India. While some museums abroad choose to hire candidates first from their

INDIAN MUSEUM, KOLKATA
Ministry of Culture, Government of India

ADVERTISEMENT FOR THE POST OF DIRECTOR
(Advertisement No. 3/2020)

Applications are invited from Indian citizens for the post of Director, Indian Museum, Kolkata in the scale of pay of Rs. 1,44,200/- - Rs. 2,18,200/- plus other allowances as per Central Government rules and as per the fixation and fitment in Level 14 of 7th CPC. Period of probation will be two years. The method of recruitment will be by direct recruitment failing which by deputation/ absorption.

The detailed advertisement and format of application are available in the website **https://indianmuseumkolkata.org/**.

Completed applications in the prescribed format should reach the Director-in-charge, Indian Museum, 27, Jawaharlal Nehru Road, Kolkata -700016 on or before 1st June, 2020.

Director-in-charge
Indian Museum, Kolkata

Figure 19.17 Newspaper advertisement (English) for the post of Director published by Indian Museum, Kolkata [21].

भारतीय संग्रहालय, कोलकाता,
संस्कृति मंत्रालय, भारत सरकार

निदेशक के पद के लिए विज्ञापन
(विज्ञापन संख्या 3/2020)

भारतीय संग्रहालय, कोलकाता में निदेशक पद के लिए भारतीय नागरिकों से आवेदन आमंत्रित किए जाते हैं वेतनमान रुपये 1,44,200/- – रुपये 2,18,200/- साथ में केंद्र सरकार के नियमों के अनुसार अन्य भत्ता अतिरीक्त एवं 7 सीपीसी के स्तर 14 में निर्देशित और निर्धारित के अनुसार। प्रोबेशन की अवधि दो साल होगी। भर्ती की विधि सीधी होगी अथवा प्रतिनियुक्ति / अवशोषण द्वारा होगी ।

विस्तृत विज्ञापन और आवेदन का प्रारूप वेबसाइट **https://indianmuseumkolkata.org/** पर उपलब्ध है।

निर्धारित प्रारूप में पूरा किया हुआ आवेदन 1 जून, 2020 तक या उससे पहले निदेशक-प्रभारी, भारतीय संग्रहालय, 27, जवाहरलाल नेहरू रोड, कोलकाता - 700016 पहुंचना चाहिए।

निदेशक-प्रभारी,
भारतीय संग्रहालय, कोलकाता

Figure 19.18 Newspaper advertisement (Hindi) for the post of Director published by Indian Museum, Kolkata [22].

associate institutions through campus hiring, a majority of museums and freelancers forego this mode of talent sourcing, especially in India owing to the lack of specializations that fit their requirements concerning qualifications expected in candidates. Since the hiring patterns in this field are largely unregulated, it is unlikely to establish a systematic form of sourcing talent, such as the campus hiring process, to fill the erratic nature of vacancies in museums.

The common method of candidate selection from the talent pool that follows all the aforementioned sourcing modes is like the selection process followed by any other industry; candidates need to apply as per the specifications mentioned in the source fill up the application form, email the profile, walk-in for interviews.

19.6 Educational Opportunities

The preparation for the right career begins with correct stream selection, education and skills training. The following educational options can be considered for entry into this field.

19.6.1 School or Pre-University Level

While the role of a curator does not mandate any specific choice of stream, passionate aspirants of the conservatorial role may opt for a science background, especially chemistry in their Senior Secondary or Preuniversity level of education.

19.6.2 University Level (Under Graduation)

The role of a curator does not mandate a specific undergraduate degree but a background in arts and history (ancient/medieval) is preferable since it gives the aspirants a better understanding of the historic and artistic significance of the garment or collection they handle.The role of a conservator mandates a background in Chemistry and Material Sciences, which can also be pursued at the undergraduate level.

19.6.3 Postgraduation

The role of the curator mandates a Masters in Ancient or Medieval History and/or Masters in Museology, which may be pursued in combination with any other undergraduate or postgraduate degree. The conservators are required to possess a Masters in Chemistry or a Bachelors in Chemistry in combination with a Masters in Fabric and Apparel Sciences or Textile Conservation or Museology. Fashion and textile studies programs drive students to learn about the historical and cultural significance of fashion interpretation, as well as the means to approach costume and textile history from a variety of traditional and innovative viewpoints while also gain practical knowledge, such as how to identify fibers and treat various types of objects, as well as how to roll a textile and dress a mannequin [23].

While Art conservators prefer Apprentice training to other means of qualifying for jobs, a graduate training program is the more widely accepted path to acquiring the theory, philosophy, skills, and credentials to be a textile conservator (though not necessarily the experience). An MA or MS degree is typically required for a museum job or advancement within a museum department. Some textile conservators have previously worked in fields such as fashion design, theatre, textile science, archaeology, or historic preservation. In this case, a certificate or advanced internship in textile conservation may be the most effective way to gain additional training—and expand job opportunities [13].

Table 19.2 relays some of the famous institutions that offer courses that are relevant to pursuing a career in museum textile [24–30].

19.7 Sample Organization Structure

Organization Structure is a chart depicting various levels and hierarchies in the organization. This shows various jobs in the organization and reporting system. This shows the growth opportunity an individual can have while working in the organization. Museums, in general, need experts for developmental work, fundraising, financial management, public relations etc. among a multitude of other specialized tasks. While organizational structures can be hierarchical, matrix or task force forms, there does not exist a single overall pattern that suits every museum in the world. These organizations often adopt structures that are suitable to their needs and evolve through their experience in managing their human resource. One can grow to take up administrative, curatorial or other functions in the organization. It will be interesting to note various sample structures here [31–33, 35]. While Figure 19.19 showcases the organization structure of a private museum (INTACH), Figure 19.20 allows a glimpse into the rather rigid and often observed organizational structure in government/ public funded museums through the organizational chart of Assam State Museum.

19.8 Limitations and Challenges in the Field of Textile Museum

While the field is abundant with opportunities for passionate and talented aspirants, the same is not true in general. The most clinching limitation in this line of work is the lack of satisfactory remuneration, especially for those working under private conservators. While men and women alike are tasked with travelling across the country, and sometimes across the globe in connection with their specific conservation projects, the obvious physical and mental discomfort attached to the job also cannot be ignored [36].

The lack of educational specializations in textile conservation in India is a serious challenge for aspiring conservators as pursuing courses abroad may not be an affordable option for many. Lack of definite job openings, especially in the Government museums is another challenge that limits the opportunities available for aspirants.

A general lack of understanding and awareness about the field of work is the biggest challenge in pursuing careers in museum textile conservation [37].

19.9 Conclusion

Textile conservation deals with restoration, preservation, storage, and display of historical textiles present in museums or with private collectors and is a field full of opportunities for those passionate enough to seek them. India has the largest textile collection but does not have enough textile conservators. The Ministry of Culture must be advised that while they do possess a lot of

Table 19.2 Courses and fellowships related to textile conservation.

Institute	Type of course	Course	Fellowship	Relevant role
Lady Irwin College, New Delhi	Graduate training program	MSc in Fabric and Apparel Sciences		Conservator/ Curator
National Museum Institute, New Delhi	Graduate training program	MA in Museology		Curator
Maharaj Sayajirao University, Baroda	Graduate training program	MA in Museology		Curator
National Museum Institute, New Delhi	Graduate training program	MA in Conservation		Conservator
University of Applied Arts, Vienna	Graduate training program	Textile Conservation	Eurasia Pacific Scholarship	Conservator
University of Glasgow	Graduate training program	M. Phil. in Textile Conservation		Conservator
University of Delaware	Graduate training program	MS in Art Conservation		
Durham University	Graduate training program	MA in Conservation of Archaeological and Museum Objects		
UCL Institute of Archaeology	Graduate training program	MA in Principles of Conservation		

(Continued)

Table 19.2 Courses and fellowships related to textile conservation. (*Continued*)

Institute	Type of course	Course	Fellowship	Relevant role
George Washington Institute for Textiles, Washington DC	Related courses	MA in Museum Studies		Conservator
Exeter Museum, London	Related courses	Textile Conservation Enrichment program		Conservator
Metropolitan Museum of Art, New York	Related courses		Andrew Mellon Foundation Fellowship	Conservator
-		-	Agnes Bedell Fellowship	Conservator
Opificio delle Pietrre Dure Museum, Florence	Related courses		Charles Wallace Trust Fellowship	Conservator
Beloit College, Wisconsin	Nondegree Training	Program in museum studies inclusive of textile conservation		Curator & Conservator
Fashion Institute of The State University of New York & The Museum at FIT	Graduate training program	MA in Fashion and Textile Studies: History, Theory, Museum Practice		Conservator & curator
University of Rhode Island	Graduate training program	MS in Textiles, Fashion Merchandising and Design		

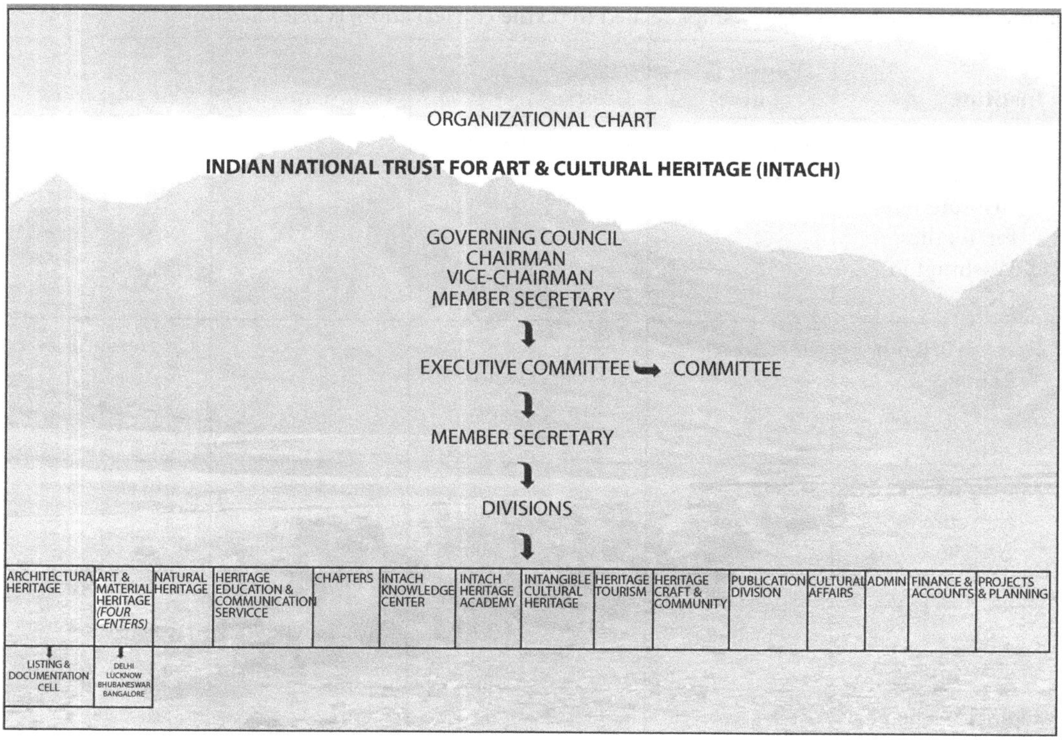

Figure 19.19 Organizational structure of INTACH (NGO) [32].

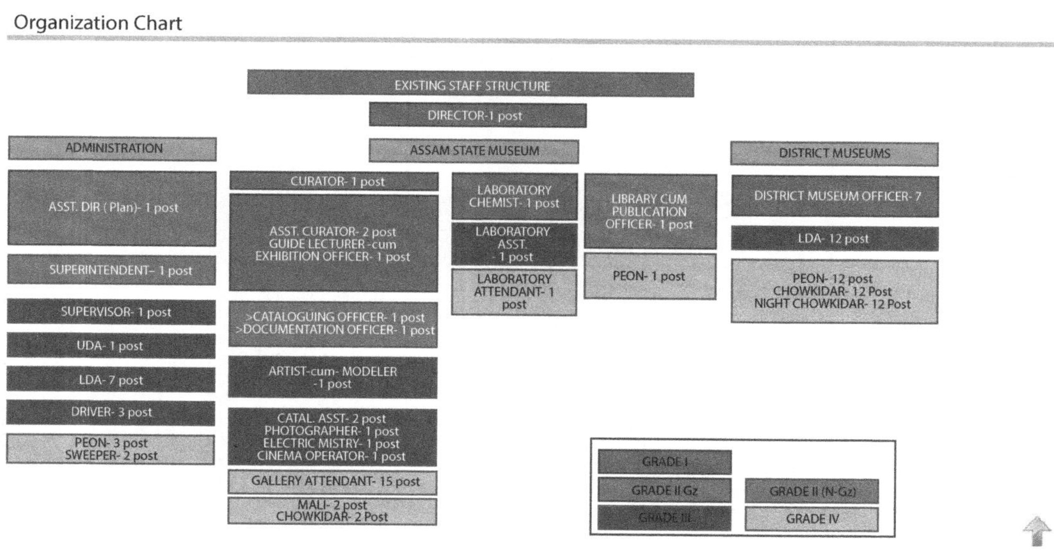

Figure 19.20 Organizational chart of Assam State Museum (Public) [35].

cultural treasures and is a repository of such rich historical artifacts, it is important to preserve them and it is important to attract people to make a career/living out of it. They need to ramp up hiring for people to look after these tangible and intangible cultural assets of the nation, open doors for people to make a flourishing career. The concept of campus placements, which is currently nonexistent in this field can also be considered and evaluated on its viability in this case.

Acknowledgment

The authors are deeply indebted to those who gave freely their time for personal interviews and agreed to be immortalized for posterity in our many recordings. By sharing their insights and advice, they have demonstrated their zeal and passion for their field of choice and the drive in them to create awareness of the abundant opportunities for flourishing careers in the field of museum textile. We would like to thank Dr Simmi Bhagat (Associate Professor in Fabric and Apparel Science, University of Delhi), Ms Radhana Raheja (teacher and PhD. Scholar, University of Delhi) and others for their patience and effort in highlighting their journey and consequently the journey of many others pursuing a career in this field. Thanks, Shweta Singh, PG student at SIMS for assisting in data collection.

References

1. Urick, J. and Keefe, A.R., Museum careers in textile conservation. In: *The Business of Textile and Fashion History. The University of Rhode Island: Seminar in Textiles and Clothing*, 2020.
2. *The Metropolitan Museum of Art, Textile Conservation* [Internet]. Metmuseum.org. *The Metropolitan Museum of Art,* 2022. Available from: https://www.metmuseum.org/about-the-met/conservation-and-scientific-research/textile-conservation.
3. Museums & Galleries Commission, *Standards in the museum care of costume and textile collections, 1998*, Museums, Libraries and Archives Council, London, 1998.
4. Wightman-Stone, D., Fashion and textile museum to highlight designers guild. *FashionUnited*, 2019. [Internet]. Available from: https://fashionunited.uk/news/culture/fashion-and-textile-museum-to-highlight-designers-guild/2019122746785.
5. Kamm, C.A., *The Museum World: Career Skills and Training for Jobs in Textile and Clothing Historic Collections*, Iowa State University, 2012, https://doi.org/10.31274/etd-180810-2280.
6. Textile Museum of Canada, *About the Textile Museum of Canada* [Internet]. Textilemuseum.ca. Textile Museum of Canada, 2022. Available from: https://textilemuseum.ca/about-the-museum/.
7. Role of textile museums and fashion museums of the world. *Fibre2Fashion*, 2014. [Internet]. Fibre2fashion.com. Available from: https://www.fibre2fashion.com/industry-article/7363/textile-history-still-alive-in-museums.
8. Labrum, B., Expanding fashion exhibition history and theory: Fashion at new Zealand's national museum since 1950. *Int. J. Fash. Stud.*, 1, 1, 97–117, 2014. https://doi.org/10.1386/infs.1.1.97_1.
9. *Sanskriti Foundation, Museum of Indian textiles* [Internet]. Sanskritifoundation.org. *Sanskriti Foundation*, 2019. Available from: http://www.sanskritifoundation.org/museum-indian-textiles.htm.
10. Dhaka, A., Khar, S.S., Dimri, N., Craft resurgence in India post independence. *Global Fashion Conference*, 2020.

11. Akosah-Twumasi, P., Emeto, T.I., Lindsay, D., Tsey, K., Malau-Aduli, B.S., A systematic review of factors that influence youths career choices—The role of culture. *Front. Educ. Front.*, 3, 58, 2018. https://doi.org/10.3389/feduc.2018.00058.
12. NCERT, Human Ecology and Family Sciences, *PART II Textbook for Class XII*. NCERT, 2020.
13. Perry, R.B., Becoming a conservator: New York City, 1960–1965, in: *The Life and Art of Felrath Hines*, pp. 52–61, Indiana University Press, 2018.
14. National Museum of Scotland. *Textile Conservator Recruitment Pack* [Internet] Nms.ac.uk. National Museum of Scotland, 2017. Available from: https://www.nms.ac.uk/media/1155030/textileconservatorrecruitmentpackaug2017.pdf.
15. The Art Institute of Chicago. *Associate Conservator Job Posting* [Internet]. Indeed.com. The Art Institute of Chicago, 2021. Available from: https://www.indeed.com/viewjob?jk=db7afbd62050f7da&tk=1f106s0ovp12n800&from=serp&vjs=3.
16. The George Washington University Museum and The Textile Museum. *Curator, Textile arts of Americas Job Posting* [Internet]. Glassdoor.com, 2021 [cited 2021 Jul 9]. Available from: https://www.glassdoor.com/Job/textile-conservator-jobs-SRCH_KO0,19.htm
17. DakshinaChitra, *Dakshinachitra Internship Posting* [Internet]. Dakshinachitra.net. DakshinaChitra Museum, 2021. Available from: https://www.dakshinachitra.net/.
18. CSMVS, *Careers*, Chhatrapati Shivaji Maharaj Vastu Sangrahalaya, 2019, [Internet] Available from: https://www.csmvs.in/careers.
19. Museum of Art and Photography. *LinkedIn-Interns (South Asian Textiles)* [Internet]. Linkedin.com. Museum of Art and Photography, 2021. Available from: https://www.linkedin.com/jobs/view/2464507050/?refId=QfgmtHqOGrRjAV6y53pPOg%3D%3D&trackingId=NbHV6JzSSKEnwZ7pDw40dg%3D%3D.
20. Durham Museum. Facebook - *The Durham Museum* [Internet]. Facebook.com. Durhan Museum, 2021. Available from: https://www.facebook.com/TheDurhamMuseum.
21. Indian Museum Kolkata. *Director Vacancy (English)* [Internet]. Indianmuseumkolkata.org. Indian Museum Kolkata, 2020. Available from: https://indianmuseumkolkata.org/resource/tenders/website%20Advertisement%201.6.20.jpg.
22. Indian Museum, Kolkata. *Director Vacancy (Hindi)* [Internet]. Indianmuseumkolkata.org. Indian Museum Kolkata, 2020. Available from: https://indianmuseumkolkata.org/resource/tenders/IM%20Appointment%20Advertisement%20size%208%20X%208%20Hindi.jpg.
23. https://www.fitnyc.edu/fashion-textile-history/index.php
24. Shri. Chanda, S., Prof. Bhaskar, P., Dr. Ray, D., *Museum Staff-Role and Responsibilities*, E-Pathshaala, MHRD India, 2015, [Internet] Available from: http://epgp.inflibnet.ac.in/epgpdata/uploads/epgp_content/S000829IC/P001546/M024022/ET/1507703696P15-M21-MuseumStaffRoleandResponsibilities-ET.pdf.
25. Beloit College, Beloit College-Textile conservation student testimonial, in: *Testimonials*, 2020, [Internet] Available from: https://www.beloit.edu/summer-programs/center-for-collection-care/testimonials/.
26. Richmond, A., *Careers and Training in Conservation in the UK*. Conservation Department, V&A, 2008.
27. Lithgow, K. and Timbrell, H., How better volunteering can improve conservation: Why we need to stop wondering whether volunteering in conservation is a good thing and just get better at doing it well. *J. Inst. Conserv.*, 37, 1, 3–14, 2014. https://doi.org/10.1080/19455224.2013.873730.
28. Lennard, F., *Textile conservation education in the United Kingdom*, University of Glasgow, United Kingdom, 2010.
29. Good, I., Archaeological textiles: A review of current research. *Ann. Rev. Anthropol.*, 30, 1, 209–226, 2001. https://doi.org/10.1146/annurev.anthro.30.1.209.

30. Foskett, S. and Thompson, K., *An Evaluation of the Teaching and Learning of Reflective Practice at the Centre for Textile Conservation*, University of Glasgow, 2017.
31. Singhal, D. and Bhagat, S., Cleaning of textile artifacts in museums. *Int. J. Appl. Home Sci.*, 2, 7-8, 237–242, 2015.
32. INTACH, *Indian National Trust for Art and Heritage Annual Report 2018–19*, 2019, [Internet]. Intach.org. Available from: http://www.intach.org/pdf/Annual-Report-2018-19.pdf.
33. The George Washington University Museum. *Work with Us* [Internet]. Gwu.edu. [cited 2021 Jun 21]. The George Washington University Museum and The Textile Museum, 2021. Available from: https://museum.gwu.edu/work-with-us.
34. Pandey, S., Pandit, P., Pandey, R., Pandey, S., Marketing Strategies for Upcycling and Recycling of Textile and Fashion, in: *Recycling from Waste in Fashion and Textiles: ASustainable and Circular Economic Approach*, pp. 253–275, 2020, https://doi.org/10.1002/9781119620532.ch12.
35. Assam State Museum, *Organization Charts*, Assam State Museum, 2021, [Internet] Available from: https://museums.assam.gov.in/about-us/detail/organization-chart-4.
36. Labhard, L.A., Textile conservation internships cal poly students at Hearst Castle. *History News, JSTOR*, 45, 6, 15–17, 1990.
37. Asokan, A., *Textile Conservation Professional* [Internet]. Theinterviewportal.com. [cited 2021 Jun 21] Offbeat, unusual, unconventional & interesting career interviews, 2020. Available from: https://theinterviewportal.com/2020/01/05/textile-conservation-professional/
38. National Museum of Scotland. *Assistant Preventive Conservator* [Internet]. Nms.ac.uk. National Museum of Scotland, 2017. Available from: https://www.nms.ac.uk/media/739292/assistant-preventative-conservator-june2015.pd

Index

Abrasion, 221
Academy of Sciences, 3, 4
Accession number, 149
Accretions, 221
Acid free cardboard, 144
Adhesive, 221
Adinkra, 239, 241, 253–255, 259
Adire, 29, 239, 241, 248–250, 259
Adivasi Academy, 7
African museums, 259–261
African textiles, 239–241, 253, 259, 261
Akwete cloth, 239, 240, 245–246
Albert Hall Museum, 10
Allucciolato, 71
Amalgamated, 109
American Association of Museums (AAM), 10
American Textile Museum, 6
Ancient armours, 209, 210
Ancient civilization, 234
Ancient costume, 19
Ancient craft, 19
Ancient textile, 19
Andrew Mellon Foundation, 12
Animal scale armours, 216
Ankara/African wax prints, 239, 241, 253, 256–257, 259
Annette, 117
Antaria, 27
Anthropologist, 320
Anti-allergic properties, 316
Antonio, 108
Applique, 221
AR/VR Technology, 16
Archaeological textiles, 153
Architectural design, 145
Ardabil, 66
Areni, 43
Armou, 209
Armours, 209, 210
  designs, 209, 215
  materials, 215
  parts, 209, 210
  uses, 209, 210
Arrest decay, 148
Art collection, 293
Artifacts, 2, 111, 114, 161, 164, 166, 221
  as primary sources, 342
  deterioration of, 341, 343
  handling of, 342, 344, 349
Artistic, 2
Ashmolean, 4
Aso-oke, 240, 242, 243, 259
Attachés, 111
Automatic cut-off valves, 156
Azarakh, 22
Azemmour, 29

Babbage, 70
Baked enamel metal, 144
Baluchari, 54
Banarasi, 54
Bandhani, 293
Barbuta, 210
Bark cloth, 21, 239, 241, 257–259
Bascinet, 210
Bashoufu, 20
Beaux, 118
Berber cloth, 239–241, 246–247
Bias-cut, 151
Biocidal, 156
Biodeterioration, 163, 296
Biological threat, 156
Bizarre silks, 69

Blanket stitch, 178
Body heat parameters, 153
Bogolan, 28
Bonnie cashin, 114
Boots, 210
Borowitz, 119
Braid loop, 228
Breathable waterproof fabric, 301
Brignole, 110
British galleries, 292
British Museum, 4
Brocade, 21, 45, 49, 52, 54, 56, 62–64, 66, 69–71, 81, 82
Brocatelle, 70–71

Calico Museum, 6, 8
Cannellé, 71
Carbon dioxide or nitrogen chambers, 157
Care, 161–164, 166, 171, 178
Career opportunities,
  assistant conservator(s), 360, 362
  associate conservator, 363–364, 369, 370
  conservator(s), 356, 357, 359–364, 366–372, 374–377
  consultant(s), 359, 362, 371
  curator(s), 357–363, 365, 366, 372, 374–377
  deputy conservator(s), 360, 361
  deputy curator, 359, 360
  freelance conservator(s), 361, 362, 371, 372
  freelance consultant(s), 362
  senior assistant(s), 359, 360
  technical assistant(s) (TAs), 359, 360
Cavalry, 210
Cellulose degradation, 221
Cellulose nitrate, 155, 158
Central Museum of Textiles, 5
Ceremonial apron, 209, 217
Cervelliere, 210
Chain mail shirt, 213
Chancay, 48
Chemical degradation, 296
Chimu, 45–46
Chiné a la Branche, 70
Chintz, 221, 293
Chronicles, 113
Clasps, 111
Cleaning, 171, 300
Climate control, 147, 185
Climatic buffering, 146
Coif, 210
Cold poultice, 302
Collection management,
  assessment, 128
  categories, 127
  conservation, 130
  object description, 128
  policy, 126–128
  technological survey, 129
Collection storage, 143–144, 149
Collections manager, 181, 184, 187, 188
Collective value, 7
Color catalogs, 321, 323
Color mordants, 315, 316
  citric acid, 316
  copper sulphate, 316
  iron sulphate, 316
  soda, 316
Colourfast, 304
Complementary warp, 45
Connective value, 6
Connoisseurship, 3
Conservation, 162, 219, 355–363, 368, 374–377
Conservation safe frames, 153
Conservation scientist, 181, 185, 186
Conservation studio, 224
Conservation techniques used in the museum, 287
  backing material used in conservation of museum textiles, 288
  methods of reinforcing the deteriorated textile antiquities, 288
  pre-treatments given to textile antiquities and display boards, 287
Conservation treatment, 229
Conservator, 294
Controlled access, 145
Coptic, 58, 60, 74
Coptic tapestry, 178
Costumes, 219, 312
Crisis management, 154
Crossed warp, 45, 48
Cubitiere, 211
Cuirass, 211
Culture, 162
Cumbi, 48
Curation, 356, 357
Curators, 162, 294, 377
Customized storage designs, 152

Damask, 49, 51, 53, 55, 56, 71
Dark storage, 148
Decomposition, 178
Degeneration, 146, 152, 158
Degradation, 144, 148, 177
Deterioration, 146, 152, 156–159
Dhaka muslin, 57
Digitized documentation,
 advantages, 137
Disaster handling, 144, 154
Disaster management, 146, 154–155, 183, 184, 191
Disaster potential management, 154
Disaster response, 154, 156
Discontinuous-warp-and-weft, 42, 45, 47, 48, 53, 57, 60, 66
Dismantled, 103
Display methods, 328–334
 display design, 330–332
 display lighting, 334–335
 draped textiles, 331
 framing of textiles, 331
 free-hanging textiles, 331
 guidelines, 329
 mannequin display, 332
 open display or display cases, 329–330
 rolled display, 332
 slanting method, 332
 tactile learning, 333
 visitor interaction, 333
 workplace method, 333
Documentalist, 181, 184, 187, 188
Documentation,
 aims, 124–125
 features, 125–126
 function, 124–126
 importance, 123
 process, 124–125
 styles of, 136–137
 types, 124, 130–135
Doeuillet, 111
Domínguez, 115
Don Bosco Museum, 7
Double cloth, double weave, 45, 46, 54–56
Double interlocking, 47
Douglas, 107
Dovetailing, 47
Draught-proofing, 147, 158
Drawloom, 49, 57, 66, 69, 70, 74
Droguets, 70
Duan, 53
Dust-proof, 149, 150, 159
Dye analysis, 300
Dye process,
 mordant, 205
 vat, 204
Dyeing, 240–241, 248–251, 254, 256, 259, 261
Dyes, 296
 brazilwood, 205
 flavonoid, 205
 indigo, 203–204
 logwood, 205
 madder, 204
 saffron, 204
 tannin, 205
 tyrian purple, 203–204

Ecological methods, 315
Economic value, 11
Educational qualification,
 ancient history, 359
 chemistry, 361, 374
 fabric and apparel sciences, 374, 376
 material sciences, 361, 374
 medieval history, 359, 360, 374
 museology, 357, 362, 374, 376
 textile conservation, 359, 361, 362, 374-377
Educative value, 10
E-gasuri, 65
Egyptian Textile Museum, 5
Electromagnetic radiation, 148
Electronic key, 154
Electronic security systems, 153
Embossed, 209
Embroidery, 241, 247–248, 254, 259, 261, 293
Emergency doors, 153
Emergency security policy, 154
Engraved, 209
Environment, 167
Environment control, 183, 184
Environmental conditions, 143–144, 146, 147, 159
Environmental control, 143–144, 148
Environmental factors, 341
Environmental monitoring, 147
Environmental requirements, 145, 152
Environmental stability, 147
Environmentally stable conditions, 145

Epaulette, 228
Equilibrium, 146, 148
Ethnobotanist, 320
Ethnological, 115–116
Exhibition protocols, 335–336
Exhibitions, 1, 2
Exhibits designer, 181, 188, 189

Fabric of India, 298
Fellow collectors, 2
Femme fatale, 114
Finishes, 296
Fire sensors, 155
Firefighting equipment, 155
Fire-resistant, 155
First Lady Michelle Obama, 14
Flood detection alarms, 156
Flood management, 156
Fly shuttle, 4
Fort rock sandals, 42, 72
Fragrance, 117
Frequently fluctuation environmental conditions, 146
Frida Khalo Museum, 13
Frock coat, 231
Fugu/smock weaves, 239–242, 259
Fungal growth, 221

Gaseous fire suppression system, 145
Gauntlet, 210, 211
Gauze, 45, 48, 49, 59, 62, 73
General information about the museums, 270–276
  acquisition of textile antiquities, 272–273
  ageing of textile articles placed in different museums, 274
  air circulation facilities in museums, 274–275
  conservation laboratories, 271–272
  details of temperature and relative humidity ranges in a variety of museums, 276
  methods used for identification of fibers, 276
  number of people who work at the museums that have been chosen, 270–271
  protective measures used to protect the windows/ventilators from sunlight and dust, 275
George Santanya, 7
Gold leaf, 301
Gothic, 108
Government museum and art gallery, 7
Gravitational pull, 149, 151
Greave, 210, 211, 214
Guggenheim Museum Bilbao, 13
Gum Arabic, 301

Hallstatt, 42, 69
Handmade textiles, 298
Hargreaves, 4
Hauberk, 213
Health/safety, 183, 184
Health promoting properties of natural dyes, 316
Helm, 210
Helmet, 210
Heraldic design, 210
Heraldry, 210
Heritage, 162
Heritage management, 181, 188
Heritage textile(s), 19, 358, 361, 362
Hieroglyphics symbol, 215
Himachal State Museum,
  traditional costume, 9
  traditional jewelry, 8, 9
Historians, 162
Historic, 2
Historical costumes, 162
Huaca Prieta, 42, 44
Humidification, 301
Humidifiers, 148, 156
Humidistats, 148
Humidity, 143–147, 157, 159
Humidity levels, 144, 147, 157
Hygroscopic, 159

ICCROM, 143, 182
ICOM-CC, 184–185
Ikat, 45, 46, 55, 57, 62, 64–66, 70
Imperativeness, 116
Importance of conservation, 88
Inca, 45, 48
Indian museum, 293
Indian textile, 291
Indian textile collection, 291
Indigenous, 239–241, 246–248, 253, 257, 259, 261
Industrialization, 4
Inert, 149–151, 159
Infrared/IR, 148
Inherent vice, 343

Inlay design, 209
Insect infestation, 229
Intangible cultural heritage, 19
Intangible cultural properties, 20
Integrated pest management (IPM), 156
International Council of Museums (ICM), 2
Interwoven Globe, 5

Jacquard, 37, 57, 70, 71, 73, 75
Jaime, 115
Jama, 225
Jamdani, 5, 54
Japanese Textile Museum,
  yukari, 6
Jardinière, 71
Jin Duan zi, 53
Jin Dynasty, 52
Jin silks, jin warp faced compound weave, 49–51, 65
Job opportunities,
  cross-sector, 361
  exhibition(s), 355, 356, 358, 360–362, 378
  freelancers, 362, 363, 371, 374
  opportunities offshores, 362
  private sector, 361
  public sector, 359, 363
Jousting, 209, 210
Jousting armours, 210
Jufti knot, 28

Kabati, 178
Kalakshetra, 236
Kalamkari, 22, 221, 297
Kancheepuram, 54
Kanga, 239, 241, 253, 255
Karamushi, 21
Kashmir shawl, 57, 74
Katazome, 25
Kayabandh, 27
Keitō Kinran, 64
Kente, 59, 61, 239–241, 243–244, 259
Kente cloth, 28
Kesa, priests robe, 62–64
Kesi, 53
Kete, 239–241, 244–245, 259
Kitenge, 239, 241, 253, 256–257, 259
Knight, 210
Kontush sash, 315, 324
Kuba raffia cloth/Kuba raffia/Kuba cloth, 239–241, 247, 248, 259

Labelling, 169, 170
Labelling tool, 165
Lacquered leather, 152
Lampas, 49, 50, 52–56, 66, 69–72, 74
Lampasette, 70
Laoguanshan loom, 51
Laotian art, 6
Latten, 215
Leak detectors, 156
Leno, 62
Linen, 177
Ling, 51
Living and learning design center, 10
Localized storage, 145
Loom,
  backstrap, 198–202
  continuous warp, 202
  horizontal, 195
  vertical, 194–196
  warp weighted, 194–197, 202
Loose knits, 151
Louver, 109

Mail, 210, 213
Mail armour, 210, 211
Mail armour shirt, 213
Mail hood, 211
Mail mitten gauntlet, 211
Maintenance, 162–164, 166–168
Mannequin mount, 232
Mannequins, 173
Manual documentation,
  advantages, 137
  risk, 137
Manual monitoring, 154
Material culture,
  frameworks, 344
Materials and methods, 269–270
  data collection, 269
  selection of locale, 269
  statistical analysis of the data, 270
    weighted mean score, 270
  tool preparation and data collection, 269
Mayan textile, 30
Mechanical ventilation equipment, 158

Medicinal plants, 316
Medieval galleries, 292
Medieval jousting, 210
Melee combat, 209
Merchandise, 118
Metal breastplate, 215
Metallic costumes, 209, 217
Metallic textiles, 209, 210
Methods of conservation,
  curative/interventive conservation, 95–96
    solvent or dry cleaning, 96
    stabilization, 96
    surface cleaning, 95
    vacuuming, 95
    wet cleaning, 95–96
  preventive conservation, 91–95
    climate, 92
    disaster, 94–95
    dust, soil, and other contaminants, 94
    insects, 93
    light, 92–93
    microorganisms, 93–94
Metropolitan Museum of Art, 5
Microbiological infections, 157
Military frock coat, 227
Minimizes, 152, 155, 158
Minimizes the risk of a fire, 155
Miyako-joufu collector, 20
Modachrome, 116
Monitoring procedures, 146
Morrell, 106
Mosi, 20
Mount making tool, 165, 172
Mud cloth, 239, 241, 251–253, 259
Mughal, 54–57
Muses, 2
Museum,
  exhibition, 161–163, 172–174, 177
  gallery, 161, 166, 172
  handling, 161–163, 167, 175, 177
  objects, 161–163, 166–172, 175
  textiles, 161, 162, 164
Museum environment, 294
Museum galleries, 312
Museum of textile industry, 5
Museum storage, 144–148, 150, 152–154, 157, 159, 337
  visitor access, 337
Muslin, 302

Naphthalene balls, 157
Napoleonic, 111
Naqshband, 66
Nasij, 52–53
National Handloom & Handicraft Museum, 7, 13, 14
National Museum, 234
National Textile Museum, 6
Natural fibres, 315
Natural plant mordants, 316
  galls, 316
  rhubarb, 316
  sumac, 316
  tree bark (oak, willow, alder), 316
Navajo, 31
Nawab Mir Yusuf Ali Khan, 3
Netherland Museum Association, 10
New approaches in conservation,
  digitization, 98–99
Nishiki, 62–63
Non-tangible, 2
NRLC, 182
Nubia, 58–59

Object retrieval, 145
Object treatment, 300
Oddy tests, 158
Olivettes, 228
Oricalco fabric,
  Chicago Museum of Science, 11
  Grado Zero Escape, 11
Ottoman, 39, 68–69, 73

Packing, 168
Paithani, 54
Paracas, 45–47
Parameters of the building works, 146
Parasols, 111
Pashmina, 54–55, 57, 70
Patka, 56, 232, 233
Patola, 21, 55, 57
Patterns,
  brocade, 194, 199–202
  ikat, 199–200
  warp-patterned, 199
  weft-patterned, 200
Pauldron, 210, 211, 213
Pazyryk, 44, 65
Perimeter alarm, 153

Pest infestation, 152, 154, 157
Pest management, 185
Pesticides, 156
Pests, 167
Philippe de Lasalle, 70
Photography and memorabilia, 336–337
Pile, 41, 42, 44, 45, 49, 54, 55, 58–61, 66–69, 71, 74
Plain weave, 48, 57, 60, 61, 304
Plants yielding blue color, 320
Plants yielding red color, 319, 320
Plants yielding yellow color, 317, 318
Pohon Budi Gallery of National Textile Museum, 8
Pointille, 215
Pollaxe, 210
Pollution, 144, 146, 157, 158
Polychrome glass, 209
Pra-Kashi silk, 220, 232
Preventative conservation, 343
Prince of Wales Museum, 5
Principles of conservation,
  conservation collection condition survey, 90
  determining the need and laying the foundation for conservation, 89
  object treatment, 90
  obtaining a conservation assessment, 89
  risk assessment and management, 90
  selecting an assessor or conservator, 89
  the assessment as a planning tool, 90
  the conservation assessment and process, 89
Printing/block printing, 241, 253–254, 256–257, 259
Professional hierarchy, 182
Proto-lampas, 50, 53, 74
Provincial, 115
Pseudomorphs, 41

Quarantine and sanitation, 156

Ra, 62
Rank insignia, 228
Reconstruction of the ancient color palette, 315
Reconstruction of the old dyeing techniques, 324
Recuay, 45–46
Reemay, 308
Refurbished, 118
Registry of saree, 235
Rehearses, 115
Relative humidity, 145, 147
Renaissance galleries, 292
RE-ORG, 143–144
Resist-dyeing, 241, 256
Restoration, 177, 357–359, 361, 362, 375
Ribbon, 114
Richard Arkwright, 4
Rignai and risa, 21
Risk assessment, 154
Ro, 62
Romanones, 115
Royal Society, 4
Rudi Gernreich, 117

Sabarmati Ashram, 5
Sabatons, 210, 211, 213, 214
Safavid, 66–67
Safety systems, 144, 153
Sakiori, 24
Salar Jung Museum, 3
Sallet, 210
Salt migration, 221
Samite, 54, 55, 66–68, 74
Sample job descriptions (JDs), 364–366
San Jose Museum of Quits & Textiles, 10
Sasanian, 66–68, 74
Satin, 37, 39, 49, 53, 54, 62, 64, 67, 70, 71, 74
Scale armours, 210
Scientific, 2
Scope of conservation,
  education and knowledge dissemination, 97–98
  lecture, seminar, workshops, and research, 98
  photography and publication, 98
Secretariat, 103
SEM microscope, 16
Semple, 70, 72
Sha, 62
Shell armours, 210
Shield, 210
Shield boss, 211
Showcase, 172
Shroud, 177
Shuka cloth, 239, 241, 247
Shweshwe, 239, 241, 253, 256
Silica gel, 147
Silk Museum, Bsous, 10

Skills,
agility, 359, 361, 366
conservation techniques, 361
creativity, 357, 360
decision-making, 360, 361
documentation, 362, 378
financial management, 375
fund-raising, 360, 375
leadership, 360
management, 360
marketing, 360, 371
sewing, 361
Sogdian, 66–68
Soiling, 300
Solvent cleaning, 301, 305
Somerset, 105
Sources of textile museum collections,
government museums, 359–361, 370, 372, 375
private museum(s), 358, 361–363, 370–372, 375
private owners, 356
public-private partnerships, 356
trust(s), 356, 361, 377, 378
Sourcing of talent,
campus hiring, 372, 374,
consultant hiring, 371
museum websites, 370
newspaper advertisement, 371, 373
private job sites, 369
social media, 370, 371
word-of-mouth, 362, 372
Space allocation, 145, 146
Spangenhelm, 210
Spectrophotometric analysis, 324, 325
Spinning Jenny, 4
Storage, 161–163
material, 164
units, 163
Storage area, 144–146, 149, 153–157
Storage design, 143–145, 148, 150–152
Storage equipment used in selected museums, 280–281
Storage facilities, 143–145
Storage location, 144, 146
Storage units, 144–145
Storage, display, and handling of museum textiles, 96–97
Sulphur, 152, 158
Supplementary warp, 45, 48, 61, 74
Supplementary weft, 46, 53, 54, 59–61, 68, 70, 71, 74
Surface cleaning, 301
Sutra, 235
Symbolism, 239–240, 245, 261

Tabby weave, 40, 41, 43, 44, 46–52, 54–56, 58–62, 65, 66, 69, 72, 74
Taffeta, 66, 70
Tangible, 2
Tapestry, 178
Tapestry weave, 44, 45, 47–49, 53–55, 57–63, 65, 74
Taqueté, 50, 58, 65–68, 74
Tarkhan dress, 164
Tasses, 211
Taxonomy of significant learning experiences, 342, 345–346
application, 347
caring, 350–351
foundational knowledge, 346–347
human dimensions, 349–350
integration, 348–349
learning how to learn, 351–352
Tehran Carpet Museum, 10
Tellem, 61
Temperature, 143–148, 152, 159
Temperature levels, 147
Temperature variations, 146
Textiles, 19, 219, 220, 312
design, 20
Textile collection, 291
Textile conservation, 219, 234
Textile conservator, 219
Textile exhibitions, 292, 328–335
case studies, 338–339
ideal exhibition, 338
Textile heritage, 220, 221
Textile museum, 1
Textile panels, 232
Textile storage, 149
The collection of textile artifacts collections in various museums of Uttarakhand, 281
details of stored textile materials, 281
Govind Ballabh Pant Museum, Almora, 281
Gurney House Museum, Nainital, 284
Jim Corbett Museum, 283–284
Kumaon Regiment Museum, Ranikhet, 282–283

Lok Sangrah, Folk Culture Museum, Bhimtal, 283
Tribal Museum, Munsyari, 282
The Fashion & Textile Museum, 6
The Textile Museum, 6
Thermal insulation, 147
Thermo-hygrograph, 147
Thermometer, 147
Thomas Stearns Eliot, 1
Toggles, 228
Transitory, 116
Transportation, 178
Triple cloth, 45, 46
Tsutsugaki, 25
Turpan Museum, Xinjiang, 41, 44
Tutankhamen, 209, 217
Twill, 39, 41–45, 48, 52–54, 62, 64, 66, 67, 69–72, 74
Twill Damask, 49, 51, 55, 56
Twill tapestry weave, 54, 55, 57
Twined, twining, 37, 39–42, 45, 48, 72, 73
Types of display techniques used for textile antiquities, 276–279
labeling methods carried out for the displayed artifacts, 279–280
Types of textile articles conserved, 90–91

Ukara cloth, 239, 250, 259
Ultrasonic humidification, 301
Ultraviolet rays, 177
Ultraviolet/UV, 148
UNESCO, 143
University collections,
access to, 343
as teaching resources, 342, 344–347, 350–351
conservation of, 343–344, 347
Iowa State University Textiles and Clothing Museum, 345, 346, 348, 350–351
University of Basil, 4
Unpacking, 168
Untreated polyester wadding, 151
Uttariya, 27
UV protection, 324, 325

V&A, 291–295, 297–299, 312
VAACHA Museum, 7
Vacuum cleaners, 157
Various methods of prevention used in various selected museums, 284–285
covering materials used for various artifacts displayed in selected museums, 285–286
special kind of lighting system in the museum to protect the textiles/garments from fading/ageing, 286–287
touching on museum antiquities, 287
Velcro, 307
Velcro fasteners, 168
Velvet, 49, 54, 55, 66, 67, 69–71, 74
Ventail, 210, 213
Vermeer, 107
Vernon, 103
Victoria and Albert Museum, 4, 8, 220, 291

War, 209
Wari, 45, 47
Warp and weft substitution, 45, 46
Warp faced, 43, 48–52, 56, 61, 63, 65, 74
Warp wrapping, 45
Water frame, 4
Water proof sheeting, 156
Weaving, 240–241, 243–246, 248, 252, 259, 261
Weaving techniques, 193–203
Weft faced, 51, 55, 56, 59–61, 63, 65–67, 74
Wet cleaning, 301, 303
Wet-pipe suppression system, 145
Wilkinson, 119
Work-life balance, 363

Yanghai, 44
Yusoku-orimono, 21
Yuzen, 25

Zhusi, 53

Also of Interest

Check out these published and forthcoming related titles from Scrivener Publishing

Handbook of Museum Textiles
Volume I: Conservation and Cultural Research
Edited by Seiko Jose, Sabu Thomas, Pintu Pandit, Ritu Pandey and Vandana Gupta
Forthcoming 2022. ISBN 978-1-119-79170-6

Handbook of Museum Textiles
Volume II: Scientific and Technological Research
Edited by Seiko Jose, Sabu Thomas, Pintu Pandit and Ritu Pandey
Forthcoming 2022. ISBN 978-1-119-98338-5

Smart Nanotextiles
Wearable and Technical Applications
Nazire Deniz Yilmaz
Published 2022. ISBN 978-1-119-65478-0

Textile Dyes and Pigments
A Green Chemistry Approach
Edited by Pintu Pandit, Kunal Singha, Subhankar Maity, and Shakeel Ahmed
Published 2022. ISBN 978-1-119-90491-5

Sustainable Practices in the Textile Industry
Edited by Luqman Jameel Rather, Mohd Shabbir and Aminoddin Haji
Published 2021. ISBN 978-1-119-81888-5

Innovative and Emerging Technologies for Textile Dyeing and Finishing
Edited by Luqman Jameel Rather, Aminoddin Haji and Mohd Shabbir
Published 2021. ISBN 978-1-119-71014-1

Recycling from Waste in Fashion and Textiles
A Sustainable & Circular Economic Approach
Edited by Pintu Pandit, Shakeel Ahmed, Kunal Singha and Sanjay Shrivastava
Published 2020. ISBN 978-1-119-62049-5

Frontiers of Textile Materials
Polymers, Nanomaterials, Enzymes, and Advanced Modification Techniques
Edited by Mohd Shabbir, Shakeel Ahmed and Javed N. Sheikh
Published 2020. ISBN 978-1-119-62037-2

Advanced Functional Textiles and Polymers
Edited by Shahid-ul-Islam and B.S. Butola
Published 2019. ISBN 978-1-119-60579-9

Textiles and Clothing
Environmental Concerns and Solutions
Edited by Mohd Shabbir
Published 2019. ISBN 978-1-119-52631-5

Smart Textiles
Wearable Nanotechnology
Edited by Nazire D. Yilmaz
Published 2019. ISBN 978-1-119-46022-0

Nanomaterials in the Wet Processing of Textiles
Edited by Shahid-ul-Islam and B.S. Butola
Published 2018. ISBN 978-1-119-45980-4

Advanced Textile Engineering Materials
Edited by Shahid-ul-Islam and B.S. Butola
Published 2018. ISBN 978-1-119-48785-2

Handbook of Renewable Materials for Coloration and Finishing
Edited by Mohd Yusuf
Published 2018. ISBN 978-1-119-40775-1

Textile Finishing: Recent Developments and Future Trends
Edited by K.L. Mittal and Thomas Bahners
Published 2017. ISBN 978-1-119-42676-9

Chemicals and Methods for Conservation and Restoration
Paintings, Textiles, Fossils, Wood, Stones, Metals, and Glass
Edited by Johannes Karl Fink
Published 2017. ISBN 978-1-119-41824-5

Printed in the USA/Agawam, MA
September 17, 2026

910661.171